Advances in Multi-Band Microstrip Filters

The first of its kind, this work offers a detailed insight into a range of design procedures for dual-band and tri-band microstrip filters, from theory to practical design. Originating from the FP7 MultiWaveS project, this comprehensive resource includes the most recent results from several well-established research groups as well as detailed coverage of competing approaches ranging from the conventional approach, to advanced multi-layer fabrication technologies, and the development and application of several novel geometries and concepts. In-depth coverage of basic theoretical foundations, detailed design procedures and rules, and comparisons of measured and simulated results enable you to select the optimal approach for your purposes and designs, making this an invaluable resource for both students and industry professionals in the field of microwave engineering.

Vesna Crnojević-Bengin is Associate Professor at the University of Novi Sad, Serbia, as well as leader of the European Microwave Association's topical group MAGEO and Associate Editor of the *International Journal of Electronics*. In 2005 she received the Yugoslav Microwave Theory and Techniques Award for Scientific Contribution.

EuMA High-frequency Technologies Series

Series Editor
Peter Russer, Technical University of Munich

Homayoun Nikookar, *Wavelet Radio*
Thomas Zwick, Werner Wiesbeck, Jens Timmermann and Grzegorz Adamiuk (Eds.), *Ultra-wideband RF System Engineering*
Er-Ping Li and Hong-Son Chu, *Plasmonic Nanoelectronics and Sensing*
Luca Roselli (Ed.), *Green RFID Systems*
Vesna Crnojevic-Bengin, *Advances in Multi-band Microstrip Filters*

Forthcoming
Peter Russer, Johannes Russer, Uwe Siart, and Andreas Cangellaris, *Interference and Noise in Electromagnetics*
Maurizio Bozzi, Apostolos Georgiadis, and Ke Wu, *Substrate Integrated Waveguides*
George Deligeorgis, *Graphene Device Engineering*
Luca Pierantoni and Fabio Coccetti, *Radiofrequency Nanoelectronics Engineering*
Alexander Yarovoy, *Introduction to UWB Wireless Technology and Applications*
Natalia Nikolova, *Introduction to Microwave Imaging*
Philippe Ferrari, Rolf Jakoby, Onur Karabey, and Gustavo Rehder, *Reconfigurable Circuits and Technologies for Smart Millimeter-Wave Systems*

Advances in Multi-Band Microstrip Filters

VESNA CRNOJEVIĆ-BENGIN
University of Novi Sad

CAMBRIDGE
UNIVERSITY PRESS

CAMBRIDGE
UNIVERSITY PRESS

University Printing House, Cambridge CB2 8BS, United Kingdom

Cambridge University Press is part of the University of Cambridge.

It furthers the University's mission by disseminating knowledge in the pursuit of education, learning and research at the highest international levels of excellence.

www.cambridge.org
Information on this title: www.cambridge.org/9781107081970

First published 2015

Printed in the United Kingdom by TJ International Ltd. Padstow Cornwall

A catalogue record for this publication is available from the British Library

ISBN 978-1-107-08197-0 Hardback

Contents

Contributors

Tobias Gerhardus Brand
Stellenbosch University

Vesna Crnojević-Bengin
University of Novi Sad

Norbert Cselyuszka
University of Novi Sad

Riana Helena Geschke
University of Cape Town

Jia-Sheng Hong
Heriot-Watt University

Nikolina Janković
University of Novi Sad

Dmitry Kholodnyak
St. Petersburg Electrotechnical University LETI

Petrie Meyer
Stellenbosch University

Jia Ni
Heriot-Watt University

Shilong Qian
Heriot-Watt University

Vasa Radonić
University of Novi Sad

Alexander Rusakov
St. Petersburg Electrotechnical University LETI

Viacheslav Turgaliev
St. Petersburg Electrotechnical University LETI

Irina Vendik
St. Petersburg Electrotechnical University LETI

Kiril Zemlyakov
St. Petersburg Electrotechnical University LETI

Abbrevations

2D:	two-dimensional
3D:	three-dimensional
AWR MWO:	Advancing the Wireless Revolution Microwave Office
BPF:	bandpass filter
BW:	bandwidth
CDMA:	code division multiple access
CLC:	capacitively loaded cavities
CRLH:	composite right/left-handed
CSRR:	complementary split ring resonator
CST:	computer simulation technology
CTE:	coefficient of thermal expansion
D-CRLH:	dual composite right/left-handed
DGS:	defected ground structure
DNG:	double-negative
DPS:	double-positive
EM:	electromagnetic
ENZ:	epsilon-near-zero
FBW:	fractional bandwidth
FCC:	Federal Communications Commission
GPR:	grounded patch resonator
GSM:	global system for mobile communications
HFSS:	high-frequency structure simulator
HTCC:	high-temperature cofired ceramics
HTS:	high-temperature superconductor
I/O:	input/output
IEEE:	Institute of Electrical and Electronics Engineers
IFS:	iterated function system
LAN:	local area network
LCP:	liquid crystal polymer
LH:	left-handed
LPF:	lowpass filter
LTCC:	low-temperature cofired ceramics
LTE:	long-term evolution
MEMS:	microelectromechanical system

MIC:	monolithic integrated circuits
MNZ:	mue-near-zero
MROP:	microstrip rectangular open-loop
NZ:	near-zero
PCB:	printed circuit board
PGPR:	perturbed grounded patch resonator
RF:	radio frequency
RFID:	radio frequency identification
RGPR:	rectangular grounded patch resonator
RH:	right-handed
SI-SLR:	stepped impedance stub-loaded resonator
SIR:	stepped impedance resonator
SLR:	stub-loaded resonator
SNG:	single-negative
SoP:	system-on-package
SRR:	split ring resonator
TBCCO:	thallium barium calcium copper oxide
TE:	transverse electric
TEM:	transverse electromagnetic
TL:	transmission line
TM:	transverse magnetic
UMTS:	universal mobile telecommunications system
UWB:	ultra-wideband
WiFi:	wireless local area network product based on the IEEE 802.11 standards
WiMAX:	worldwide interoperability for microwave access
WLAN:	wireless local area network
YBCO:	yttrium barium copper oxide
$\lambda/2$:	half-wavelength
$\lambda/4$:	quarter-wavelength

1 Introduction

Vesna Crnojević-Bengin

Filters play an irreplaceable role in virtually any type of radio frequency (RF)/microwave system today. With the recent rapid development and widespread use of various wireless communication systems, ever-more stringent requirements are posed on RF/microwave filters – smaller size, higher performance, and lower cost are all simultaneously required today. Even more recently, multi-band operation is considered necessary to solve the challenges of insufficient capacity of the various wireless systems. To that end, high-performance and compact microwave filters that operate at two or more non-harmonically related frequencies are greatly needed, and several different approaches to their design are being investigated today.

While waveguide architecture has been traditionally used to design filters with high quality factors and capable of handling high power, their bulky construction proved to be a limiting factor for applications where miniaturization is a primary concern. To that end, various planar solutions were developed, based on transmission electron microscopy (TEM) or quasi-TEM transmission lines. Besides compact dimensions, the main advantages of planar filters are easier integration in the circuit or module environment, and well-developed and automated manufacture procedures that typically require no manual tuning.

In this book, we present the most recent results in the development of multi-band planar filters. We focus on the microstrip architecture, but note that most of the filters presented here could easily be redesigned in other architectures, such as coplanar waveguide, or stripline.

This edited book presents the results of the research efforts of four well-established research groups, all working in the field of multi-band filter design, performed within the FP7 project MultiWaveS funded by the European Commission under Framework Programme 7, grant no. PIRSES-GA-2009-247532. The book presents in detail six competing approaches and concepts which are used to solve a common problem: the design of compact microstrip multi-band filters with high performance. The approaches presented range from the conventional one, to the application of advanced multi-layer fabrication technologies, and to the development and application of several novel resonator geometries and filter concepts. The book consists of eight chapters, where each chapter starts with an introduction and ends with conclusions and a list of references.

Chapter 1, the Introduction, is followed in Chapter 2, by a comprehensive overview of the existing methods for the design of dual-band and tri-band filters, together with representative solutions from the literature for all types of filters. Specifically, Sections 2.1–2.4 provide a detailed discussion of the existing planar solutions, while Section 2.5 addresses other advanced multi-band filters, including tunable, non-planar (3-D), superconductive, multi-layer, and other solutions. The existing design methods for multi-band bandpass filters are divided into four groups, and the most representative from each group are presented and analyzed. Apart from dual-band and tri-band bandpass filters, this chapter also presents multi-band bandstop filters proposed in recent years.

Chapter 3 is specifically devoted to the theoretical foundations, where recent developments in algorithms for the construction of multi-band transfer functions are described. Two approaches developed by the authors are presented in detail: one an optimized approach, where poles and zeros of the filter function are systematically positioned and fine-tuned to achieve equi-ripple passbands and/or stopbands, and the other a rigorous approach where the reactance function of a passive liquid crystal (LC)-one-port is used as frequency mapping function. In both cases, the mathematical theory is discussed, examples are presented, and both techniques have proved to produce excellent results.

The remaining chapters are devoted to detailed presentation of specific solutions for the design of multi-band filters developed by the authors.

One of the most widely used approaches in the design of multi-band bandpass filters is to combine two single-band filters to obtain a dual-band response, and to combine a single-band and a dual-band filter to obtain a tri-band response. However, although it allows independent control of the passbands, this approach most often results in filters with increased size. On the other hand, filter miniaturization can be achieved using conventional quarter-wavelength resonators instead of their half-wavelength counterparts which are two times longer. The size of such single-band filters can further be reduced by using traditional techniques, such as folding the resonators of which a filter is composed. This approach has been used in Chapter 4, where dual-band and tri-band bandpass filters based on quarter-wavelength resonators and developed by the authors are presented. A detailed description of their design procedures is given and the filters are compared with other similar solutions. The proposed filters outperform all other recently published configurations since they exhibit high selectivity and compactness, excellent transmission and return-loss characteristics, and, at the same time, allow independent control of all passbands.

A number of multi-band filters have recently been published based on the use of various innovative geometric designs. Nevertheless, grounded patch resonators or mushroom structures are also known, and widely exploited in antenna design, but have not been extensively adopted for multi-band filter applications, despite a number of advantages over other commonly used planar resonators. Chapter 5 focuses on the design of compact dual-band filters using the grounded patch resonator (GPR) and its modifications. Since the square GPR is inherently a non-degenerate dual-mode resonator, it is shown how to independently control the two fundamental modes by changing the geometrical parameters of the resonator. Introduction of a slot perturbation in the rectangular GPR gives the perturbed grounded patch resonator (PGPR) and the behavior of such a structure is analyzed in detail. It is demonstrated that the inter-resonator couplings of the PGPR at two centre frequencies can be adjusted separately, which allows the design of dual-band filters with independent passband centre frequencies and respective bandwidths. Unlike many other approaches, this method allows passbands to be spaced further apart than is typically possible with narrow-band coupled-resonator methods.

Chapter 6 presents novel dual-band filters based on the use of fractal curves – space-filling curves that, theoretically, allow the design of infinite-length lines on a finite substrate area. Due to their space-filling ability, fractal curves have great potential for miniaturization of passive microwave circuits. Special attention has been given to the Hilbert fractal curve, which is very convenient for application in resonators and filters due to its configuration. In the first part of the chapter, two dual-band bandpass filters, based on the Hilbert fractal curve and proposed recently by the authors are shown, which exhibit small size and good performance in terms of insertion loss and selectivity, and are comparable to other recently published solutions. In addition, the proposed filters allow independent control of the passbands and their fabrication is very simple since the circuits are fully planar and do not require vias or multi-layer structures. In the second part of the chapter, the Hilbert fractal curve is also used to design a complex resonator that consists of two dual-mode resonators. The proposed resonator, which exhibits many degrees of design freedom, has been used to design one tri-band bandpass filter and one tri-band bandstop filter. Owing to a specific configuration and various couplings that exist in the structure, the proposed bandpass filter simultaneously exhibits very good performance and very compact dimensions. In comparison to other tri-band bandpass configurations, the proposed structure is the most compact one and it has significantly better selectivity, as well as the possibility to independently control the passbands. The proposed tri-band bandstop filter is also characterized by good performance and it represents the first tri-band bandstop filter operating at 2.4, 3.5, and 5.2 GHz.

All filters presented so far have followed one of the typical approaches to the design of multi-band filters, such as the superposition of single-band filters or the use of complex multi-mode resonators. Although allowing various degrees of design freedom in terms of the independent control of passbands, and exhibiting different relative weaknesses and strengths, all of these approaches have one thing in common: resonators which form filters are regarded as electronic components, and then suitably coupled to result in a desired filtering response. In Chapter 7, we propose an alternative view to the design of electrical circuits, multi-band filters in particular, where the resonator is not considered as a component, but rather as a piece of material tailored to exhibits specified filtering properties. Namely, we show how artificial electromagnetic materials, also called metamaterials, can be used to design multi-band filters with novel properties – small size, low losses, and reduced group delay. To achieve such behavior, we focus on a specific sub-class of metamaterials, called near-zero (NZ) metamaterials, which have recently been shown to support propagation of electromagnetic waves with the zero propagation constant, giving rise to various interesting phenomena. The chapter focuses on NZ propagation of quasi-TEM modes, needed for the design of NZ microstrip filters. It examines both the permittivity NZ and permeability NZ propagation of quasi-TEM waves and demonstrates how they can be used to design single-band NZ filters. Finally, this idea is extended to multi-band filters, and a dual-band microstrip NZ filter is presented and analyzed. Owning to its flat phase response, such filters are promising candidates for applications where flat group delay is required.

Chapter 8 presents recent advances in the field of multi-layer microwave filters, as multi-layer technologies provide three-dimensional flexible designs with the integration of microwave components, circuits, andsub-systems. They have been recognized as promising for the integration of multi-standard or multi-band functions in a single device, with the aim to reduce the complexity and size of the system. In this chapter, recent results of the authors are presented, including filters realized using low-temperature co-fired ceramics (LTCC), and liquid crystal polymer (LCP) technologies. The chapter demonstrates a novel approach to filter design which relies on non-traditional resonators based on a combination of right-handed and left-handed transmission line sections and capacitively loaded cavities. In this way, filter characteristics can be drastically improved and their functionalities enlarged. Among the presented devices are ultra-wideband (UWB) filters, dual-band filters, and tunable single-band and dual-band filters. In all cases, the use of multi-layer technology has resulted in miniature filters, highly demanded today by various wireless applications.

The chapters present a detailed overview of the competing approaches, as well as providing full insight into the design procedures of some of the best dual-band and tri-band microstrip filters today. Apart from simulation and measurement results, which are compared for each presented filter, real-world fabrication-related details are discussed, such as process and/or material tolerances.

This book, which comprises theoretical foundations, detailed design rules, comparisons of measured and simulated results, as well as discussion of the influence of fabrication tolerances, is aimed at a wide audience, ranging from students to the industry professionals in the field of microwave engineering.

2 Design methods of multi-band filters

Nikolina Janković, Vesna Crnojević-Bengin, Petrie Meyer, and Jia-Sheng Hong

2.1 Introduction

Development of a number of wireless communication standards and devices imposed a requirement for components to simultaneously operate at two or more frequencies that correspond to the standards such as IEEE 806.16, IEEE 806.11, GSM, CDMA, etc.

Dual-band bandpass components were the first multi-band circuits to answer this requirement. The first dual-band filter was proposed almost two decades ago [1], and since then there has been a growing interest in dual-band filters.

In comparison to dual-band filters, design of bandpass configurations with three or more bands represents a greater challenge when it comes to filter performance and compactness, because good characteristics have to be achieved in three or more closely positioned passbands. In addition to possible signal crosstalk in closely positioned passbands, design of these filters is a demanding task since it requires a steep slope of transfer function. What is more, almost all bands that are commercially used are closely positioned – for instance WiFi, WiMAX, and GSM systems that operate at 0.9/1.8 GHz, 2.4/2.45 GHz, 3.5 GHz, and 5.2/5.25 GHz.

A number of multi-band bandpass filters and design methods can be found in the literature; in this chapter, an overview of those methods will be given and comparisons made. Although great attention has been paid to multi-band filter development in recent years, significantly fewer papers have been published on configurations with three or more bands in comparison to those devoted to dual-band structures.

Broadly speaking, there are four different approaches to the design of multi-band bandpass filters, each of which can be divided into sub-groups.

The first approach employs classical filter design theory that includes basic concepts such as filter transfer functions, low-pass prototype filters and elements, and frequency and element transformations. This theory is a universal technique that can be applied to any type of filters – waveguide, microstrip, or dielectric. Also, it is an analytical method and thus a very powerful synthesis technique for single-band filters. Since the theory has been widely investigated and applied in single-band filter circuits, it has also been applied in multi-band filter design. In the second section, some classical-theory-based methods for multi-band filter design will be presented and discussed.

The second design approach is based on introducing transmission zeros in the responses of a wideband single-band filter and this method will be presented in the third section.

The third approach employs multi-mode resonators. Since multi-mode structures support several resonant modes, they represent good candidates for miniature multi-band filters. Although the positions of their resonant modes cannot be independently controlled to the full extent, the third approach allows more design freedom in comparison with the previous methods. The fourth section is devoted to multi-band structures that were realized using multi-mode resonators.

Combining two single-band filters to obtain a dual-band response, and single-band and dual-band filters to obtain a response with three or more bands represents the fourth approach to the design of multi-band filters. Since such filters comprise several independent structures, this method gives the most freedom in design in terms of the independent control of passbands. This method will be presented and discussed in the fifth section.

The four methods include design of planar multi-mode bandpass filters. However, there have also been proposed multi-band bandpass filters that are realized in a different manner – using multi-layer architecture or fabrication technologies other than PCB, or using advanced theoretical concepts such as metamaterials. A brief overview of these methods will be given in the sixth section.

Unlike multi-band bandpass filters, multi-band bandstop filters have not been widely explored. Whilst there are four principal and several advanced design methods for multi-band bandpass filters, multi-band bandstop filters have been designed using only some of those approaches. It is also noticeable that most multi-band bandstop filters have been designed to operate at two bands, whilst there have been only a few tri-band bandstop filters. In the last section of this chapter, an overview of multi-band bandstop filters will be given.

2.2 Design based on the classical filter design theory

Filter design theory is an analytical method, and has proven to be a universal and very powerful technique for single-band filter design [2–6]. It has been extensively investigated, and numerous filter transfer functions and topologies have been proposed in the literature. Generally speaking, this design technique includes three major steps:

- filter order and transfer function determination in accordance with predefined specifications,
- network and coupling matrix synthesis,
- design and physical realization of filter.

Behavior of every filtering network can be defined using transfer and characteristic functions which can be defined as

$$H(s) = \frac{E(s)}{P(s)}, \tag{2.1}$$

$$K(s) = \frac{F(s)}{P(s)}, \tag{2.2}$$

where s represents complex frequency and $F(s)$, $P(s)$, and $E(s)$ characteristic polynomials. In low-pass prototypes, $F(s)$ is a polynomial with real coefficients whose roots are positioned on an imaginary axis. Its roots are the frequencies at which there is no reflection and they represents reflection zeros. $P(s)$ is an even polynomial with real coefficients whose roots are the frequencies at which there is no transmission and they represent transmission zeros, whilst $E(s)$ is Hurwitz polynomial.

The most well-known filter functions are Butterworth, Chebyshev, elliptic, and Bessel functions.

Coupling matrices are N x N matrices whose elements represent coupling coefficients between the filter resonators. The main diagonal elements are self-couplings, $M_{i,i+1}$ elements are couplings between adjacent resonators, whilst all other elements are cross-couplings, i.e. couplings between non-adjacent resonators.

The physical layout of the filter is formed in accordance to the coupling matrix. Namely, the coupling matrix determines the filter order as well as the distance between the resonators and ultimately the filter configuration.

The wide application of classical filter design theory has been the main motive to apply it in multi-band filter design.

So far, there have been several design procedures for dual-band and multi-band filters, each of which is characterized by a different approach to transfer function synthesis and the coupling matrix. Some of the procedures are fully analytical yet applicable only to some filter types and topologies, whilst others require demanding optimization methods but at the same time are more universal in terms of filter topology. The common characteristic of the most proposed procedures is that the starting point is the synthesis of a single-band wideband filter. Afterwards, transmission zeros are introduced in the filter response, and passbands are formed. Thus, the passbands cannot be independently formed, which is one of the main drawback of this multi-band filter design approach. In addition, filters designed using this approach have exceedingly large dimensions.

An overview of design procedures for multi-band filters and analysis of their properties are given below.

The first dual-band bandpass filter was proposed in [7], and was realized using classical filter design theory. The proposed procedure allows design of symmetrical dual-band filters, i.e. filters whose passbands are symmetrically positioned around frequency and have equal insertion loss, return loss, and bandwidths.

According to dual-band filter specifications related to in-band and out-of-band characteristics, a low-pass prototype is obtained, Figure 2.1, i.e. the normalized frequencies Ω_a and Ω_s are determined. These frequencies together with number of poles and zeros in bandpass and bandstop regions represent the starting point in the formation of characteristic polynomials $P(s)$, $F(s)$, and $E(s)$. After the characteristic polynomials are synthesized, a coupling matrix is formed, and geometrical parameters of the final filter are determined.

Although the final filter in [7] is realized in the waveguide architecture that does not diminish the applicability of the procedure because the classical theory approach is a universal technique that can be applied to any filter architecture.

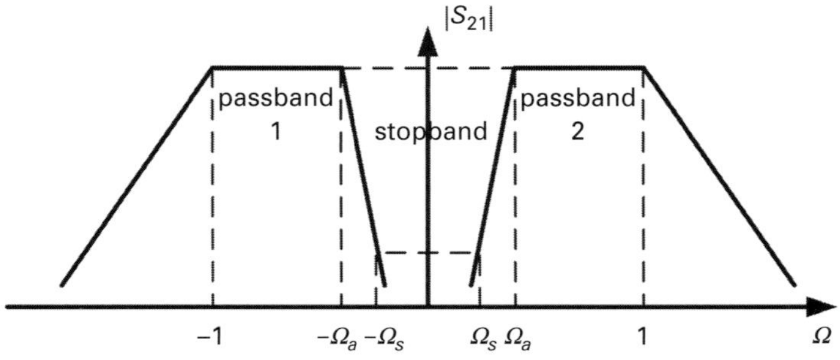

Figure 2.1 Low-pass prototype of the symmetrical dual-band filter.

An advantage of this procedure is the fact that it represents an analytical approach, i.e. an efficient and precise method, as well as the fact that this procedure gives a unique coupling matrix solution. However, its main drawback is the fact that it is applicable only to dual-band filters with symmetrically positioned passbands. In addition, the procedure cannot be readily applied to all filter topologies.

Based on this procedure, in [8] a fully analytical yet simpler method was presented that enables the design of dual-band filters with asymmetrically positioned passbands. However, the two passbands cannot be independently positioned, which is the major drawback. Also, this method cannot be applied to all filter topologies.

Another procedure for synthesis and design of asymmetrical dual-band bandpass filters was presented in [9]. The initial transfer function is formed using the transfer functions of two single-band filters and coupling matrix using optimization procedures. The first step involves an exact and exhaustive synthesis yielding a list of equivalent coupling matrices. In a second step, the proposed approach takes advantage of the multiple solution property by providing some rules for selecting a coupling matrix to be used as the starting point for an approximate synthesis procedure. The approximate synthesis then allows some simplifications of the initial coupling topology by cancelling one or several weak couplings between resonators, which ultimately makes the hardware implementation easier.

Relying on optimization procedures is the major drawback of this procedure, since the convergence time of such procedures is very sensitive to the initial guess. Also, due to the applicability of the method on various filter topologies, optimization methods do not give a unique solution for the coupling matrix.

The three previous procedures can be only used for dual-band filter design, which limits their applicability. In the following, procedures that can be used in the design of filters with a theoretically arbitrary number of passbands will be presented.

A procedure for dual-band and tri-band filter design that can be applied to all filter topologies was presented in [10]. The first step in the procedure is the design of a wideband response which encompasses all passbands of dual-band or tri-band filters. The response is realized by the standard Chebyshev function. Afterwards, the number of transmission zeros is chosen, and they are introduced in the wideband response so as to obtain a dual-band/tri-band response.

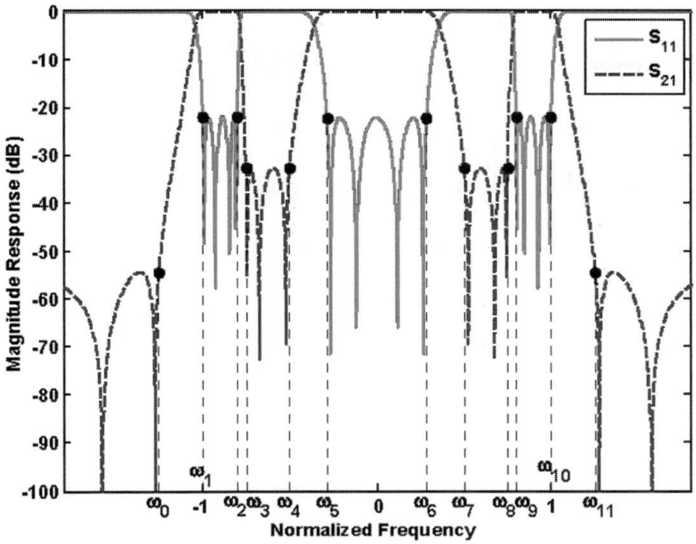

Figure 2.2 Normalized prototype of multi-band filter.

In order to form the coupling matrix, the transfer function needs to be synthesized, which is conventionally achieved using predefined poles and zeros. In this procedure, the transfer function is formed using a different approach. Namely, the transfer function is formed using the transfer functions of two or three single-band filters whose passbands correspond to those of the desired dual-band/tri-band filter. Afterwards, an initial coupling matrix is formed whose non-zero elements are distributed so that it corresponds to the final filter topology. However, the matrix needs to be optimized to obtain the final values that correspond to the filter specifications. The need for optimization, which is sensitive to the initial guess, is the main disadvantage of this procedure. What is more, the optimization may not lead to an optimal response in terms of filter zeros and poles.

The method presented in [11] is the first analytical method for synthesis of symmetrical and asymmetrical multi-band responses with an arbitrary number of passbands and stopbands. Also, it allows synthesis of the transfer function that has both real and imaginary zeros.

According to this method, each passband in the response is defined by two frequencies – ω_1 and ω_2 for the first passband, ω_3 and ω_4 for the second passband, etc., whilst each stopband is defined by one frequency, Figure 2.2.

The transfer function is formed so that it has equal ripple in the passbands, and the first step in its synthesis is to form the initial characteristic function in accordance to passband and stopband specifications. In other words, an initial guess is assigned to zero and pole values in passbands and stopbands.

Figure 2.3 shows an example with two passbands and one stopband, denoted by p_i and z_i, respectively. Each zero and each pole are positioned between two transfer function extrema. For instance, the pole p_2 is positioned between α_1 and α_2. Based on the initial guess of a pole (zero), it is examined whether two extrema in the vicinity of

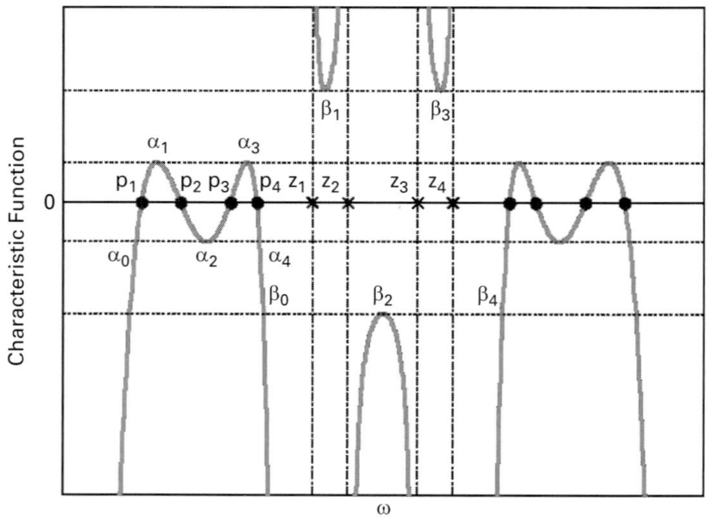

Figure 2.3 Typical filter function with critical frequencies.

the pole (zero) have equal values – if not, a new initial guess is assumed, otherwise the next pole (zero) is determined.

Several tens of iterations are needed to arrive at the final transfer function, which means that this method also requires a certain level of optimization. Yet, the optimization procedures are less demanding than those in [9] and [10].

After the transfer function is obtained, the coupling matrix is formed, and subsequently the geometrical parameters of the filter are determined.

Using this procedure, fourth-order and eighth-order dual-band filters in cul-de-sac topology were designed [12]. Their configurations and responses are shown in Figure 2.4.

The resonators used in the filter configuration were realized using conventional ring-shaped $\lambda/2$ resonators. In both responses, the central frequencies of the passbands are positioned close to each other, and the stopband region has poor performance which is due to the fact that the dual-band response is formed by introducing transmission zeros into the wideband response.

Also, both filters have large overall dimensions, $0.41\lambda_g$ x $0.35\lambda_g$ and $1.18\lambda_g$ x $0.28\lambda_g$, where λ_g represents the guided wavelength at the central frequency of the first passband. Large overall size is a common property of all filters based on classical filter design theory. Namely, the resonators usually employed in filter configuration are conventional $\lambda/2$ resonators and the quality factor and mutual coupling factor are readily determined. In this manner, multiple couplings between resonators are avoided, which is of great importance since the coupling matrix usually foresees only direct couplings and several cross-couplings. On the other hand, such resonators occupy a large area, and thus even low-order filters have large overall size.

In [13], another dual-band filter based on the previous method was proposed, and its configuration and response is shown in Figure 2.5. In this case, dual-mode resonators

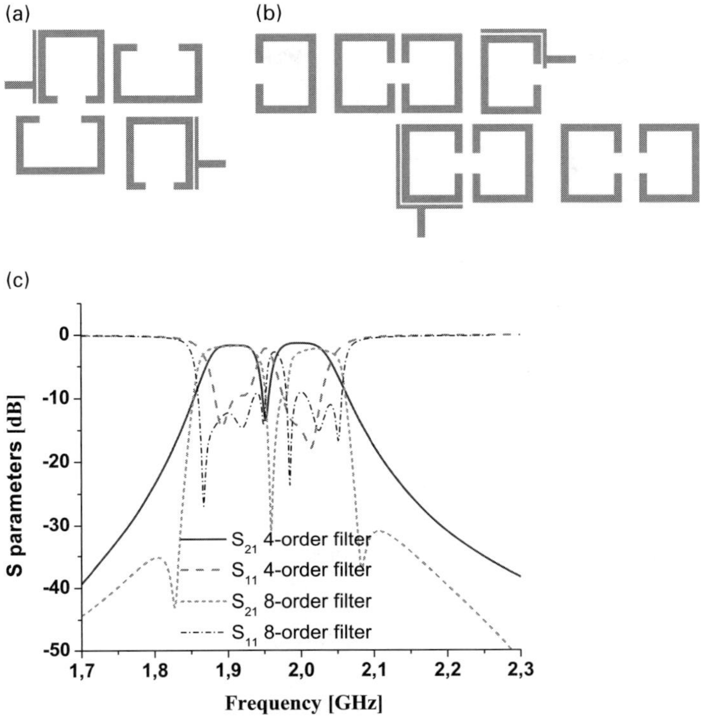

Figure 2.4 (a) Configuration of the fourth-order filter proposed in [12]. (b) Configuration of the eighth-order filter proposed in (c). (c) Filter responses [12].

Figure 2.5 Configuration of the filter proposed in [13].

with perturbation were used in filter configuration. The filter response is characterized by poor performance in the stopband region between the two passbands, which can be seen in Figure 2.6. In spite of using dual-mode resonators, which reduces the number of resonators used in filter configuration, the circuit has large dimensions – $0.44\lambda_g$ x $0.2\lambda_g$.

The proposed procedure was also used to design bandpass filters with three and four passbands [14]. The filters are comprised of E-shaped resonators, which represent stub-loaded resonators. To implement the tri-band filter, a dual-mode dual-band filter was used to govern the first two passbands, and a single bandpass filter was used to govern the third passband. In the case of the quad-band filter, two dual-mode dual-band filters

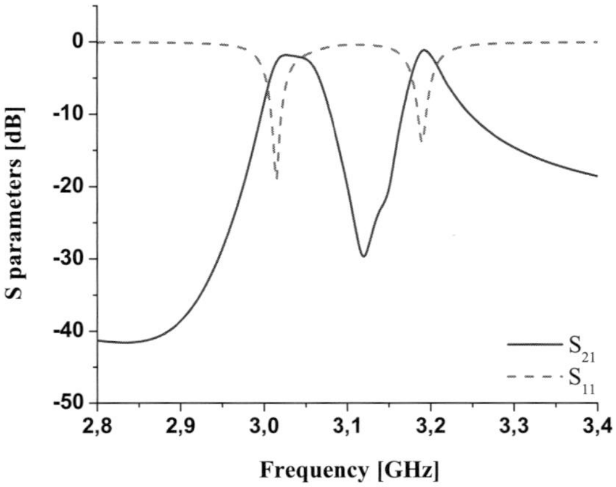

Figure 2.6 Response of the filter proposed in [13].

were used to obtain the four passbands. The filters are characterized by closely positioned passbands with relatively good performance in stopband regions. Also, filter dimensions are very large – $0.31\lambda_g$ x $0.3\lambda_g$, and $0.62\lambda_g$ x $0.48\lambda_g$, respectively. The configuration and response of the proposed tri-band filter are shown in Figure 2.7.

Apart from this method, another analytical method has been proposed in [15]. The transfer function is generated using corresponding single-band filter functions that are obtained using classical filter theory. According to the transfer function, scattering parameters are analytically obtained and afterwards used for coupling matrix synthesis. This method allows the realization of real and imaginary zeros as well as an arbitrary number of passbands with arbitrary bandwidths.

The transfer function obtained by this method is a generalized Chebyshev function, which can be expressed as

$$C_N(\omega) = \frac{F_N(\omega)}{P_N(\omega)}, \tag{2.3}$$

where $F_N(\omega)$ and $P_N(\omega)$ are characteristic polynomials.

The dual-band filter function with a generalized Chebyshev function can be expressed using two single-band filter functions whose passbands correspond to the dual-band filter passbands

$$\begin{aligned} C_N(\omega) &= \frac{1}{\dfrac{1}{C_{N1}(\omega)} + \dfrac{1}{C_{N2}(\omega)}} = \frac{C_{N1}(\omega)C_{N2}(\omega)}{C_{N1}(\omega) + C_{N2}(\omega)} \\ &= \frac{1}{\dfrac{P_{N1}(\omega)}{F_{N1}(\omega)} + \dfrac{P_{N2}(\omega)}{F_{N2}(\omega)}} = \frac{F_{N1}(\omega)F_{N2}(\omega)}{P_{N1}(\omega)F_{N2}(\omega) + P_{N2}(\omega)F_{N1}(\omega)} \\ &= \frac{F_N(\omega)}{P_N(\omega)}, \end{aligned} \tag{2.4}$$

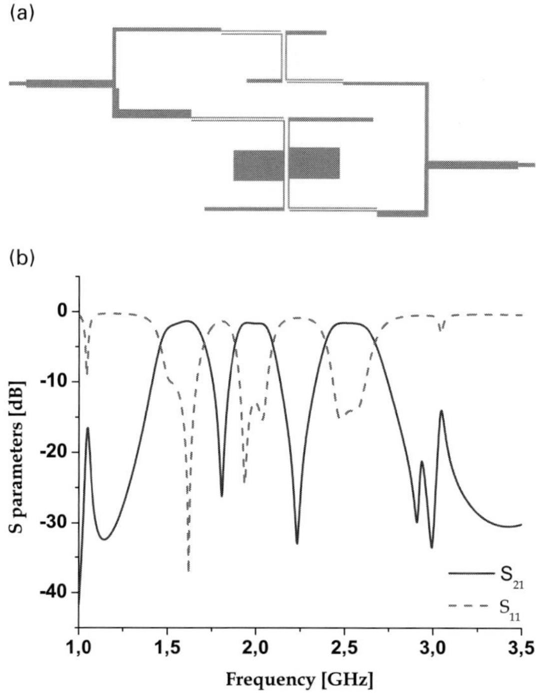

Figure 2.7 (a) Configuration of the filter proposed in [14]. (b) Filter response.

where $F_{N1}(\omega)$, $P_{N1}(\omega)$, $F_{N2}(\omega)$, and $P_{N2}(\omega)$ are formed by shifting the frequencies of the original filter function

$$\begin{aligned} F_{Ni}(\omega) &= F'_{Ni}(\omega - \omega_{si}), \\ P_{Ni}(\omega) &= P'_{Ni}(\omega - \omega_{si}), \end{aligned} \tag{2.5}$$

where i =1 or 2, and ω_{si} the central frequency of the ith passband.

If the passbands to be realized have different bandwidths, then the polynomials in the previous expression need to be modified in the following manner

$$\begin{aligned} P_{Ni}(\omega) &= \prod_{j=1}^{NP_i}(BW_i(\omega - \omega_{si}) - p_{i,j}), \\ F_{Ni}(\omega) &= \prod_{j=1}^{NF_i}(BW_i(\omega - \omega_{si}) - f_{i,j}), \end{aligned} \tag{2.6}$$

where $p_{i,j}$ is the jth root of P_{Ni}, $f_{i,j}$ the jth root of F_{Ni}, and N_{Pi} and N_{Fi} the number of roots of P_{Ni} and F_{Ni}.

In the case of multi-band filter design, the filtering function can be expressed as follows

$$C_N(\omega) = \frac{1}{\dfrac{1}{C_{N1}(\omega)} + \dfrac{1}{C_{N2}(\omega)} + \ldots + \dfrac{1}{C_{Nm}(\omega)}} = \frac{F_{N1}(\omega)F_{N2}(\omega)\ldots F_{Nm}(\omega)}{\sum_{i=1}^{m}\left(P_{Ni}(\omega)\prod_{\substack{j=1 \\ j \neq i}}^{m} F_{Nj}(\omega)\right)}, \tag{2.7}$$

where m is the number of the passbands, C_{Ni} the filtering functions, P_{Ni} the numerator, and F_{Ni} the denominator of C_{Ni}. Each passband has its own order and number of transmission zeros.

Finally, the transfer function is transformed into the bandpass prototype using the following expression

$$\omega_{LP} = \frac{\omega_{C,BP}}{\omega_{H,BP} - \omega_{L,BP}} \left(\frac{\omega_{BP}}{\omega_{C,BP}} - \frac{\omega_{C,BP}}{\omega_{BP}} \right) = \frac{1}{\Delta} \left(\frac{\omega_{BP}}{\omega_{C,BP}} - \frac{\omega_{C,BP}}{\omega_{BP}} \right), \tag{2.8}$$

where ω_{LP} is a low-pass prototype frequency, ω_{BP} a bandpass prototype frequency, and $\omega_{C,BP}$, $\omega_{H,BP}$, and $\omega_{L,BP}$ the central frequency and upper and lower passband limit, respectively. Δ represents the bandwidth.

After the transfer function is obtained, the coupling matrix is formed and subsequently the geometrical parameters of the filter are determined. When the coupling matrix is formed, the frequencies may be shifted. This requires a certain level of optimization, which is the main disadvantage of this procedure.

Fourth-order and sixth-order dual-band bandpass filters designed by the proposed method were presented in [16]. Stub-loaded resonators were used in the filter configuration, as shown in Figure 2.8. Similarly, to the previous cases, the filter response reveals that the passbands are closely positioned. At the same time, the stopband region has slightly better performance in comparison to the filter in [12]. The filter dimensions are large – $0.35\lambda_g$ x $0.25\lambda_g$ in the case of the fourth-order filter and $0.5\lambda_g$ x $0.44\lambda_g$ in the case of the sixth-order filter.

Finally, two optimization methods for the generation of multi-band responses will be noted. The method presented in [17] is an efficient optimization method for generating multi-band characteristics, which guarantees that for given critical specifications in the passbands and stopbands, both optimal transmission and reflection zeros are computed which was not the case in [9] and [10]. The computational part consists of a differential correction-like algorithm proven to be convergent, and which guarantees the optimality of the response.

A similar optimization method was presented in [18] and represents an efficient optimization method for generating asymmetric dual- or multi-band characteristics, which guarantees that certain critical specifications in the passbands and stopbands are met. It is the exact generalization of the classical Chebyshev procedure used for single band synthesis, but obtained for the case of a multi-band characteristic. However, similar to other optimization methods this method also requires large-scale computational resources.

From the previous analysis of filter design methods, it can be concluded that their main advantage is their applicability to any type of filter in terms of architecture, as well as the fact that they provide efficient and relatively simple design since they represent analytical and semi-analytical methods.

On the other hand, these methods have two significant drawbacks. Firstly, their passbands cannot be independently formed since they are formed from one wideband response by introducing transmission zeros or by "mapping" one single-band response to several bands. Secondly, filters designed using classical filter design theory inevitably

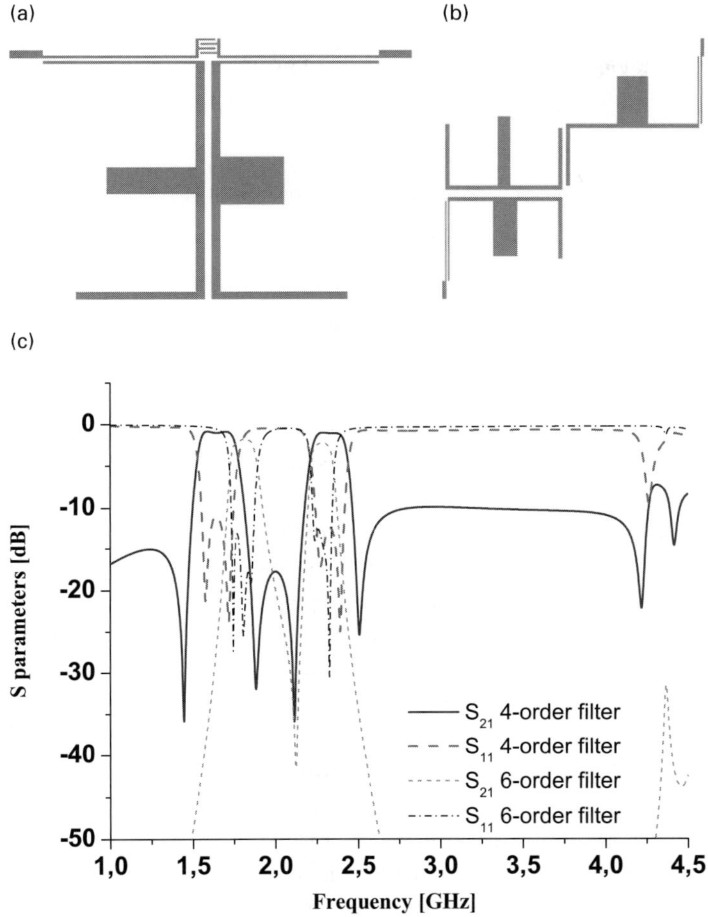

Figure 2.8 (a) Configuration of the fourth-order filter proposed in [16]. (b) Configuration of the sixth-order filter proposed in [16]. (c) Filter responses.

have large dimensions due to the fact that the resonators usually employed in filter configuration are conventional $\lambda/2$ resonators. In applications where other filter characteristics have priority, this drawback can be neglected, but in the design of filters for miniature wireless communication devices, this drawback makes the method almost unusable.

Apart from these disadvantages, it should be noted that the responses of filters designed by this method exhibit poor performance in stopband regions.

2.3 Design by introducing transmission zeros

Introducing transmission zeros into the transfer function of a single-band filter represents the simplest method for multi-band filter design. In the previous section, it was

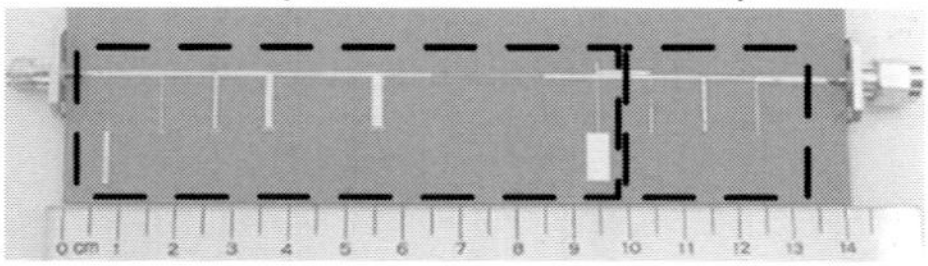

Figure 2.9 Fabricated circuit of the dual-band filter proposed in [19].

shown that multi-band filters designed using classical filter design theory are realized by introducing transmission zeros into a previously synthesized wideband response. In other words, it can be said that the previous method and the method presented here share a common design approach. However, there is a key difference between the two approaches.

Namely, in the previous method transmission zeros are due to couplings between non-adjacent resonators. In other words, the structure that forms wide passband also forms transmission zeros. On the other hand, in the method presented here two separate structures are used to obtain passband and stopband, i.e. bandpass and bandstop filters are separately designed. In addition, the approach presented in the previous section is based on analytical and optimization procedures, whilst the method presented here is significantly simpler.

Although it is very simple and offers straightforward and efficient realization, this method has been rarely used for multi-band filter design and only dual-band and tri-band filters have been realized using this approach. The reason for this lies in the fact that most aspects do not make this method very favorable in comparison to others.

One of the first proposed dual-band filters is the structure presented in [19]. It consists of a cascade connection of a bandpass filter and bandstop filter. Both filters are designed using ideal filter prototypes in the z-domain.

Firstly, the bandpass filter with type II Chebyshev response and central frequency of 3.8 GHz is designed. Afterwards, a bandstop filter with the same central frequency yet with narrower bandwidth is realized. The fabricated circuit is shown in Figure 2.9 in which the cascade connection of the two filters can be noted. The bandpass and bandstop filters are realized using open and short-circuited stubs, respectively. The filter response reveals that there are two passbands at 2.4 and 5.2 GHz, which are characterized by good performance, Figure 2.10. However, due to the cascade connection the final dimensions of the filter are equal to $2.15\lambda_g$ x $0.35\lambda_g$, which is unacceptably large for miniature communication devices.

A somewhat different structure was presented in [20]. It represents a combination of bandpass and bandstop filters that are comprised in one circuit, rather than separately as in the previous case. Short-circuited $\lambda/4$ shunt stubs, which are separated by transmission line sections $\lambda/4$ long, comprise the bandpass filter, Figure 2.11. Introducing serial LC circuits in between two $\lambda/4$ stubs, the bandstop filter with the same central frequency but with the narrower bandwidth is formed. LC circuits are realized using stubs and capacitors, and their values are used to control the stopband bandwidth.

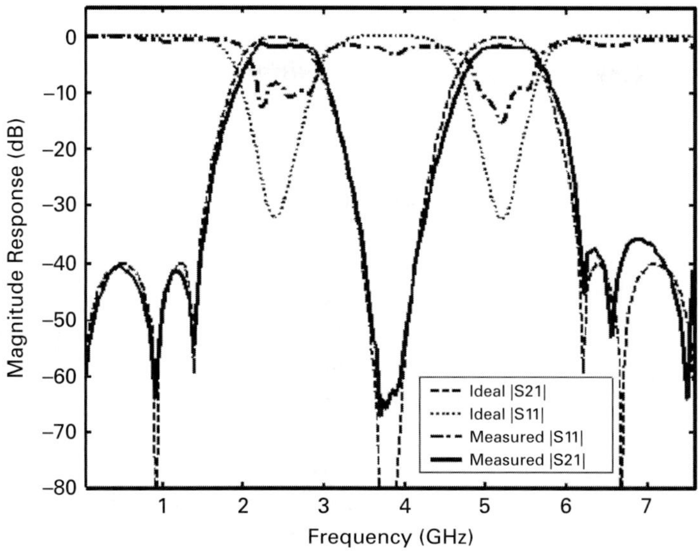

Figure 2.10 Response of the dual-band filter proposed in [19].

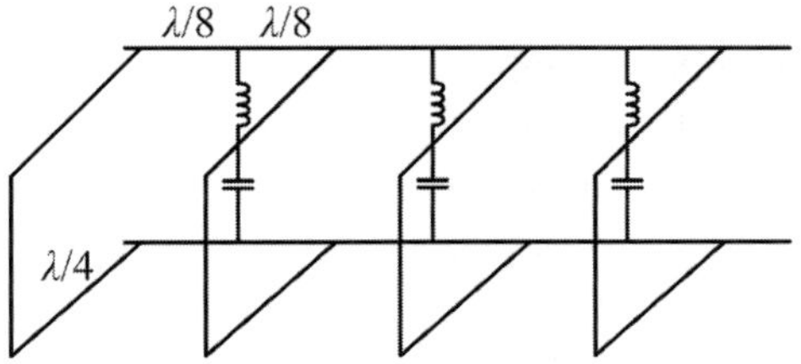

Figure 2.11 Schematic of the dual-band filter proposed in [20].

Although the cascade connection is avoided in this structure, the final circuit has large overall dimensions due to employment of a serial connection of $\lambda/4$ transmission lines.

Similarly to dual-band filters, there have been proposed only a few tri-band filters based on the method of introducing transmission zeros in a wideband response. However, tri-band filters require more elements than dual-band filters, which makes the design slightly more demanding.

The first tri-band filter based on this approach was proposed in [21], and it was realized using a very simple idea. The proposed structure consists of only bandstop filters that provide transmission zeros at specified frequencies. The filters are realized by open shunt stubs whose lengths and impedances are easily calculated once the positions of the transmission zeros are known.

The configuration and response of the presented filter are shown in Figure 2.12. The main advantage of the filter lies in its utter simplicity. On the other hand, there are several significant drawbacks. Namely, the response clearly reveals that the structure cannot have good characteristics in stopband regions. In addition, in order to avoid a

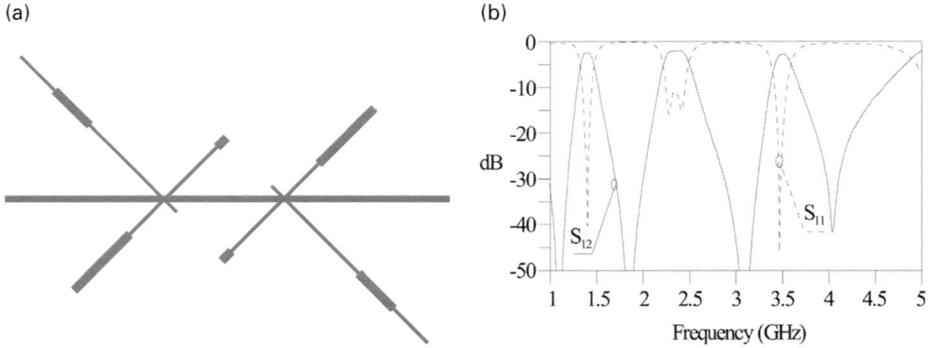

Figure 2.12 (a) Configuration of the filter proposed in [21]. (b) Filter response.

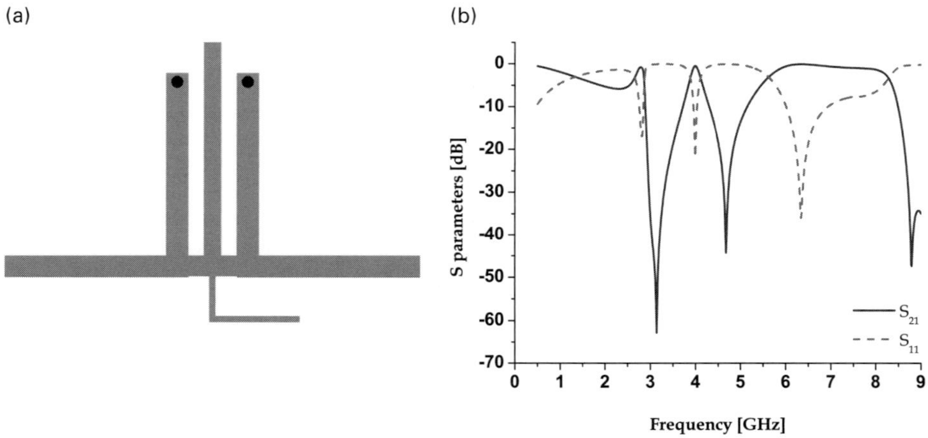

Figure 2.13 (a) Configuration of the filter proposed in [22]. (b) Filter response.

strong coupling between adjacent filters, they need to be distantly positioned, which leads to the large overall size of the filter.

Even though this type of filter seemingly allows arbitrary positioning of passbands, this is not the case. Namely, due to periodicity of the response of open stubs, i.e. due to periodicity of the transmission zeros in the response, the positions of the passbands cannot be distant. Another disadvantage is the fact that bandwidths cannot be independently controlled since one zero is "shared" between two passbands and thus two bandwidths cannot be simultaneously controlled.

Other filters based on this method, rely on the idea presented in [20]. The idea employs a wideband filter realized with $\lambda/4$ short-circuited stubs and introduction of transmission zeros using shunt stubs.

In [22], a tri-band filter was realized by introducing two transmission zeros into a wideband response, Figure 2.13. The transmission zeros are obtained using open shunt stubs whose resonances correspond to the positions of the transmission zeros, whilst the wideband response is based on $\lambda/4$ stubs. As the response reveals, this filter has poor performance in stopband regions. What is more, the overall size of the filter is very large.

A similar configuration was proposed in [23]. In this case, a transmission line $\lambda/4$ long is loaded with the shunt open and short-circuited stepped-impedance resonator (SIR) stubs. This structure provides three passbands, two of which represent higher-order harmonics, and three transmission zeros, the third being the result of the periodicity of the second transmission zero.

The first step in the design of the tri-band filter in [24] is the design of a dual-band response. Afterwards, in one of its passbands a transmission zero is introduced and a tri-band response is obtained. The dual-band filter is realized using shunt SIR stubs, which are connected by a common transmission line. The bandstop filter is designed with shunt stubs that are loaded with capacitors. The fact that the dual-band filter is realized using SIRs implies that independent control of the passbands is limited. What is more, independent control is additionally limited by the fact that the third passband is formed by the introduction of a transmission zero. In addition, the overall dimension of $0.42\lambda_g$ x $0.37\lambda_g$ makes this structure very large.

In addition to being based on the same very simple idea , the filters in [21–24] have in common that their stopband regions have poor performance and their passbands cannot be independently controlled.

The previous examples show that the main advantage of this approach lies in its simplicity. However, since the filters realized using this method, are comprised of bandstop and bandpass filters they inevitably have large dimensions. In addition, passbands are formed from one wideband passband, which implies that passband performance and positions cannot be independently controlled.

In other words, although this method offers straightforward and easy realization, all other critical aspects, such as independent control of passbands, passband performance, stopband characteristic, as well as compactness, do not make this approach very favorable in comparison to others.

2.4 Design by using multi-mode resonators

In the two previous sections, two filter design approaches were presented and they both have multi-band filters whose passbands cannot be independently controlled and whose dimensions are very large. In this section, a design method that allows more design freedom and filter size reduction will be presented. Filters presented in this section are based on multi-mode resonators.

Multi-mode resonators are the structures that support simultaneous propagation of two or more resonant modes whose resonant frequencies are not harmonically related. Therefore, they have been widely used in the design of both single-band and multi-band filters.

In single-band filters, the resonators used are usually dual-mode, which reduces the number of resonators needed for the realization of the specified filter function. Namely, each dual-mode resonator in a filter provides two closely positioned modes, i.e. transmission poles, and in that manner the number of resonators is reduced by half.

As for multi-band filters, multi-mode resonators are used since they can provide passbands or stopbands at frequencies that are not harmonically related. In other words,

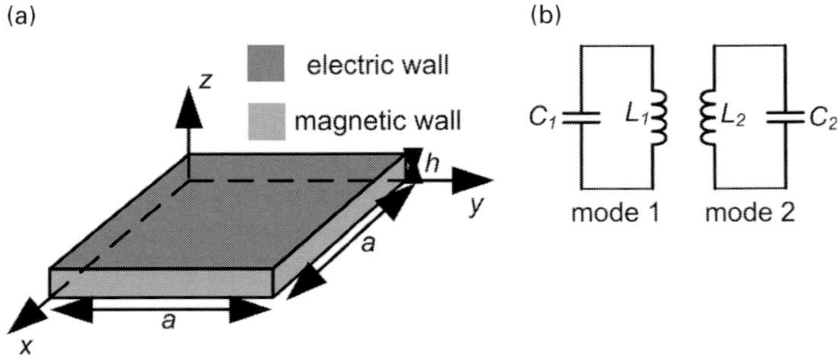

Figure 2.14 (a) Dual-mode microstrip resonator modeled as a cavity. (b) Equivalent circuit of the resonator.

multi-mode resonators offer more design freedom when it comes to independent control of passbands. In addition, since multi-mode resonators support several modes, the size of multi-band filters can be significantly reduced.

Dual-mode resonators have been widely used in dual-band filters. As for tri-band filters, the most used structures in their design are tri-mode resonators. There have also been proposed quad-mode resonators that have been used in quad-band filter design. However, it will be shown that dual-mode resonators can be used in tri-band and quad-band filter design.

The main types of multi-mode resonators are resonators with perturbation, stepped-impedance resonators, and stub-loaded resonators. In the following sections, these resonators will be presented and analyzed, and an overview of multi-band filters based on them will be given.

2.4.1 Multi-band filters with dual-mode resonators with perturbation

Dual-mode resonators with perturbation are the structures that have two-dimensional symmetry and they support propagation of two degenerative modes, i.e. two modes that have the same resonant frequency but orthogonal field distributions. To illustrate this phenomenon, a microstrip square patch resonator can be analyzed and modeled as a resonant cavity [25], Figure 2.14(a). The upper side and lower side of the cavity are modeled as the electrical wall, whilst its lateral sides are modeled as magnetic walls. The electromagnetic field in the cavity is defined by $TM_{mn0}{}^{z}$ modes

$$\begin{aligned}
E_z &= \sum_{m=0}^{\infty}\sum_{n=0}^{\infty} A_{mn}\cos\left(\frac{m\pi}{a}x\right)\cos\left(\frac{n\pi}{a}y\right),\\
H_x &= \left(\frac{j\omega\varepsilon_{eff}}{k_c^2}\right)\left(\frac{\partial E_z}{\partial y}\right),\\
H_y &= -\left(\frac{j\omega\varepsilon_{eff}}{k_c^2}\right)\left(\frac{\partial E_z}{\partial x}\right),\\
k_c^2 &= \left(\frac{m\pi}{a}\right)^2+\left(\frac{n\pi}{a}\right)^2,
\end{aligned} \tag{2.9}$$

where A_{mn} is the magnitude, ω frequency, a effective width, and ε_{eff} permittivity. The resonant frequency of the cavity can be expressed in the following manner

$$f_{mn0} = \frac{1}{2\pi\sqrt{\mu\varepsilon_{eff}}}\sqrt{\left(\frac{m\pi}{a}\right)^2 + \left(\frac{n\pi}{a}\right)^2}. \tag{2.10}$$

It can be seen that there is an infinite number of resonant frequencies with different field distributions. The modes that have the same resonant frequencies are degenerative modes and two fundamental modes $TM_{100}z$ and $TM_{010}z$ are a pair of degenerative modes since

$$f_{100} = f_{010} = \frac{1}{2a\sqrt{\mu\varepsilon_{eff}}}. \tag{2.11}$$

It can be also seen that the two modes are orthogonal to each other. Although the resonator supports two fundamental modes, in its response only one resonance would be seen since the degenerative modes have the same resonant frequency. When the two modes are coupled, the response reveals two resonances since the two modes act as two coupled resonators. Coupling between the modes is achieved when a perturbation is introduced in the resonator symmetry.

Microstrip dual-mode resonators do not have to be of a square shape but they have two-dimensional symmetry. According to its physical configuration, this type of resonator can be divided into two groups – resonators with patch configuration and resonators with ring configuration, Figure 2.15. In the first case, the perturbations are corner cuts, whilst in the second case the perturbations have the shape of a patch. The behavior of the two resonators is the same, with the difference that the mode analysis of patch resonators is more complex.

Dual-mode resonators with perturbation were first applied in single-band filters. As was stated previously, a resonator with coupled resonances acts as a structure with two coupled resonators, i.e. as a filter of the second order. In other words, by mode coupling a structure with two transmission poles is achieved which provides good performance of the filter in terms of insertion loss and quality factor. An example of a single-band filter with a dual-mode resonator with perturbation is presented in [26]. One meander-shaped ring resonator with perturbation d x d comprises the filter that exhibits a passband with two transmission poles, which implies dual-mode behavior of the resonator.

Recently, dual-mode resonators have been used in the design of multi-band filters. Dual-mode behavior of resonators with perturbation can be used in several manners to obtain dual-band response. The first way is to use two coupled degenerative modes as sources of two separate passbands. This has been applied only in a few configurations since the positioning of two degenerative modes on relatively distant frequencies requires a very strong coupling between the modes which makes the filter design more complex.

The configuration presented in [27] employs a dual-mode square patch resonator with a perturbation with the shape of a cross, Figure 2.16(a). By varying different geometrical parameters of the cross slot. the resonant frequency of one mode can be changed independently of the resonant frequency of the other mode. Also, the configuration of

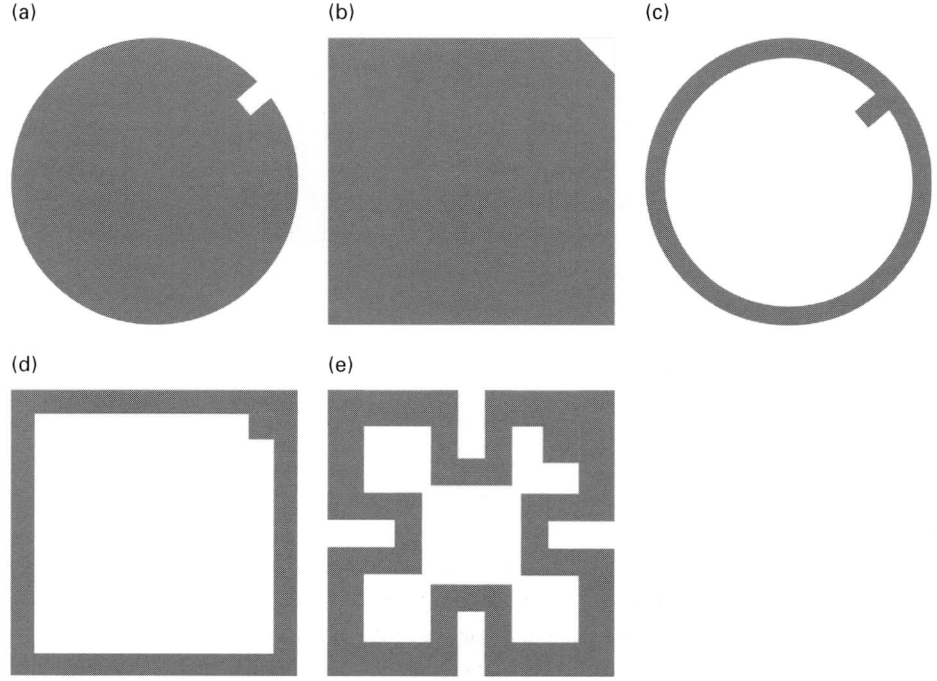

Figure 2.15 Typical dual-mode resonators with perturbation: (a) Circular patch. (b) Square patch. (c) Circular ring. (d) Square ring. (e) Meandered ring.

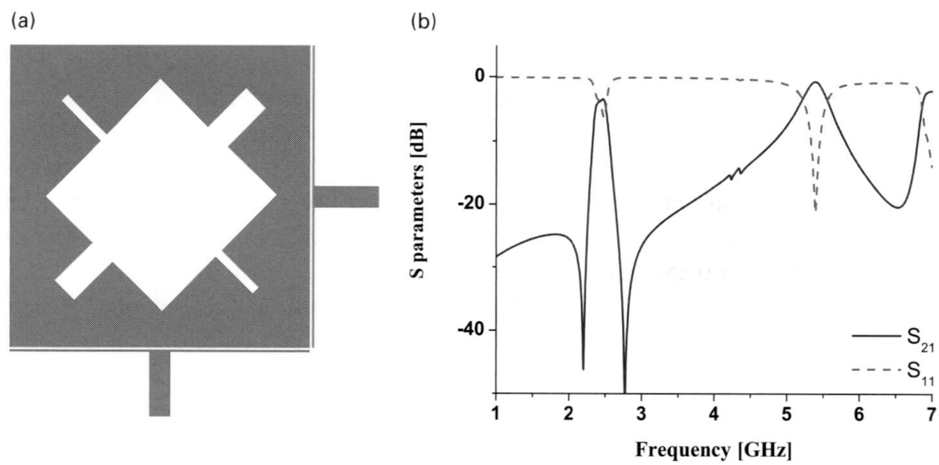

Figure 2.16 (a) Configuration of the filter proposed in [27]. (b) Filter response.

the feeding lines allows transmission zeros to occur. Although this filter is a very smart dual-band solution, which enables independent control of one passband, it is also characterized by poor selectivity in the second passband as well as by relatively large overall dimensions, Figure 2.16(b).

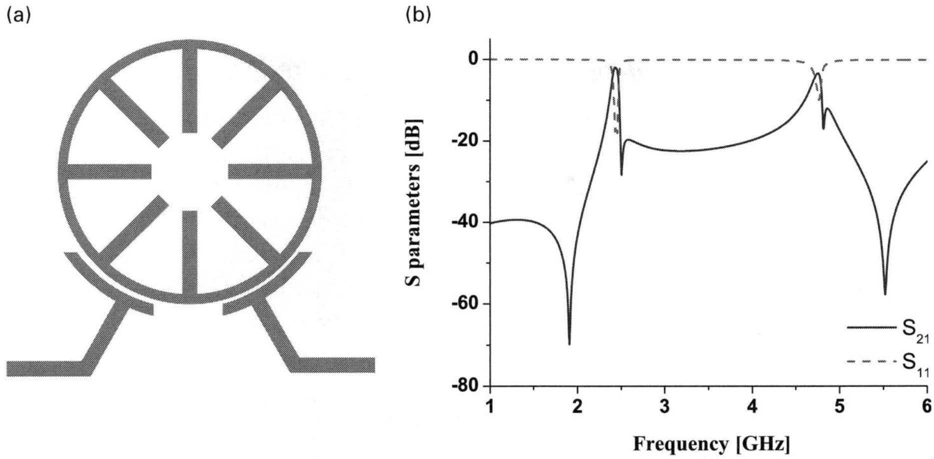

Figure 2.17 (a) Configuration of the filter proposed in [30]. (b) Filter response.

A structure similar to the previous one in which the problem of poor selectivity was solved was proposed in [28]. By embedding a pair of slits in the square patch, the dual-band response is realized and one transmission zero is introduced. Slits introduced at the edge of the resonator provide another two transmission zeros, which improve selectivity of the passbands. However, the problem of large dimensions and independent control of the second passband remain unsolved.

This design approach was also applied in [29]. The proposed dual-band filters consist of square-shaped ring resonators with a very large perturbation that has the shape of a patch. In that way, a very strong coupling between the two modes is achieved and they are positioned at distant frequencies. The realized dual-band filters are the second-order filters, and they are characterized by good in-band performance. However, the selectivity of the passbands is poor.

A disadvantage of these three configurations is the fact that positioning of passbands is determined by the coupling between the two modes and is thus very limited. Also, poor selectivity as well as relatively large dimensions that cannot be reduced below $0.25\lambda_g$ x $0.25\lambda_g$ represent further drawbacks.

A second way of applying dual-mode behavior is the realization of two passbands using higher-order modes. Namely, every resonator exhibits not only a fundamental resonance but also higher-order modes that occur periodically in the spectrum. Also, each higher-order mode supports degenerative modes. In case of dual-mode resonators, a perturbation causes coupling of degenerative modes at fundamental and higher-order resonances. Although a dual-band response can be achieved using the higher-order modes of conventional resonators, an advantage of dual-mode resonators lies in the fact that they enable two transmission poles to occur in each passband, which makes filter performance significantly better in comparison to conventional resonators.

This idea was employed for the first time in closed ring-shaped resonators [30]. A symmetrical ring resonator without perturbations is excited by asymmetrically positioned feeding lines, Figure 2.17(a). Two excited ports are placed at 135° separated positions along the ring resonator and coupled with the ring via parallel-coupled lines, leading to synchronous excitation of two transmission poles in dual passbands. Namely, apart from introducing a perturbation, another way to excite two degenerative modes is to asymmetrically position feeding lines [31]. Such feeding lines also enable transmission zeros, i.e. good selectivity of passbands, Figure 2.17(b). In addition, the resonator is loaded with short stubs, which to some extent allows control of the position of the passbands.

Very similar dual-band configurations were proposed in [32–34]. In [32], a dual-band ring-resonator bandpass filter is proposed. Two capacitors are placed at two non-orthogonally excited positions along a ring resonator in order to excite the two first-order degenerate modes and create multiple transmission zeros. A set of perturbations is introduced on the ring to excite the two second-order degenerate modes.

In [33], two coupled-line sections are simply installed on a single ring resonator at the two ports with 90° separation. By increasing the length of the loaded open-circuited stub, the two first-order degenerate modes are excited and they make the first passband, whilst one of the third-order degenerate modes moves downward and forms the second passband together with a second-order degenerate mode. In addition, three transmission zeros occur and they provide good selectivity.

The configuration proposed in [34] is a dual-mode dual-band ring resonator bandpass filter, which is implemented by a cascade of several microwave coupled transmission line sections that have non-linear phase shift properties in frequency, and are suitable for the development of dual-band devices. In that manner, the filter occupies less than 30% of the area of a traditional ring resonator filter.

These structures are characterized by very good passband performance in terms of insertion loss and selectivity. Nevertheless, their major disadvantage is a very limited passband control. Although the position of the second passband can be controlled to some extent by change in some geometrical parameters, the position is predominantly determined by the frequency of the second harmonic. Also, these filters have relatively large dimensions.

The idea was also applied in the filters with patch resonators [35–36]. Whilst the filter in [35] is based on a patch resonator with a cross slot and two different groups of loaded stubs, the configuration in [36] is comprised of a right-angled triangular patch resonator with fractal perturbation. In spite of having different configurations from those in [30–34], these filters also exhibit the same advantages and drawbacks as filters with ring-shaped resonators.

In [37], different modes of a circular patch resonator were used in a specific way to obtain a dual-band response. Namely, the resonant frequencies of TM_{21} and TM_{01} mode are positioned close to each other by circular cuts so they form the second passband with two transmission poles, whilst rectangular cuts are used to couple TM_{11} mode to the degenerative mode and in that way the first passband with two transmission poles is formed, Figure 2.18.

Although very smart, this idea also has drawbacks in terms of the resonator dimensions and passband positioning. Namely, although the resonant frequencies of TM_{21} and TM_{01} mode are closely positioned their control is very limited since control of circular cut dimensions is also limited.

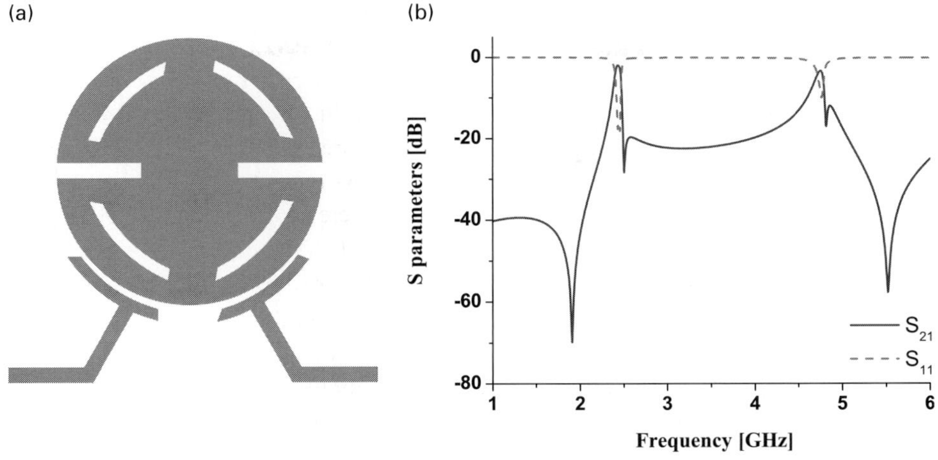

Figure 2.18 (a) Configuration of the filter proposed in [37]. (b) Filter response.

We can conclude that dual-mode resonators with perturbation enable the design of dual-band filters with good passband performance. Nevertheless, it was shown that their main drawbacks are limited passband control and relatively large dimensions, which although smaller than in the filters designed using classical theory, they cannot be reduced below $0.25\lambda_g$ x $0.25\lambda_g$.

As for filters with three or more passbands, dual-mode resonators with perturbation have been implemented in only a few tri-band configurations. In this section, several ways to design dual-band filters using dual-mode resonators with perturbation were described. In the case of tri-band filters, the only way that can be implemented is to form passbands using the fundamental and higher-order harmonics since such resonators support only two modes with the same resonant frequency, i.e. it is not possible to design a tri-mode structure using perturbation.

The structure proposed in [30] was also used to realize a tri-band filter [38]. A very similar structure was presented in [39]. The proposed filter consists of a ring-shaped resonator that is excited by asymmetrically positioned feeding lines, Figure 2.19(a). Three pairs of degenerative modes form three passbands with a small insertion loss and good selectivity in all passbands, Figure 2.19(b).

However, as in the case of a dual-band filter, the major drawback of the structure is the fact that the passbands are formed by the harmonics. Although there are mechanisms for control of the passbands, the passband positions are predominantly determined by the resonant frequencies of the harmonics. In addition, a common disadvantage of such filters can be noted in this example – large overall dimensions which are equal to $0.31\lambda_g$ x $0.31\lambda_g$.

2.4.2 Multi-band filters with stepped-impedance multi-mode resonators

Dual-mode behavior is the simultaneous propagation of two modes whose resonant frequencies are not harmonically related. Strictly speaking, dual-mode resonators are

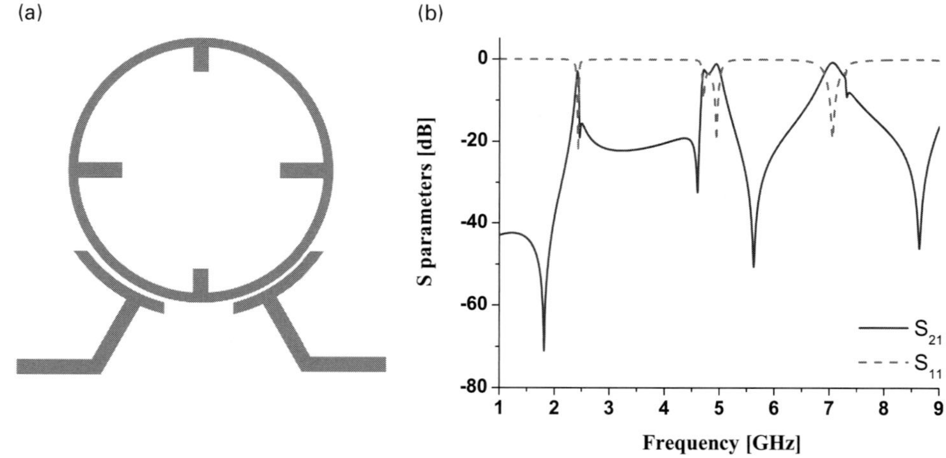

Figure 2.19 (a) Configuration of the filter proposed in. [38]. (b) Filter response.

only those resonators that support two modes that have the same resonant frequencies but different field distributions. However, dual-mode behavior can be also achieved in the structures that do not support degenerative modes. In that case, the two modes do not stem from coupled degenerative modes but they are formed by the fundamental and higher-order mode whose resonant frequencies are not harmonically related as in conventional structures.

In other words, dual-mode behavior occurs in configurations that do not represent classical dual-mode resonators but they have perturbations that cause a change in the ratio of resonant frequencies.

An example of such configuration is the stepped-impedance resonator (SIR), Figure 2.20, which was proposed for the first time in [40]. It can be said that the SIR originates from the conventional microstrip resonator, in which a stepped-impedance perturbation is introduced.

Due to its symmetry, SIR can be analyzed using even-/odd-mode analysis. The even and odd equivalent circuits are shown in Figure 2.21 and the expressions for the even and odd input impedances are the following

$$\begin{aligned} Z_{ine} &= jZ_2 \frac{Z_2 \tan\theta_1 \tan\theta_2 - Z_1}{Z_1 \tan\theta_2 + Z_2 \tan\theta_1} = jZ_2 \frac{K \tan\theta_1 \tan\theta_2 - 1}{\tan\theta_2 + K \tan\theta_1}, \\ Z_{ino} &= jZ_2 \frac{Z_1 \tan\theta_1 + Z_2 \tan\theta_2}{Z_2 - Z_1 \tan\theta_1 \tan\theta_2} = jZ_2 \frac{\tan\theta_1 + K \tan\theta_2}{K - \tan\theta_1 \tan\theta_2}, \end{aligned} \tag{2.12}$$

where Z_1 and Z_2 are the characteristic impedances, θ_1 and θ_2 their electrical lengths, and K the ratio Z_2/Z_1.

From the resonant condition $1/Z_{in} = 0$, we have that the fundamental resonance occurs when

$$K = \tan\theta_1 \tan\theta_2. \tag{2.13}$$

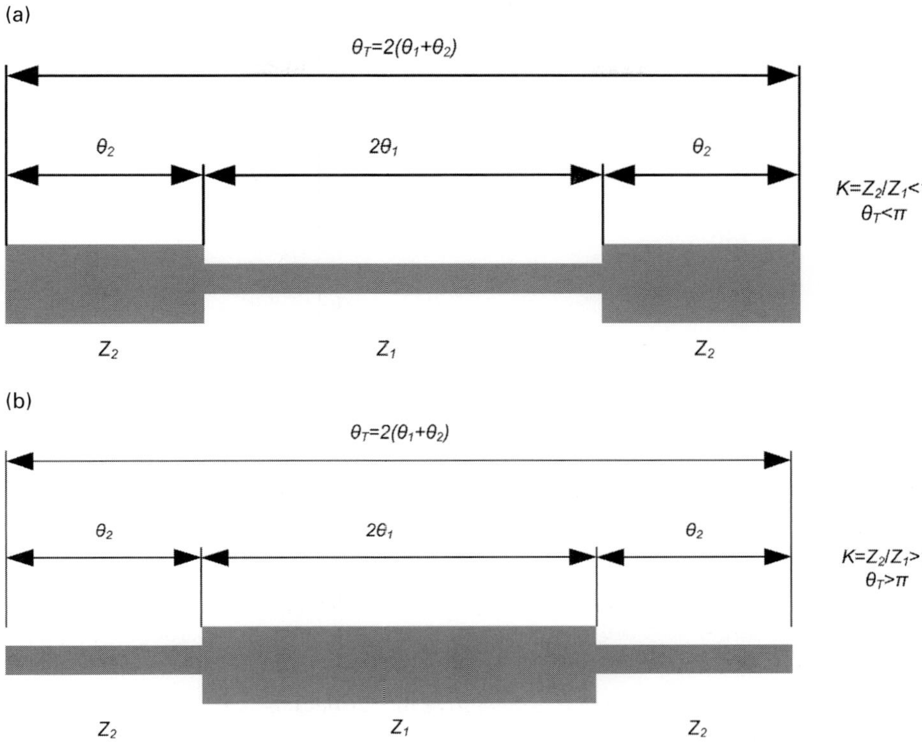

Figure 2.20 Configuration of stepped-impedance resonator.

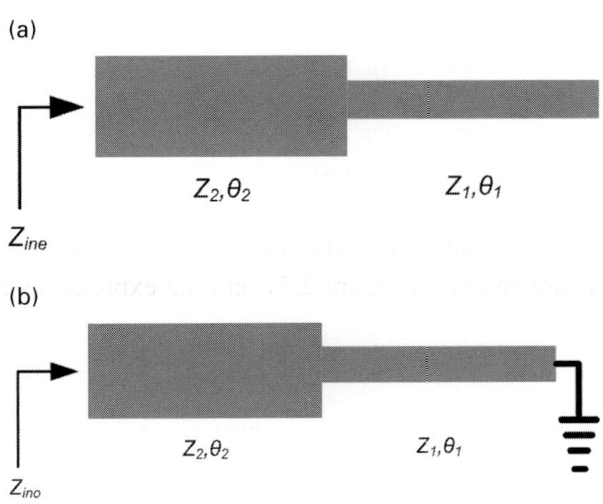

Figure 2.21 Equivalent circuit for: (a) Even mode. (b) Odd mode.

From the previous expression, we can derive the ratio θ_T/θ_1

$$\begin{array}{ll} \tan\dfrac{\theta_T}{2} = \tan(\theta_1 + \theta_2) = \dfrac{\tan\theta_1 + \tan\theta_2}{1 - \tan\theta_1\tan\theta_2} = \dfrac{1}{1-K}\left(\dfrac{K}{\tan\theta_1} + \tan\theta_1\right), & K \neq 1 \\ \theta_T = \pi, & K = 1 \end{array}. \tag{2.14}$$

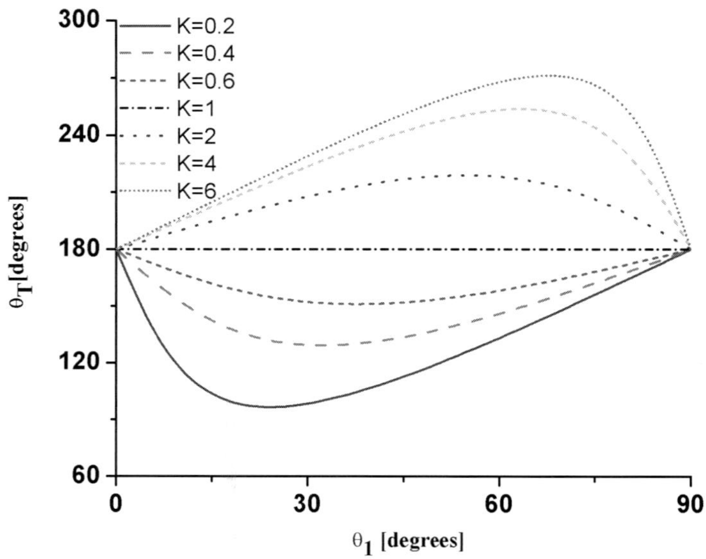

Figure 2.22 Dependence of θ_T of θ_1 for different values of K.

When $K = 1$, the structure has a uniform impedance and fundamental resonance occurs for $\theta_T = \pi$, which is in accordance with the conventional $\lambda/2$ resonator.

Figure 2.22 shows how θ_T changes with θ_1 for different values of K, and it can be seen that θ_T has minimal values for $0<K<1$, and maximal values when $K>1$.

In order to analyze the ratio of fundamental and higher-order modes resonances, we can analyze the case when $\theta_1 = \theta_2 = \theta$. In that case, the expressions (2.12) have a simpler form

$$Z_{ine} = jZ_2 \frac{K\tan^2\theta - 1}{\tan\theta(1+K)},$$
$$Z_{ino} = jZ_2 \frac{\tan\theta(1+K)}{K - \tan^2\theta}, \tag{2.15}$$

and the resonant conditions are the following

$$\tan^2\theta = K, \quad \tan\theta \to \infty, \quad \tan\theta = 0. \tag{2.16}$$

Using the expression (2.16), the ratios of the resonant frequencies are derived

$$\frac{f_1}{f_0} = \frac{\pi}{2\tan^{-1}\sqrt{K}},$$
$$\frac{f_2}{f_0} = 2\frac{f_1}{f_0} - 1, \tag{2.17}$$
$$\frac{f_3}{f_0} = 2\frac{f_1}{f_0},$$

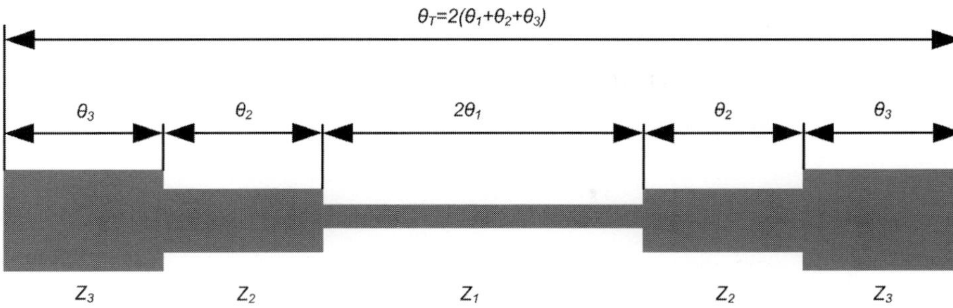

Figure 2.23 Configuration of tri-stage stepped-impedance resonator.

where f_0 is the resonant frequency of the fundamental mode, and f_1, f_2, and f_3 are the resonant frequencies of the higher modes. It can be seen that the ratio f_1/f_0 does not represent a conventional ratio of the fundamental and the first higher mode, whilst f_2 and f_3 are conventional harmonics in comparison to f_0 and f_1. In other words, the mode at f_1 is not conventional harmonic but together with the fundamental mode comprises two resonant modes. Hence, this SIR structure represents a dual-mode resonator.

Also, it can be seen that $f_1/f_0 = 2$ for $K = 1$, which is expected since SIR is a conventional $\lambda/2$ resonator when $K = 1$. The resonant frequencies f_1 and f_0 move away as K decreases, and approach each other as K increases. However, it should be noted that only the fundamental mode is influenced by K, whilst the second mode is a function of θ, which implies that only the fundamental mode can be independently controlled, whilst control of the second mode inevitably affects the first resonant mode.

By the same token, the tri-mode stepped-impedance resonator can be realized, i.e. a structure that support three resonant modes. Figure 2.23 shows a stepped-impedance resonator with three segments with different impedances. As in the previous case, even-/odd-mode analysis can be applied to analyze the behavior of the structure. The equivalent circuits are shown in Figure 2.24 and the expressions are given in the following form

$$
\begin{aligned}
Z_{ine} &= jZ_3 \frac{\tan^2\theta(K_1 + K_2 + K_1K_2) - 1}{\tan\theta(K_1 + K_1K_2 + 1 - K_2\tan^2\theta)}, \\
Z_{ino} &= jZ_3 \frac{\tan\theta(1 + K_2 + K_1K_2 - K_1\tan^2\theta)}{K_1K_2 - \tan^2\theta(K_1 + K_2 + 1)},
\end{aligned}
\tag{2.18}
$$

where K_1 is the ratio Z_3/Z_2, K_2 the ratio Z_2/Z_1, Z_1, Z_2, and Z_3 the characteristic impedances of the segments, and θ their electrical lengths. For the sake of simplicity, it is considered that $\theta_1 = \theta_2 = \theta_3 = \theta$.

The fundamental and two higher resonances occur when the following conditions are met

$$
\tan^2\theta = \frac{K_1K_2}{K_1 + K_2 + 1}, \quad \tan^2\theta = \frac{K_1 + K_1K_2 + 1}{K_2}, \quad \tan\theta \to \infty. \tag{2.19}
$$

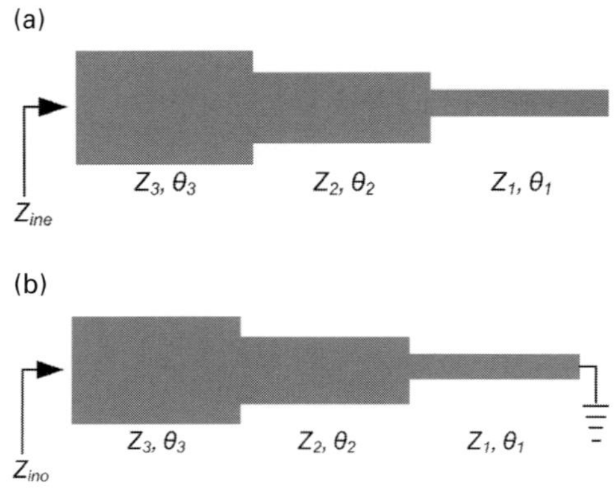

Figure 2.24 Equivalent circuit for: (a) Even mode. (b) Odd mode.

The ratios of the modes can be expressed in the following manner

$$\frac{f_1}{f_0} = \frac{\tan^{-1}\sqrt{\dfrac{K_1 + K_1K_2 + 1}{K_2}}}{\tan^{-1}\sqrt{\dfrac{K_1K_2}{K_1 + K_2 + 1}}}, \qquad \frac{f_2}{f_0} = \frac{\pi}{2\tan^{-1}\sqrt{\dfrac{K_1K_2}{K_1 + K_2 + 1}}}, \tag{2.20}$$

where f_0 is the resonant frequency of the fundamental mode, and f_1 and f_2 the resonant frequencies of higher modes. Similar to dual-mode SIR, the ratios f_1/f_0 and f_2/f_0 do not represent conventional ratios of the fundamental and higher modes, but two higher harmonics together with the fundamental one comprise three resonant modes. Hence, the SIR structure with three segments with different impedances represents a tri-mode resonator.

It should be noted that only the first two modes can be controlled by K_1 and K_2, whilst the third mode is influenced only by θ. Also, the first two modes cannot be independently controlled since both modes are functions of K_1 and K_2.

We can conclude that dual-mode and tri-mode SIR structures are good candidates for multi-band filters since they can provide resonances at non-harmonically related frequencies. Due to the possibility to control the positions of the fundamental and higher-order harmonics by a simple mechanism such as stepped impedance, stepped-impedance resonators have attracted significant attention and there have been a number of multi-mode filters based on SIRs. It was shown that the resonant modes could not be independently controlled to the full extent; but in comparison to resonators with perturbations, this type of resonator offers significantly greater design freedom in terms of passband control. Also, the behavior of SIR structures makes them good candidates for miniaturization of filters.

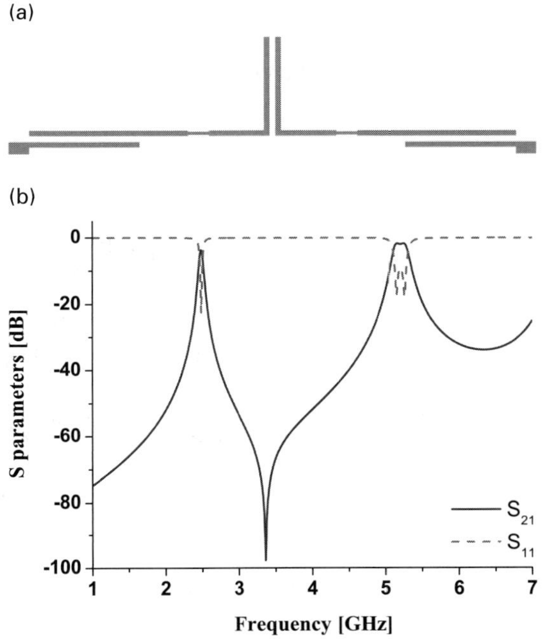

Figure 2.25 (a) Configuration of the filter proposed in [43]. (b) Filter response.

The first dual-band filters based on SIR structures were published in [41–44] and the considerable potential of SIR for high-performance dual-band filter design was demonstrated. The filter configuration proposed in [43] is shown in Figure 2.25, together with the response of the filter. Despite being based on a novel idea, the proposed filters exhibit drawbacks, such as large dimensions and the fact that all these filters operate at 2.4/5.2 GHz.

Poor selectivity is predominantly caused by the fact that the SIR structure itself cannot provide transmission zeros, but has to be generated in a different manner. One way to provide transmission zero is the employment of a greater number of resonators and specific couplings, which, however, increases overall size.

In the years following, considerable attention was paid to improving the performance of filters based on SIRs. In [45], a parallel doubly coupled stepped-impedance resonator structure is proposed to build a 2.4/5.2 GHz dual-band bandpass filter for a WLAN application, Figure 2.26. The prototype filter is a three-stage design composed of four SIRs in conjunction with the tapped-line used in the I/O ports to create transmission zeros, Figure 2.26(b). The proposed filter has the advantage of a much wider fractional bandwidth in both passbands without sacrificing the passband insertion losses. Although the problem of selectivity is solved in this configuration, this filter is not a good solution in terms of circuit size. Filters with similar configurations to this one were presented in [46–47]. They are also characterized by good selectivity and relatively large overall size.

Another way to achieve good selectivity was proposed in [48–51] in which transmission zeros are provided by source-load coupling. Namely, when there is a direct

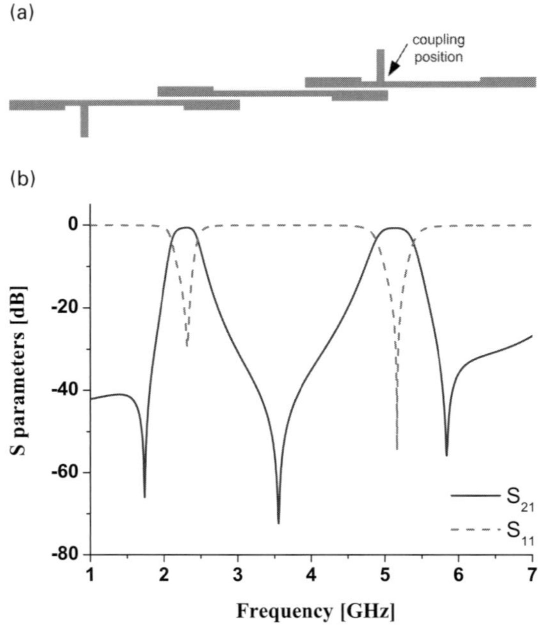

Figure 2.26 (a) Configuration of the filter proposed in [45]. (b) Filter response.

coupling between the feeding lines, an additional signal path is provided. That path provides a transmission zero at the frequency at which it is out-of-phase with the signal from the main path. However, the solution with source-load coupling usually leads to increases in filter dimensions, and thus the possibility of filter miniaturization is limited.

Filters with smaller dimensions and good selectivity can be realized by folding of the coupled resonators [52–54]. The configuration and response of the filter published in [52] are shown in Figure 2.27, and a very good dual-band response can be seen. Pseudo-inter-digital stepped-impedance resonators were used to design the bandpass filter with dual-band response. By tuning the impedance ratio and physical length of the SIRs, the filter has good dual-band performance at 2.4/5.2 GHz and high isolation between the two passbands. Additionally, the overall size of the filter is only $0.21\lambda_g$ x $0.11\lambda_g$.

The structures in [53] and [54] employ very similar resonators but they have a different feeding scheme. Nevertheless, they are both characterized by good performance.

Good characteristics can be also achieved with asymmetrically fed SIR. Namely, asymmetrical feeding provides two signal paths with different electrical length, which cancel each other out at some frequencies.

Examples of such filters were presented in [55–57], whilst the filter published in [58] is one of the best dual-band filters in terms of insertion loss, selectivity, and compactness. Its configuration and response are shown in Figure 2.28. Two coupled lines and two transmission lines are used for the filter. In such a configuration, a transmission zero appears in the middle of two equally separated passbands. Moreover, the unequal even- and odd-mode phase velocities decrease the filter rejection level. However, the

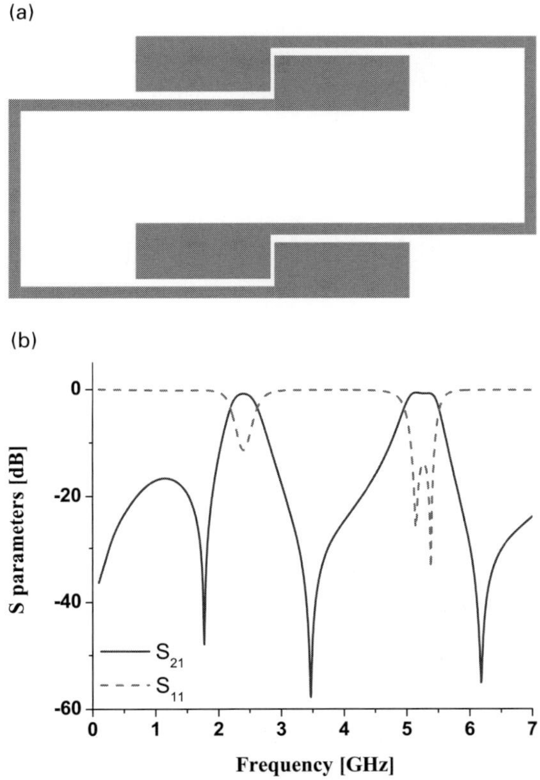

Figure 2.27 (a) Configuration of the filter proposed in [52]. (b) Filter response.

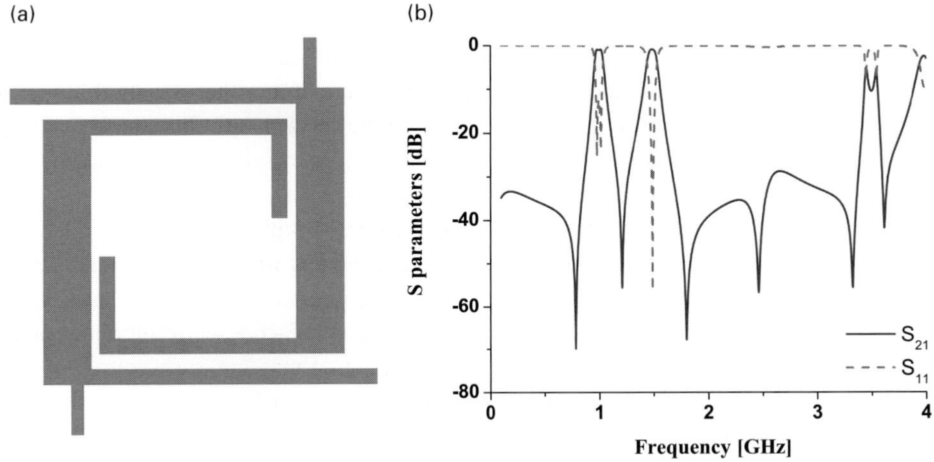

Figure 2.28 (a) Configuration of the filter proposed in [58]. (b) Filter response.

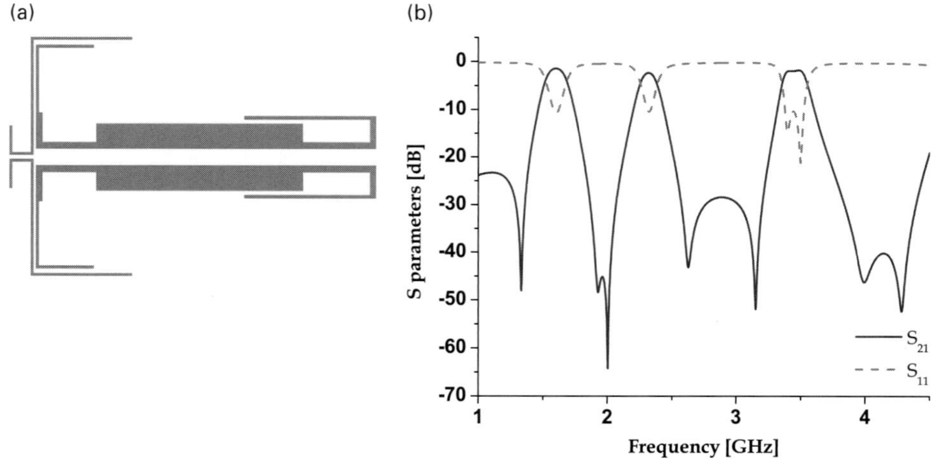

Figure 2.29 (a) Configuration of the filter proposed in [60]. (b) Filter response.

passbands cannot be independently controlled to the full extent, since the filter is based on SIRs. What is more, this configuration can be used only if two passbands are closely positioned.

As for tri-band filters, they are based on tri-mode stepped-impedance resonators, which were shown to have great potential for application in multi-band filters. Since a tri-mode stepped-impedance resonator consists of three segments with different impedances, this resonator has more degrees of freedom for the control of its response in comparison to dual-mode stepped-impedance resonators. Although this fact allows more flexible filter design, it implies that design of a filter with tri-mode SIRs is more complex due to the number of parameters that should be taken into consideration. In addition, the analysis of tri-mode SIRs implies that independent control of the passbands is very limited.

By utilization of tri-mode SIRs, tri-band filters were designed and proposed in [59–62]. The configurations in [59] and [61] are very similar and they consist of two folded tri-mode stepped-impedance resonators that are capacitively coupled to the feeding lines. They show very poor performance in terms of circuit size and especially in passband selectivity. Namely, good selectivity cannot be achieved since only edge-coupling between the resonators exists.

A filter with slightly better characteristics was proposed in [63]. The proposed filter consists of improved quarter-wavelength stepped-impedance resonators that are inductively coupled to the feeding lines. Relatively good characteristics in terms of selectivity and in-band performance were obtained at the expense of circuit dimensions, which are exceedingly large in the proposed filter.

However, the tri-band filter presented in [60], whose configuration and response are shown in Figure 2.29, represents one of the best tri-band solutions in terms of response characteristics. Similarly to dual-band filters based on SIRs, good selectivity in tri-band filters with SIRs can be achieved by creating certain couplings in the filter circuit. In this

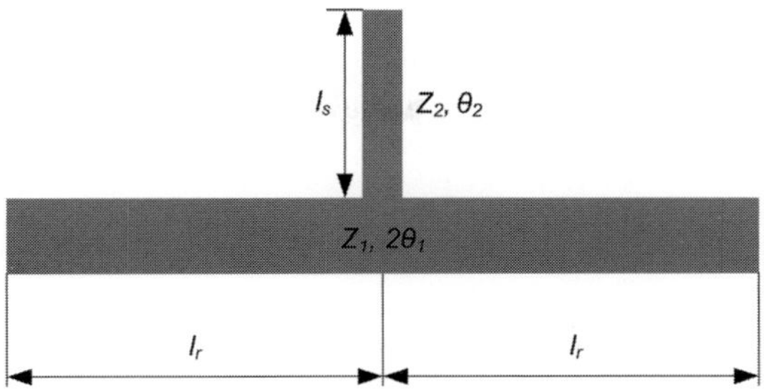

Figure 2.30 Stub-loaded resonator.

case, there is a coupling between the feeding lines, as well as a strong coupling between the two stepped-impedance resonators. On the other hand, the filter exhibits certain disadvantages. Besides the limited possibility for passband control, the large dimensions of $0.34\lambda_g$ x $0.24\lambda_g$ also represent a drawback of this filter.

SIR configuration was also used to achieve quad-band response [64]. An unconventional stepped-impedance resonator that exhibits quad-mode behavior was proposed, and it was realized by embedding short-circuit stubs into the low-impedance sections of the conventional two-sided open-circuit stepped-impedance resonator. Using two proposed resonators, a quad-band filter was designed and it exhibits passbands at 1.9, 2.8, 4.3, and 5.2 GHz. Also, it is characterized by good performance, good selectivity, and compact size. However, independent control of the passbands is limited due to the configuration of the resonator.

In the previous analysis, a number of multi-band filters based on stepped-impedance filters have been presented and it can be concluded that SIRs enable design of filters with good performance and compact size.

Nevertheless, it should be noted that most dual-band filters are aimed to operate at 2.4 and 5.2 GHz or at two closely positioned frequencies. The reason for this lies in the fact that the change in SIR segment impedance enables either significant approach or significant separation of the resonances.

In other words, filters based on SIRs are limited predominantly in terms of positioning of passbands. Also, independent control of passbands is also limited, which may, however, be improved by the introduction of certain couplings in the filtering structure.

2.4.3 Multi-band filters with stub-loaded resonators

Apart from the stepped-impedance resonator, another configuration that provides dual-mode and tri-mode behavior is a stub-loaded resonator (SLR). Such a resonator was used for the first time in the bandpass filter proposed in [65].

The structure of the simplest SLR is shown in Figure 2.30 and it can be seen that the resonator consists of a conventional transmission line with uniform impedance, which is loaded with a stub. Due to its symmetry, the resonator can be analyzed using

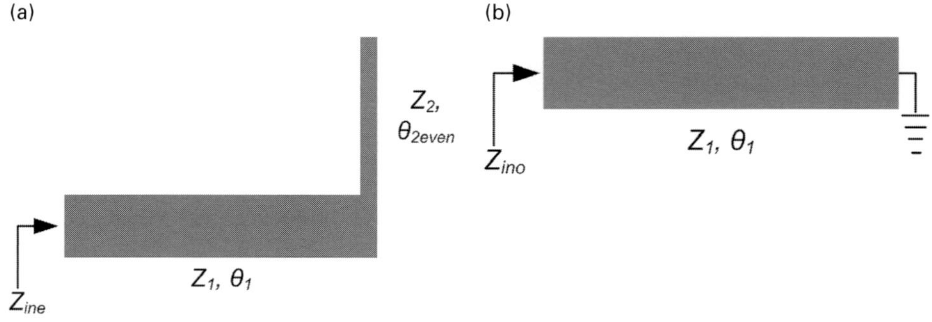

Figure 2.31 Equivalent circuit for (a) Even mode. (b) Odd mode.

even-/odd-mode analysis. The equivalent circuits are shown in Figure 2.31, whilst the even and odd input impedances are expressed in the following manner

$$Z_{ine} = jZ_2 \frac{Z_{2even} \tan \theta_1 \tan \theta_2 - Z_1}{Z_1 \tan \theta_2 + Z_{2even} \tan \theta_1}, \qquad (2.21)$$

$$Z_{ino} = jZ_1 \tan \theta_1.$$

From the resonant condition for the even mode $1/Z_{ine} = 0$ the following expression is derived

$$Z_1 \tan \theta_2 = Z_2 \tan \theta_1, \qquad (2.22)$$

whilst the resonant condition for the odd mode $1/Z_{ino} = 0$ gives

$$\tan \theta_1 \to \infty. \qquad (2.23)$$

From the expression (2.22), it can be seen that in an extreme case $\theta_2 = 0$ (no stub) the even mode corresponds to the even mode of the conventional $\lambda/2$ resonator. Also, the expressions (2.22) and (2.23) imply that the odd mode is a function of only one parameter – θ_1, whilst the even mode is influenced by the impedances Z_1 and Z_2 and the electrical lengths θ_1 and θ_2. Thus, the position of the even mode can be controlled independently of the odd mode by a change in the parameters Z_1, Z_2, and θ_2. On the other hand, the position of the odd mode cannot be changed independently, since both modes are functions of the parameter θ_1. In other words, SLR supports propagation of two modes, one of which can be independently controlled.

A tri-mode stub-loaded resonator can also be realized. Namely, if a conventional $\lambda/2$ resonator is loaded with two stubs of different lengths or by a square ring, a structure that exhibits three non-harmonically related resonances can be achieved. Also, stubs can be grounded by a via which represents an additional degree of freedom in filter design.

Although SLRs have different configurations in comparison to SIRs, these two resonator types have several common properties – resonant frequency control is significantly more flexible than in the case of dual-mode resonators with perturbation, but at the same time control is limited. Also, SLR itself is not a structure that can provide a transmission zero.

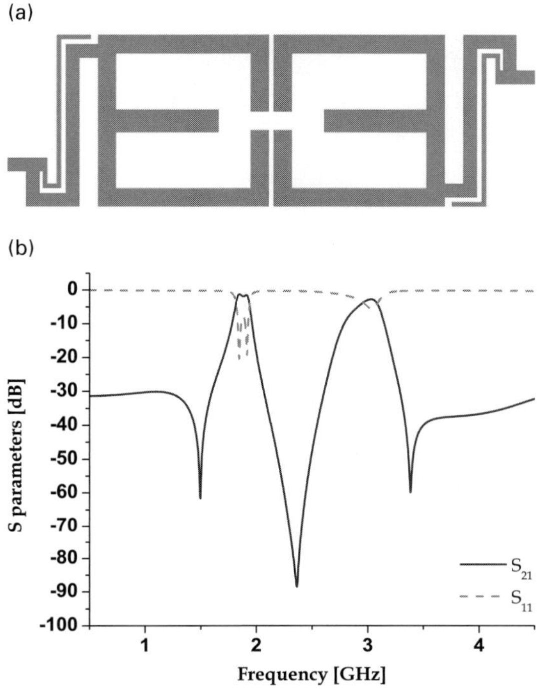

Figure 2.32 (a) Configuration of the filter proposed in [65]. (b) Filter response.

The first dual-band filter with SLR was proposed in [66]. Two SLRs are coupled through an electric field and asymmetrically fed to provide transmission zeros, i.e. better selectivity of the passbands, Figure 2.32(a). In order to reduce the size of the filter, the resonators are folded to form ring-shaped structures. Thus, the final dimensions of the filter are $0.28\lambda_g$ x $0.14\lambda_g$. The response of the filter implies that the filter has good characteristics in terms of insertion loss and selectivity, Figure 2.32(b).

Similar filters with comparable performance were proposed in [67–70]. They are all comprised of ring-shaped stub-loaded resonators but employ different schemes to provide transmission zeros. Whilst the structures in [67] and [68] employ asymmetric feeding, the filter in [69] uses inductive cross-coupling, and the structure in [70] source-load coupling.

Somewhat different structures were presented in [71] and [72] in which resonators have the configuration of a closed ring-resonator loaded with stubs. In [71], the ring resonator is loaded with two stubs that are used to control the second passband, whilst the ring resonator in [72] is loaded with four stubs, which enables filter size reduction and control of the passbands. However, these filters do not exhibit better performance in comparison to previously stated ones, and they have significantly larger dimensions.

The examples of the filters with SIRs and SLRs showed that an efficient way to reduce filter dimensions is the folding of resonators. However, in some filters based on SLRs another way of reducing size was demonstrated – the stubs were grounded by a via. Namely, it is well known that in the case of $\lambda/2$ and $\lambda/4$ the ungrounded one with

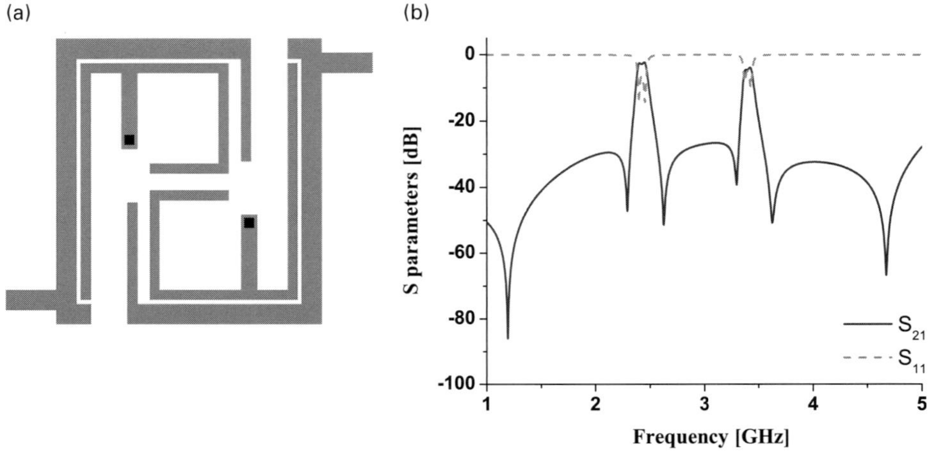

Figure 2.33 (a) Configuration of the filter proposed in [75]. (b) Filter response.

twice as long dimensions. The price that is paid for this size reduction is a more complex fabrication that requires the use of vias. By the same token, grounded stubs in SLRs enable significant size reduction in comparison to SLRs with non-grounded stubs.

Dual-band filters with SLR structures with grounded stubs were proposed in [73–78]. In [74], a dual-band bandpass filter using both an open-stub loaded resonator and a short stub-loaded resonator was presented. Apart from size reduction, use of a short stub-loaded resonator enables a transmission zero. Owing to the filter configuration and source-load coupling, a dual-band filter is implemented with four transmission zeros and it exhibits high skirt selectivity.

The configuration presented in [76] consists of two coupled grounded SLRs, and owing to the source-load coupling this filter is characterized not only by small size, $0.19\lambda_g$ x $0.19\lambda_g$, but also by very good performance in terms of insertion loss and passband selectivity, Figure 2.33.

It was stated previously in this section that if a conventional $\lambda/2$ resonator is loaded with two stubs of different lengths or by a square ring, a tri-mode SLR can be formed and used in tri-band filter design.

Tri-band filters with tri-mode SLRs were proposed in [79–81]. In [79], the proposed resonator comprises a half-wavelength resonator, a short stub, and an open stub. The tri-band characteristics are realized by properly controlling the lengths of the half-wavelength resonator and stubs. A pseudo-inter-digital structure is used to realize the compact size of the filter operating at 1.57, 2.4, and 3.5 GHz and to provide good selectivity.

The filter presented in [81] is comprised of two coupled tri-mode SLRs and its selectivity has been improved owing to the coupling between the feeding lines, Figure 2.34. This tri-band configuration that operates at 2.4, 3.5, and 5.2 GHz represents one of the best tri-band filters since it is characterized by low insertion loss, good selectivity in all passbands, as well as by small dimensions of $0.27\lambda_g$ x $0.19\lambda_g$.

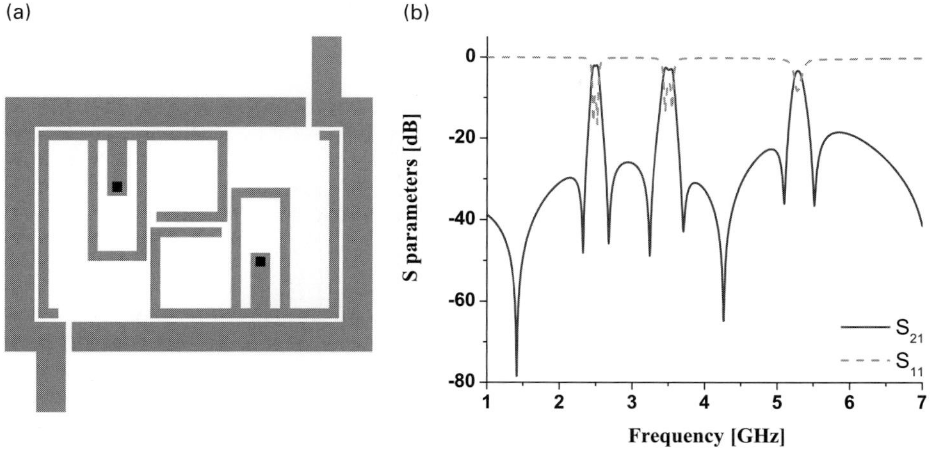

Figure 2.34 (a) Configuration of the filter proposed in [80]. (b) Filter response.

A stub-loaded resonator was also used to design a quad-band filter [82]. The two fundamental and two higher-order even and odd resonances of a dual-mode SLR were used to provide four passbands at 1.5, 2.5, 3.5, and 4.5 GHz. The proposed quad-band filter is characterized by good selectivity and small overall dimensions of $0.3\lambda_g$ x $0.3\lambda_g$. Nevertheless, employment of a dual-mode SLR for a quad-mode filter limits control of the passbands' positions.

SLR structures enable the design of filters with good performance and small dimensions. Although they have similar behavior as SIRs, filters with SLRs are not as limited in terms of the independent control of passbands. Namely, in the case of the SIR structure the position of one resonant mode can be controlled independently of the other by a change of SIR segment impedance. However, the segment impedance ratio is limited by fabrication technology, as well as by the fact that extreme values of impedance inevitably lead to high losses.

In the case of SLR, one mode can be controlled independently from the other by a change in the stub length, which is not limited by fabrication technology or by losses. However, the dimensions of the filter, which increase with stub length, should be taken into consideration. On the other hand, this problem can be diminished by variation of the filter geometry parameters.

A disadvantage of filters based on SLRs is the fact that two passbands stem from the resonant modes of the same structure, and thus it is not possible to independently control passband performance to the full extent.

2.4.4 Multi-band filters with stepped-impedance, stub-loaded resonators

Superposition of stepped-impedance and stub-loaded resonators gives a structure that exhibits multi-mode behavior and which has more parameters that can be used for control of the resonant modes. Therefore, such a structure can be used for the design of multi-band filters. Also, design of filters based on this structure offers more degrees of

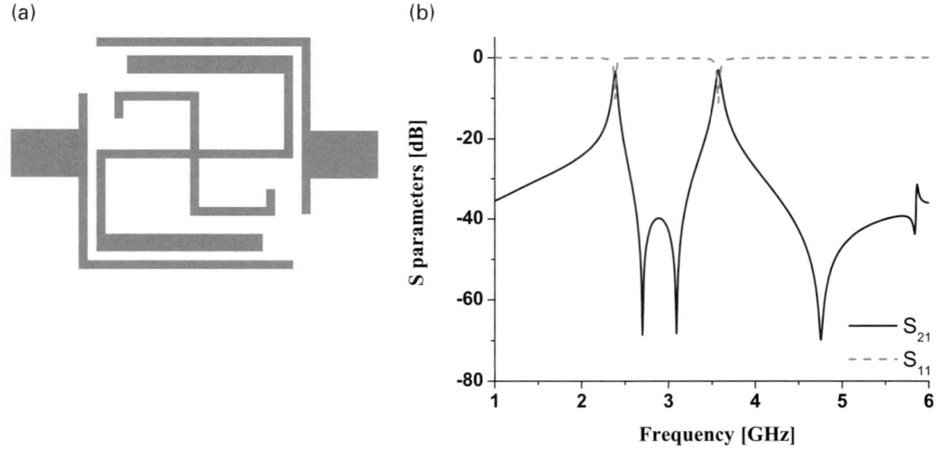

Figure 2.35 (a) Configuration of the filter proposed in [81]. (b) Filter response.

freedom, but, at the same time, more degrees of freedom imply more parameters that should be taken into consideration, i.e. a more complex design procedure.

Although SI-SLR unites the advantages of SIR and SLR structures, which implies the possibility of the design of high-performance, compact multi-band filters, configurations with SI-SLR published so far exhibit either poor performance in terms of insertion loss and selectivity or large circuit size.

The configuration and response of a dual-band filter with an SI-SLR structure are shown in Figure 2.35 [83]. It is comprised of an S-shaped stepped-impedance resonator loaded by the two same open stubs that are used to adjust the second passband in a wide range. Although it is compact, $0.18\lambda_g$ x $0.13\lambda_g$, this filter exhibits relatively poor performance in terms of insertion loss and passband selectivity.

Some other dual-band configurations with SI-SLR resonators were proposed in [84–91]. For instance, the structure presented in [91] consists of a grounded stepped-impedance resonator that is loaded with folded stubs symmetrically positioned around the midpoint of the resonator, whilst the structure in [84] is comprised of inductively fed hairpin stepped-impedance resonators that are loaded with stubs in the midpoint.

However, as was stated previously, none of the proposed configuration simultaneously exhibits good performance and compactness.

Similarly, the potential of SI-SLRs has not been fully exploited in tri-band filters. The proposed tri-band configurations have good performance but none of them simultaneously exhibits compactness [92–97].

The configuration and response of the filter proposed in [92] are shown in Figure 2.36. The proposed filter consists of two asymmetrically fed stepped-impedance resonators that are loaded with stubs in the midpoint. Even though all three passbands have good characteristics, the main disadvantage is the large overall size of the filter.

A similar structure with a different feeding scheme was proposed in [93] and it also exhibits good in-band and out-of-band performance, but the filter has large overall size.

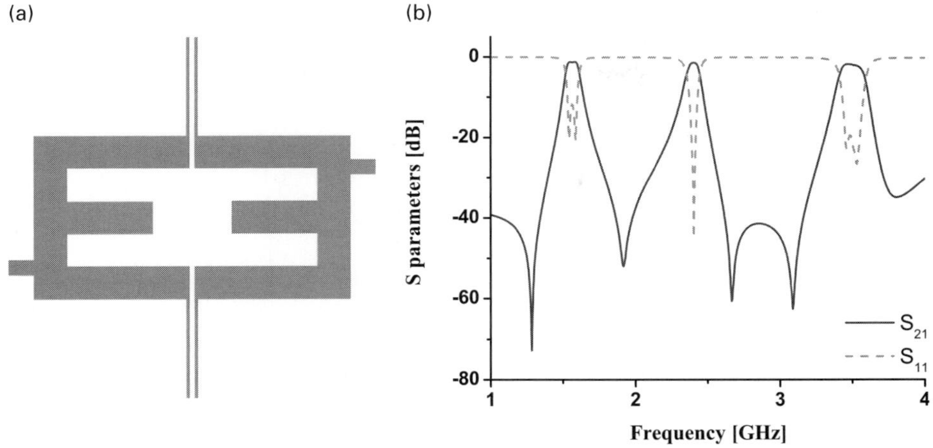

Figure 2.36 (a) Configuration of the filter proposed in [90]. (b) Filter response.

Finally, SI-SLR has also been used in the design of a quad-band filter [98]. Nevertheless, the proposed filter based on single a SI-SLR resonator exhibits compact size, but poor selectivity and in-band performance.

2.5 Design by cascading independent single-band or dual-band filters

Combining two single-band filters to obtain a dual-band response, and a single-band and a dual-band filter to obtain a response with three or more bands represents the fourth approach to the design of multi-band filters. Since such filters comprise several independent structures, this method gives the most freedom in design in terms of independent control of passbands.

Although the configuration, which consists of several structures, may imply large dimensions of filters, some of the most compact multi-band filter circuits with excellent performance have been designed by this method.

Design of dual-band filters includes separate design of two single-band filters, which are afterwards put together to form a dual-band filter. Although the two filters are independent, they influence each other's performance, and thus the final step in the design is optimization of the geometrical parameters of both single-band filters so as to obtain the desired filter response.

Resonators that are used in the design of dual-band filters include conventional $\lambda/2$ and $\lambda/4$ resonators, as well as more complex structures such as dual-mode resonators with perturbation, SIR, and SLR structures.

Dual-band filters designed by this approach can be divided into two groups. The first group comprises the filters in which the resonators of the same type but with different resonant frequencies are employed. The second group comprises filters that are realized with the resonators of different type.

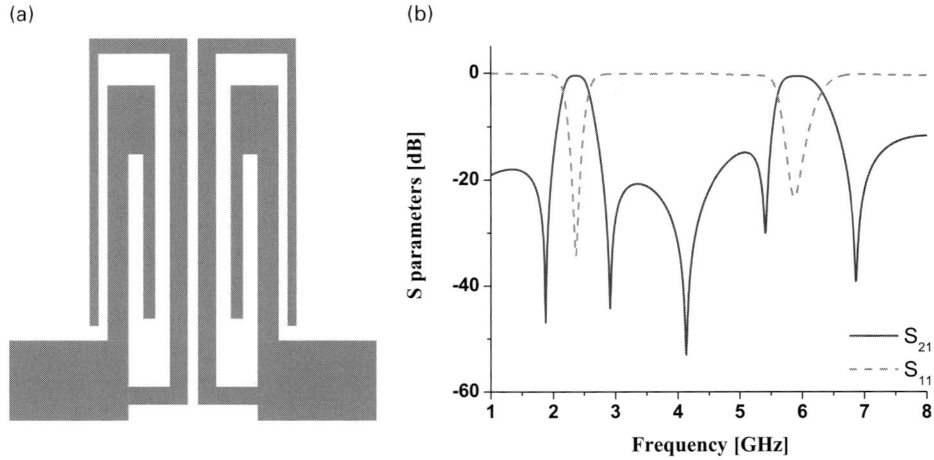

Figure 2.37 (a) Configuration of the filter proposed in [98]. (b) Filter response.

One of the simplest ways to realize a dual-band filter is to employ a pair of conventional $\lambda/2$ resonators as was done in [99] and [100]. The filter in [99] is comprised of two pairs of open-loop ring resonators, which are placed between two microstrip lines, whilst the filter in [100] has a different feeding scheme, one pair of resonators is inductively fed and the other is capacitively fed. Although the proposed filters are easy to realize and have good performance, they are not compact since they employ open-loop $\lambda/2$ resonators. Nevertheless, in [101] a compact, high-performance dual-band filter with coupled folded $\lambda/2$ resonators was proposed, Figure 2.37.

Using $\lambda/4$ resonators which are more compact but also more demanding in terms of fabrication, dual-band filters with good performance and significantly reduced size were designed [102–107].

More demanding filters in terms of design are those that employ dual-mode resonators with different resonant frequencies. In the previous section, it was shown that dual-band filters can be realized using only one dual-mode resonator. Thus, we can pose the question, why we need two dual-mode resonators with different resonant frequencies to design a dual-band filter. In that manner, better filter performance is achieved since there are two transmission poles in each passband. At the same time, the positions of the passbands can be almost arbitrarily chosen since each resonator forms one passband.

In [108–111], filters that employ two different dual-mode resonators with perturbation were proposed. For instance, in the structure presented in [110] two microstrip loop resonators, one within the other, are employed to obtain dual-mode dual-band filter responses. Each loop resonator uses two loading elements to excite the dual modes.

In all filters, the passbands are formed in such a way that each resonator provides one passband with two transmission poles. Although such structures provide a very good performance, they inevitably have large size because they are comprised of two resonators with relatively large dimensions.

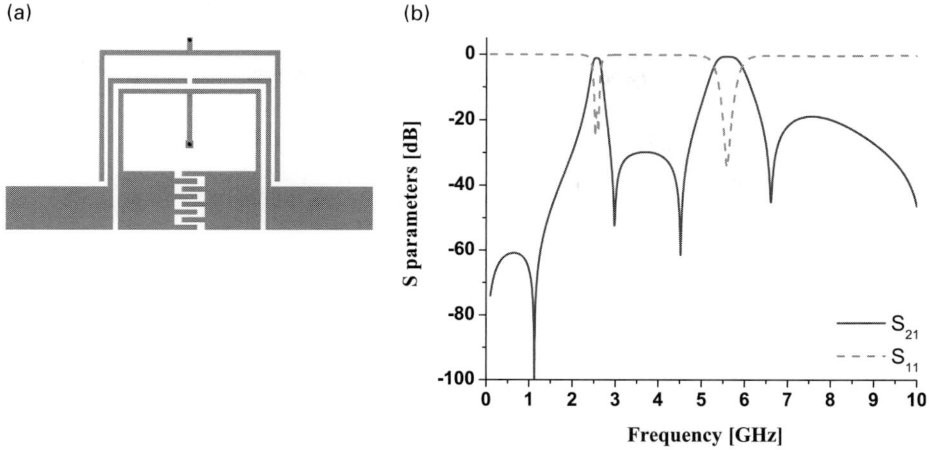

Figure 2.38 (a) Configuration of the filter proposed in [115]. (b) Filter response.

The configurations presented in [112–114] are based on SIR and SLR structures. In these filters, two different resonators have very close yet different resonant frequencies, so the first mode of the resonators are used to form the first passband, whilst the second modes are used to form the second passband. In that manner, both passbands have two transmission poles. Although the proposed filters are based on SIR and SLR structures that show great potential for miniaturization, the filters have good performance, but they are not compact.

In [115] and [116], filters are designed in a similar manner to that in [99] with the difference that asymmetrical SIR resonators are used instead of $\lambda/2$ resonators. SIRs have structure such that their modes are significantly separated and the first mode of one SIR forms the first passband and the first mode of the other SIR forms the second passband. In that manner, good filter performance is achieved whilst the circuit size is somewhat reduced in comparison to those in [99] and [100].

Very compact filters with good performance that employ SI-SLR and SLR resonators were presented in [117] and [118], respectively. In these cases, the modes of each resonator are closely positioned so that each resonator forms one passband with two transmission poles. The configuration and response of the structure presented in [118] are shown in Figure 2.38. A compact dual-band filter for 2.4/5.2 GHz WLAN application with good frequency selectivity and independently controllable bandwidth for each passband is proposed using an inter-digital capacitor loop stepped-impedance resonator and a short-circuited stub-loaded resonator.

Despite having excellent performance, both structures are designed to operate at 2.4/5.2 GHz, which significantly eases the design and reduces circuit size.

As was stated previously, the second group of dual-band filters comprises filters realized with different types of resonators. Such filters were proposed in [119–124] and they are characterized by good performance and relatively compact size.

A miniature filter with good characteristics was proposed in [121] and it is based on grounded and non-grounded stepped-impedance resonators, Figure 2.39.

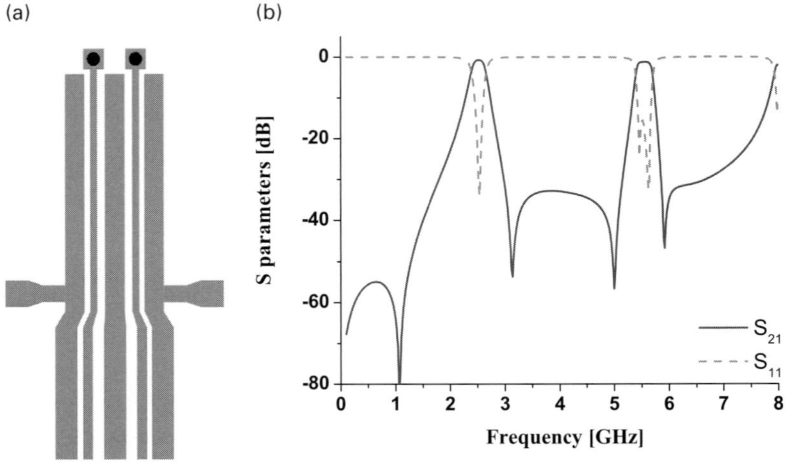

Figure 2.39 (a) Configuration of the filter proposed in [118]. (b) Filter response.

The impedance perturbation in each SIR is very small so the modes are very distant. Therefore, each SIR provides one passband – a grounded SIR at 2.4 GHz and a non-grounded one at 5.2 GHz. Similar to the previous example, it should be noted that the filter operates at 2.4 and 5.2 GHz, which eases the design and significantly reduces the filter dimensions.

Dual-path dual-band filters as well as *embedded* configuration can be treated as specific types of dual-band filters based on the fourth approach.

In the dual-path configuration, the signal branches into two paths which have different electrical lengths and in that way two passbands are formed at the output. Also, a transmission zero can occur if the two paths have such characteristics that they can provide destructive interference at the output. Dual-band filters based on dual-path configuration were proposed in [125–128]. Even though this method is very simple, it cannot meet all the requirements of high-performance dual-band filters – compactness, small insertion loss, and good passband selectivity.

Filters with *embedded* resonators are structures in which two different resonators "share" a part of a signal path. Configurations with embedded resonators were proposed in [129–133], and are characterized by good performance and small dimensions.

The configuration and response of the filter presented in [130] are shown in Figure 2.40. The proposed filters use embedded spiral resonators. Each resonator consists of two sets of spiral lines embedded in a microstrip rectangular open-loop (MROP). The resonator has two functions in the filter – dual-resonance operation mainly provided by the spiral lines is employed to generate two passbands simultaneously, and MROP acting as the dual-feeding structure is introduced to achieve good input/output (I/O) coupling for both passbands. With a proper tapping scheme, additional transmission zeros are obtained to improve passband selectivity. Excellent performance as well as very compact dimensions, $0.1\lambda_g$ x $0.1\lambda_g$, are proof of the great potential of these structures.

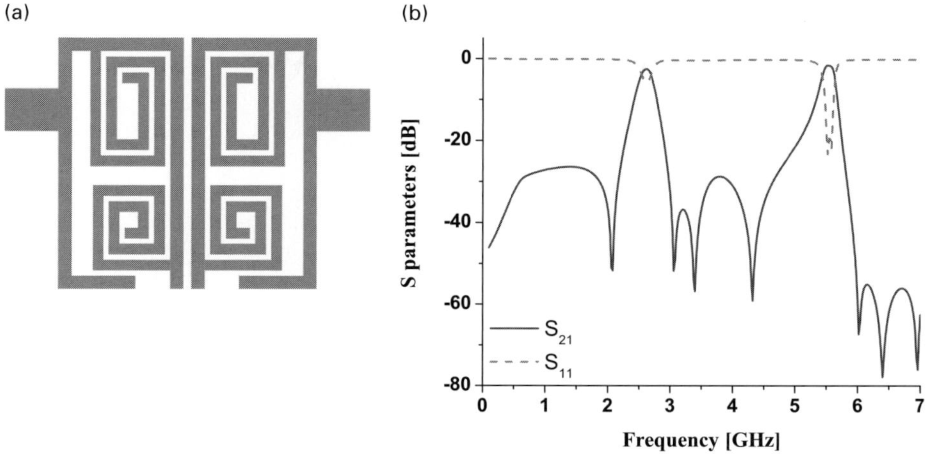

Figure 2.40 (a) Configuration of the filter proposed in [127]. (b) Filter response.

A disadvantage of these filters is the fact that the resonators "share" a part of the signal path, i.e. they are not as independent as the resonators in other filters based on this approach. Also, one should note that none of the filters is designed to operate at relatively close frequencies such as 2.4 and 3.5 GHz, but rather at 2.4 and 5.2 GHz, which eases the design procedure and decreases circuit size.

From the previous analysis, it can be concluded that numerous dual-band filters have been designed using the fourth approach, i.e. by combining two single-band filters. Application of different types of resonators as well as the fact that the two single-band filters are almost independent implies that this approach offers considerable design freedom. Against expectations, some of the most compact filters that exhibit excellent performance have been realized using this approach.

As for tri-band filters realized using this method, they are based on combining dual-band and single-band filters. Since there is a number of design methods for single-band and dual-band filters, this approach offers greater freedom in filter design.

Tri-band filters realized using this approach were presented in [134–139]. With the exception of the tri-band filter, which consists of three single-band filters [136], all configurations are comprised of one single-band and one dual-band filter. The dual-band filters have been realized using dual-mode SIR and SLR structures, whilst the single-band filters have been realized using conventional $\lambda/2$ and $\lambda/4$ resonators. These tri-band configurations are characterized by good performance and relatively compact dimensions.

The structure of the filter proposed in [138] is shown in Figure 2.41, together with the response of the filter. The configuration is comprised of a grounded SLR and $\lambda/2$ resonator and it provides not only three passbands but also good selectivity. The proposed structure is a very compact tri-band filter that also exhibits very good characteristics, i.e. low insertion loss and good performance in the stopband regions.

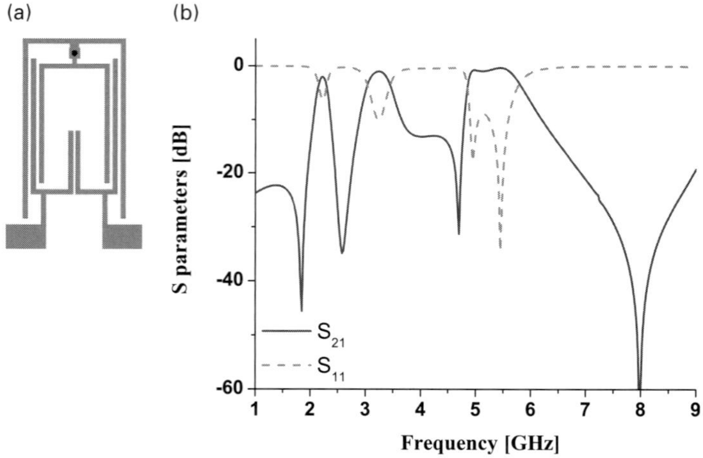

Figure 2.41 (a) Configuration of the filter proposed in [135]. (b) Filter response.

This approach has also been shown to be very applicable for the design of a filter with more than three passbands. Namely, the majority of quad-band filters introduced so far have been realized by applying this method [140–149]. Although the realization of four passbands gives greater design freedom in terms of combining single-band, dual-band, and even tri-band filters, all proposed quad-band filters except for [140] employ two dual-band filters.

The filter in [140] employs a stepped-impedance resonator and a conventional half-wavelength resonator. The first passband is formed only by the fundamental mode and the half-wavelength resonator, whilst the other three passbands have two transmission poles, one originating from the SIR, and the other originating from the half-wavelength resonator. The filter is characterized by good performance, but relatively large dimensions.

The filters proposed in [141–144] consist of two dual-band filters of the same type. Whilst the filter in [141] employs two dual-mode ring-shaped resonators each providing two passbands, the filters in [142–144] relay on dual-band filters based on stepped-impedance resonators. In [143], the filter consists of two pairs of meandered stepped-impedance resonators that provide four passbands at 1.5, 2.5, 3.5, and 5.5 GHz with excellent in-band performance. Owing to the configuration, the filter also exhibits excellent selectivity and very compact dimensions of only $0.2\lambda_g$ x $0.18\lambda_g$.

On the other hand, the filters in [145–149] employ dual-band filters with different types of resonators, most of which are stepped-impedance and stub-loaded resonators, as well as their variations. The quad-band filter proposed in [149] is based on a grounded conventional half-wavelength resonator that provides two closely positioned passbands at 2.75 and 3.3 GHz, and a grounded stepped-impedance resonator that provides another two passbands at 1.525 and 4.5 GHz. The filter has compact size and satisfactory out-of-band rejection between each band owing to transmission zeros introduced by source-load coupling and a specific coupling scheme.

Another proof of the applicability of this method is the fact that one of the very few quint-band filters proposed so far is based on this method [150]. The proposed configuration comprises five tri-mode stub-loaded stepped-impedance resonators each of which provides a passband with three transmission poles. Also, specific configuration of input lines and resonators provides a transmission zero at the upper side of each passband. The filter operates at 0.6, 0.9, 1.2, 1.5, and 1.8 GHz and exhibits good performance and overall dimensions of $0.52\lambda_g$ x $0.045\lambda_g$.

2.6 Other advanced solutions of multi-band filters

In the four previous sections, microstrip multi-band filters realized as planar structures in conventional printed-circuit board technology have been analyzed. From the fabrication complexity point of view, PCB technology represents the least demanding and the most cost effective procedure for the realization of filter circuits.

However, there have also been proposed multi-band filters that have been realized in a different manner – using multi-layer architecture or fabrication technologies other than PCB or applying advanced theoretical concepts in electromagnetic wave propagation. In the following, an overview of multi-band filters designed using unconventional methods will be given.

One of the critical aspects of filter circuit is its size and in the previous sections, some of the methods for circuit size reduction were discussed. Another method to make a configuration more compact is to realize it as a multi-layer structure. Multi-layer structure allows the elements of filter circuit to be stacked vertically in several layers and in that manner to reduce the overall dimensions.

The simplest multi-layer structures are realized in printed circuit board technology as defected ground structures (DGS). Apart from transmission lines and resonators that are realized in a standard conductive layer, there are also structures realized in the ground layer.

DGS structures can be divided into two groups. The first group comprises DGS structures in which resonators are realized in both the conductive and ground layer. On the other hand, the second group includes DGS filters in which resonators have been realized only in the ground layer.

The first type of filter aims to reduce circuit size as well as to increase the coupling between the resonators since resonators realized in two layers are coupled over the whole area and not only through the edges. Dual-band filters with such DGS configurations were proposed in [151–154]. However, none of the structures simultaneously exhibits good performance and compactness.

Defected ground structures, in which resonators are positioned in the conductive and ground layer, have also been used for the realization of several tri-band filters [155–157] and a quad-band filter [158]. The structure proposed in [155] consists of two pairs of stepped-impedance resonators in the conductive layer and one pair of stepped-impedance resonators realized in the ground layer. The filter is characterized by good performance and small overall dimensions of $0.26\lambda_g$ x $0.2\lambda_g$. The quad-band filter employs half-wavelength and SIR resonators to obtain four passbands, and it is characterized by good performance, selectivity, and compact size.

However, the DGS structure is not a guarantee of high-performance, compact filter, and this is confirmed by the structures proposed in [156] and [157], which do not exhibit good performance or compactness.

The second type of DGS structures aims to increase the coupling between the feeding lines themselves, as well as between feeding lines and resonators. In that manner, insertion losses can be decreased, but also transmission zeros can occur which improves filter selectivity. This approach has been applied in the filters published in [159–162], but similar to the structures from the first group, these filters usually do not simultaneously exhibit good performance and compactness.

Somewhat more complex structures are those designed in two separate conductive layers that are both positioned above the ground layer. Such structures employ standard dielectric layers, and examples of dual-band filters based on this idea were published in [163] and [164]. In both structures, two dual-mode resonators with perturbation are realized in separate layers and each resonator provides one passband. Even though these filters have good responses, they are not characterized by compactness, which is due to dual-mode resonators with perturbation that have relatively large size.

The same architecture has also been exploited for the design of quad-band [165–166] and quint-band filters [167]. The quad-band filter proposed in [166] consists of grounded and non-grounded stepped-impedance resonators and they provide passbands at 1.5, 2.5, 3.5, and 5.5 GHz. The filter is characterized by excellent in-band performance and selectivity, as well as by compact dimensions of only $0.14\lambda_g$ x $0.08\lambda_g$. The quint-band filter was realized using a very similar configuration and it is also characterized by excellent performance and compactness.

The most complex multi-layer filters consist of three or more layers fabricated using more demanding and expensive technology than PCB. At the same time, such technologies allow more design freedom. One of the most attractive multi-layer technologies is low temperature co-fired ceramics (LTCC).

Employing this technology, the most compact dual-band filters with linear dimensions less than $0.08\lambda_g$ have been realized [168–171]. By utilization of LTCC technology, the tri-band filters presented in [172] and [173] were designed. Similarly to the dual-band filters realized in LTCC technology, these tri-band filters are the most compact filters which, at the same time, have good selectivity and low insertion loss.

Multi-layer structures have great potential for miniaturization, but also a potential for improvement of certain filter characteristics. Nevertheless, the major drawback of multi-layer structures in comparison to planar ones is the necessity for more complex PCB fabrication or expensive technologies such as LTCC.

In the literature, one can find filters that employ conductive materials other than copper. Superconductors are materials that exhibit zero resistance at temperatures lower than a certain value. The temperature at which a material abruptly changes its properties is called the critical temperature. First superconductors were discovered in the year of 1911. However, it was only in the eighties that superconductors with a critical temperature higher than 77 K (boiling point of nitrogen) were found. The materials whose critical temperature is higher than 77 K are called high-temperature superconductors (HTS) and over recent decades they have found wide application in microwave devices.

The most often used HTS material in microwave applications is yttrium barium copper oxide (YBCO) and thallium barium calcium copper oxide (TBCCO). The main advantage of these materials is the high values of their conductivity, i.e. small conductor attenuation, which makes superconductors good candidates for applications in which small losses are of critical importance. However, despite being commercially available HTS materials are far more expensive than conventional materials. In addition, the realization of filters with superconductors is more demanding in comparison to conventional fabrication procedures.

Dual-band and tri-band filters with HTS materials were proposed in [174–176] and [177], respectively, and their responses are characterized by very small insertion losses in the passbands. Since other characteristics of HTS filters do not depend on HTS materials, they are not minature whilst their responses are very good.

Another unconventional method for multi-band filter design is based on using lumped elements. They are used together with transmission lines and other distributed elements, and their employment aims to obtain resonators with inductances and capacitances that cannot be achieved if only transmission lines are used. They are also used to improve couplings in a circuit as well as to reduce circuit size. Examples of dual-band filters with lumped elements were proposed in [178] and [179]. Additionally, a quad-band filter with lumped elements was presented in [180].

The last group of unconventional filters comprises configurations based on composite right/left-handed (CRLH) transmission lines, i.e. structures based on metamaterials.

Metamaterials are artificially made materials that exhibit extreme values of effective permittivity and permeability, which makes them very attractive for application in miniaturization of microwave circuits. Especially interesting are double-negative metamaterials that simultaneously exhibit negative permittivity and permeability over frequency range. Due to antiparallel phase and group velocities, i.e. due to the "left-handed" system of vectors E, H, and k, double-negative metamaterials are also called left-handed (LH) metamaterials.

Transmission lines that are based on the LH concept represent LH transmission lines. A pure LH transmission line does not exist, but only a composite structure – the CRLH transmission line.

Utilization of CRLH transmission lines enabled the design of dual-band filters that are very compact in comparison to conventional ones [181–184]. On the other hand, circuits based on metamaterials are also characterized by higher insertion losses in passbands, which is the major drawback of these structures.

Tri-band filters based on metamaterials have been proposed in [185] and [186]. Both configurations are very compact, but the major limitation in the application of these filters is significant insertion loss in the passbands. Finally, the CRLH concept was also used in the design of a quad-band filter [187], which does not exhibit large insertion loss in the passbands.

2.7 Multi-band bandstop filters

Unlike multi-band bandpass filters, multi-band bandstop filters have not been as widely explored. Whilst there are four principal and several advanced design methods for multi-band bandpass filters, multi-band bandstop filters have been designed using only some of those approaches.

It should be also noted that the majority of multi-band bandstop filters are designed to operate at two different frequencies, whilst only a few bandstop filters operating at three or more bands have been proposed so far.

The first multi-band bandstop filter was proposed in [188] in which the authors presented an analytically designed dual-band bandstop filter in cul-de-sac topology with directly coupled resonators which make the proposed technique similar to those used in bandpass filter design.

Generally speaking, the most widely applied approach for the design of dual-band bandstop filters has been the employment of multi-mode resonators. Dual-band filters based on stepped-impedance resonators were proposed in [189–192].

The filter presented in [189] consists of three parallel-coupled stepped-impedance microstrip structures. Two structures comprise stepped-impedance coupled open-end stubs, whilst the third one represents a single stepped-impedance microstrip line. Three signal paths and coupling between the structures provide two wide stopbands with good performance. However, the filter has a large overall size of $0.38\lambda_g$ x $0.32\lambda_g$.

In [190–191], filters are based on two-section stepped-impedance open stubs. The structure in [190] consists of a $\lambda/2$ microstrip stepped-impedance open-stub unit and two series segments that are connected to the feeding lines. By proper choice of the electrical length and impedance of stepped-impedance and serial segments two stopbands are achieved, and the filter has good performance but relatively large size. On the other hand, the structure in [191] comprises two open stubs and one serial segment, yet it has very similar properties to the one presented in [190].

In [192], more design freedom was achieved through the employment of a tri-section stepped-impedance open stub. A configuration with two open stubs and one serial segment provided two passbands at 2.4 and 5.2 GHz, and it is characterized by good performance yet large overall size.

Another type of multi-mode resonator that has been employed in dual-band bandstop filter design is the stub-loaded resonator [193–195]. The filter in [193] consists of the main transmission line and three stub-loaded resonators coupled to the main line. In other words, this is a third-order filter, consisting of three coupled-line sections and two admittance inverters, which provide two stopbands and have good characteristics in terms of stopband insertion loss. However, it has very large dimensions.

Similar structures were proposed in [194], one of which comprises a pair of non-grounded stub-loaded resonators, whilst the other comprises a pair of grounded stub-loaded resonators. The resonators are coupled to the main transmission line that is divided into two different paths. Also, they are folded into loops to enhance the loading effect. Because of the symmetrical feeding scheme, the odd-mode frequencies are almost suppressed and the harmonic resonances are utilized as the second stopband. Besides good return and insertion loss, these filters also exhibit very compact size.

In [195], a dual-band stopband filter that comprises a common microstrip uniform impedance resonator and two open-circuited stubs was proposed. The open-circuited stubs are realized as stub-loaded resonators, and in that manner two stopbands with relatively good performance were provided. Although the stubs were folded in order to reduce the filter size, its overall dimensions are still large – $0.26\lambda_g$ x $0.18\lambda_g$.

Beside methods based on multi-mode resonators, the technique based on a combination of two single-band filters has also been utilized for dual-band bandstop filter design [196–197]. The dual-band filter presented in [196] consists of two single-band filters of the fourth order. The single-band filters employ conventional open-loop $\lambda/2$ resonators and they are coupled to the main transmission line. Although filters of the fourth order provide good performance in terms of return and insertion loss, they also have exceedingly large dimensions, especially because they are realized using $\lambda/2$ resonators.

A slightly more compact dual-band filter with good rejection characteristics was presented in [197]. It is comprised of two microstrip square loop dual-mode resonators having different sizes that are located at different sides of the folded feeding line. Therefore, each resonator provides one stopband in the filter response and, at the same time, the resonators do not influence each other's performance.

A non-resonating concept for the design of a dual-band bandstop filter was proposed in [198]. The filter comprises two transmission lines that are connected in parallel and the input signal is divided into two components. Owing to the phase difference between the two transmission lines, the signals combine with each other at the output port and when out-of-phase they provide transmission zeros. Although this idea is very smart and easily applicable, its main drawbacks are that the range of the two stopband frequencies is limited as well as the fact that the overall dimensions of the filter are very large.

As for the advanced theoretical concepts, DGS and metamaterial concepts were used in [199–201] and [181], respectively, to realize dual-band bandstop responses.

A DGS bandstop filter with all resonators realized in the ground layer was presented in [199]. The proposed structure comprises two cascaded DGS SIR structures resonating at different frequencies to obtain stopbands at 2.4 and 6.7 GHz. Both stopbands are very narrow yet below 25 dB, and the overall size of the circuit is compact.

DGS dual-band structures with resonators realized in both the conductive and ground layer were proposed in [200–201]. The filter in [200] employs a pair of open-loop $\lambda/2$ resonators in the conductive layer and a pair of complementary split ring resonators in the ground layer. The filter in [201] is comprised of a T-shaped defected microstrip structure and a U-shaped defected ground structure. Both filters have very compact dimensions. However, only the latter one simultaneously exhibits good performance in terms of insertion and return loss.

The metamaterial transmission line concept used to design the dual-band bandpass response in [181] was also applied to realize the dual-band bandstop response. The conventional bandpass and bandstop filter were first designed using quarter-wave short-circuited and open-circuited stubs, respectively, and then the dual-band filter with arbitrary dual operating bands was easily implemented by replacing the microstrip lines with the dual-band CRLH TLs. Due to the non-linear phase response of the CRLH TLs, the dual-band filters with two arbitrary passbands/stopbands can be easily obtained. Although very compact, the bandstop filter does not exhibit very good performance.

As for bandstop filters with three or more stopbands, there have only been a few publications on tri-band [202–204] and quad-band structures [205–206].

The first tri-band filter was the circuit presented in [202]. It consists of a spur line and a section of quarter-wavelength coupling structure, and it operates at 6.4, 14.5, and 21.5 GHz. The spur line structure behaves as a shunt series open circuited resonator inducing transmission zeros. However, the filter has large overall dimensions and extremely poor stopband characteristics.

In [203], a filter designed using inter-digital hairpin resonators with three pairs of folded fingers was presented. The bands can be independently controlled by modifying the parameters of each corresponding pair of folded fingers. The filter operates at 6.4, 10.7, and 15 GHz, and although it has relatively compact size performance is very poor.

The tri-band filter in [204] consists of a ladder-shaped defected-microstrip structure, i.e. a ladder-shaped perturbation made in the main signal line and it operates at 1.55, 4.59, and 7.59 GHz. Although this structure is more compact and has better stopband characteristics in comparison to the previous tri-band structures, there is still plenty of room for improvement in the performance of tri-band bandstop filters.

As for quad-band structures, in [205] a multi-band bandstop response was proposed that was achieved using the analytical approach, i.e. using a transversal coupling network comprising resonators that have the same or similar resonant frequencies. A theoretical approach was also used to design a quad-band bandstop filter in a cul-de-sac configuration in [206]. The proposed procedure comprises evaluation of the poles and zeros of the filtering function using frequency transformation and afterwards the generation of a coupling matrix for the filtering function. The procedure is advantageous to some other theoretical approaches in terms of complexity of determination of the transfer function.

2.8 Conclusion

Design methods for multi-band bandpass filters can be divided into four groups each of which can be divided into several sub-groups.

The first approach relies on classical filter design theory. Its main advantage is its applicability to different types of filters as well as the fact that it comprises analytical and semi-analytical procedures, which provides efficient and relatively simple filter design.

On the other hand, this method has several significant drawbacks. Passbands cannot be independently formed and controlled since they are formed from one wideband response by introducing transmission zeros, or by “mapping” one single-band response to several bands. Also, due to the employment of simple resonating structures such as

$\lambda/2$ resonators, this approach leads to filter circuits with large dimensions. In addition, it should be noted that filter responses have poor performance in stopband regions.

The second approach relies on introducing transmission zeros into the transfer function of a single-band filter and it represents the simplest method for multi-band filter design. The main advantage of this approach lies in its simplicity. However, all other aspects do not make this approach very favorable in comparison to others. Since the filters realized using this method are comprised of bandstop and bandpass filters, they inevitably have large dimensions. In addition, passbands are formed from one wideband passband which implies that the performance and positions of passbands cannot be independently controlled.

The third approach is based on multi-mode resonators, and it offers significantly more freedom in multi-band bandpass filter design.

Dual-mode resonators with perturbation allow the realization of filters with relatively good performance of passbands. However, limited control of passbands as well as relatively large dimensions that cannot be reduced below $0.25\lambda_g$ x $0.25\lambda_g$ represent their drawbacks.

SIR structures allow the design of filters with good performance and compact circuits. Nevertheless, due to the behavior of SIRs, filters based on SIRs are limited in terms of positioning of the passbands.

SLR structures enable the design of filters with good performance and small dimensions. Although they behave similarly to SIRs, filters with SLRs are not as limited in terms of the independent control of passbands. In the case of SLRs, one mode can be controlled independently from the other by a change in the stub length which is not limited by fabrication technology or by losses. A disadvantage of filters based on SLRs is the fact that two passbands stem from the resonant modes of the same structure, and thus it is not possible to independently control passband performance to the full extent.

Superposition of stepped-impedance and stub-loaded resonators gives a structure that exhibits multi-mode behavior and which has more parameters than can be used for control of the resonant modes. Although SI-SLR unites the advantages of SIR and SLR structures, which implies the possibility of design of high-performance, compact multi-band filters, configurations with SI-SLR published so far exhibit either poor performance in terms of insertion loss and selectivity, or large circuit size.

The fourth approach is based on two single-band filters to obtain dual-band response, and single-band and dual-band filters to obtain response with three or more bands. Resonators used in the design of multi-band filters include conventional $\lambda/2$ and $\lambda/4$ resonators as well as more complex structures such as dual-mode resonators with perturbation, SIR, and SLR structures. Application of different types of resonators as well as the fact that the comprising filters are almost independent implies that this approach offers greater design freedom. Against expectations, some of the most compact multi-band filters that simultaneously exhibit excellent performance have been realized using this approach.

Apart from bandpass filters based on these four approaches, there have also been proposed multi-band bandpass filters that have been realized in a different manner – using multi-layer architecture or fabrication technologies other than PCB, or applying advanced theoretical concepts in electromagnetic wave propagation. Although some of these methods allow design of filters with better characteristics in comparison to the four conventional methods, their main drawback is a more complex and more expensive fabrication procedure.

Although multi-band bandstop filters have not attracted as much attention as multi-band bandpass filters, they have also been explored in recent years. It was shown that the majority of multi-band bandstop filters were designed to operate at two different frequencies, whilst only few bandstop filters operating at three or more bands have been proposed so far.

Also, most dual-band bandstop filters have been designed using multi-mode stepped-impedance and stub-loaded resonators, whilst several of them have been designed using the technique based on a combination of two single-band filters and advanced techniques such as DGS and metamaterial concepts.

As for bandstop filters with three or more bands, only a few structures have been proposed so far and thus there is still a plenty of room for improvement in their performance.

References

[1] H. Miyake, S. Kitazawa, T. Ishizaki, T. Yamada, and Y. Nagatomi, "A miniaturized monolithic dual band filter using ceramic lamination technique for dual mode portable telephones," *Proceedings of IEEE MTT-S International Microwave Symposium Digest*, 1997, pp. 789–792.

[2] R. Cameron, C. Kudsia, and R. Mansour, *Microwave Filters for Communication Systems: Fundamentals, Design, and Applications*, John Wiley & Sons, New Jersey, 2007.

[3] G. Macchiarella, "Accurate synthesis of inline prototype filters using cascaded triplet and quadruplet sections," *IEEE Transactions on Microwave Theory and Techniques*, vol. 50, no. 7, pp. 1779–1783, July 2002.

[4] U. Rosenberg and S. Amari, "Novel coupling schemes for microwave resonator filters," *IEEE Transactions on Microwave Theory and Techniques*, vol. 50, no. 12, pp. 2896–2902, Dec. 2002.

[5] U. Rosenberg and S. Amari, "Novel coupling schemes for microwave resonator filters," *IEEE Transactions on Microwave Theory and Techniques*, vol. 50, no. 12, pp. 2896–2902, Dec. 2002.

[6] R. J. Cameron, "Advanced coupling matrix synthesis techniques for microwave filters," *IEEE Transactions on Microwave Theory and Techniques*, vol. 51, no. 1, pp. 1–10, Jan. 2003.

[7] G. Macchiarella and S. Tamiazzo, "A design technique for symmetric dualband filters," *Proceedings of IEEE MTT-S International Microwave Symposium Digest*, 2005, pp. 12–17.

[8] A. Garcia Lamperez, "Analytical syntesis algorithm of dual-band filters with asymmetric passbands and generalized topology," *Proceedings of IEEE/MTT-S International Microwave Symposium*, 2007, pp. 909–912.

[9] P. Lenoir, S. Bila, F. Seyfert, D. Baillargeat, and S. Verdeyme, "Synthesis and design of asymmetrical dual-band bandpass filters based on equivalent network simplification," *IEEE Transactions on Microwave Theory and Techniques*, vol. 54, no. 7, pp. 3090–3097, July 2006.

[10] M. Mokhtaari, J. Bornemann, K. Rambabu, and S. Amari, "Coupling-matrix design of dual and triple passband filters," *IEEE Transactions on Microwave Theory and Techniques*, vol. 54, no. 11, pp. 3940–3946, Nov. 2006.

[11] Z. Yunchi, K. A. Zaki, J. A. Ruiz-Cruz, and A. E. Atia, "Analytical synthesis of generalized multi-band microwave filters," *Proceedings of IEEE/MTT-S International Microwave Symposium*, 2007, pp. 1273–1276.

[12] E. Corrales, O. Menéndez, P. de Paco, M. Ramirez, and J. Verdú, "Microstrip dual-band bandpass filter based on the cul-de-sac topology," *Proceedings of 40th European Microwave Conference*, 2010, pp. 549–552.

[13] Z. Chunxia, G. Yongxin, and Y. Shaolin, "A compact dual-band filter based on dual-mode resonators," *Proceedings of Asia-Pacific Microwave Conference*, 2011, pp. 1278–1281.

[14] Y. Kuo, J. Li, S. Cheng, and C. Chang, "Tri-band and quad-band filter design using E-shaped resonators," *Proceedings of Asia-Pacific Microwave Conference*, 2011, pp. 1270–1273.

[15] Y. Kuo, J. Lu, C. Liao, and C. Chang, "New multiband coupling matrix synthesis technique and its microstrip implementation," *IEEE Transactions on Microwave Theory and Techniques*, vol. 58, no. 7, pp. 1840–1850, July 2010.

[16] Y. Kuo and C. Chang, "Analytical design of two-mode dual-band filters using E-shaped resonators," *IEEE Transactions on Microwave Theory and Techniques*, vol. 60, no. 2, pp. 250–260, Feb. 2012.

[17] V. Lunot, S. Bila, and F. Seyfert, "Optimal synthesis for multi-band microwave filters," *Proceedings of IEEE/MTT-S International Microwave Symposium*, 2007, pp. 115–118.

[18] S. Bila, R. J. Cameron, P. Lenoir, V. Lunot, and F. Seyfert, "Chebyshev synthesis for multi-band microwave filters," *Proceedings of IEEE/MTT-S International Microwave Symposium Digest*, 2006, pp. 1221–1224.

[19] L. Tsai and C. Hsue, "Dual-band bandpass filters using equal-length coupled-serial-shunted lines and Z-transform technique," *IEEE Transactions on Microwave Theory and Techniques*, vol. 52, no. 4, pp. 1111–1117, April 2004.

[20] Y. Liu and W. Dou, "A dual-band filter realized by alternately connecting the main transmission-line with shunt stubs and shunt serial resonators," *IEEE Microwave and Wireless Components Letters*, vol. 19, no. 5, pp. 296–298, May 2009.

[21] C. Quendo, E. Rius, A. Manchec, Y. Clavet, B. Potelon, J. Favennec, and C. Person, "Planar tri-band filter based on dual behavior resonator (DBR)," *Proceedings of 35th European Microwave Conference*, 2005, pp. 1–4.

[22] W. Feng, M. Doan, and W. Che, "Compact tri-band bandpass filter based on short stubs and crossed open stubs," *Proceedings of International Conference on Advanced Technologies for Communications*, 2010, pp. 213–216.

[23] C. Law and K. Cheng, "A new, compact, low-loss, microstrip filter design for tri-band applications," *Proceedings of Asia-Pacific Microwave Conference*, 2008, pp. 1–5.

[24] Y. Liu, Y. Zhao, Y. Zhou, and Z. Niu, "Integrated dual-band BPF and single-band BSF for tri-band filter design," *Journal of Electromagnetic Waves and Applications*, vol. 25, no. 17, pp. 2420–2428, Nov. 2011.

[25] J. Hong and M. J. Lancaster, *Microstrip Filters for RF/Microwave Applications*, John Wiley & Sons, New Jersey, 2001.

[26] J. Hong and M. J. Lancaster, "Microstrip bandpass filter using degenerate modes of a novel meander loop resonator," *IEEE Microwave and Guided Wave Letters*, vol. 5, no. 11, pp. 371–372, Nov. 1995.

[27] M. Weng, S. Wu, S. Jhong, Y. Chang, and M. Lee, "A novel compact dual-mode filter using cross-slotted patch resonator for dual-band applications," *Proceedings of IEEE/MTT-S International Microwave Symposium*, 2007, pp. 921–924.

[28] Y. Sung, "Dual-mode dual-band filter with band notch structures," *IEEE Microwave and Wireless Components Letters*, vol. 20, no. 2, pp. 73–75, Feb. 2010.

[29] S. Fu, B. Wu, J. Chen, S. Sun, and C. Liang, "Novel second-order dual-mode dual-band filters using capacitance loaded square loop resonator," *IEEE Transactions on Microwave Theory and Techniques*, vol. 60, no. 3, pp. 477–483, March 2012.

[30] S. Luo and L. Zhu, "A novel dual-mode dual-band bandpass filter based on a single ring resonator," *IEEE Microwave and Wireless Components Letters*, vol. 19, no. 8, pp. 497–499, Aug. 2009.

[31] I. Wolff, "Microwave bandpass filter using degenerate mode of a microstrip ring," *IET Electronics Letters*, vol. 8, no. 12, pp. 302–303, June 1972.

[32] S. Luo, L. Zhu, and S. Sun, "A dual-band ring-resonator bandpass filter based on two pairs of degenerate modes," *IEEE Transactions on Microwave Theory and Techniques*, vol. 58, no. 12, pp. 3427–3423, Dec. 2010.

[33] S. Sun, "A dual-band bandpass filter using a single dual-mode ring resonator," *IEEE Microwave and Wireless Components Letters*, vol. 21, no. 6, pp. 298–300, June 2011.

[34] Y. Chiou, C. Wu, and J. Kuo, "New miniaturized dual-mode dual-band ring resonator bandpass filter with microwave C-sections," *IEEE Microwave and Wireless Components Letters*, vol. 20, no. 2, pp. 67–69, Feb. 2010.

[35] Y. Li, H. Wong, and Q. Xue, "Dual-mode dual-band bandpass filter based on a stub-loaded patch resonator," *IEEE Microwave and Wireless Components Letters*, vol. 21, no. 10, pp. 525–527, Oct. 2011.

[36] J. Xiao and X. Zu, "Dual-band bandpass filter using right-angled triangular resonator," *Proceedings of Cross Strait Quad-Regional Radio Science and Wireless Technology Conference (CSQRWC)*, 2011, pp. 690–695.

[37] R. Zhang, L. Zhu, and S. Luo, "Dual-mode dual-band bandpass filter using a single slotted circular patch resonator," *IEEE Microwave and Wireless Components Letters*, vol. 22, no. 5, pp. 233–235, May 2012.

[38] S. Luo, L. Zhu, and S. Sun, "Compact dual-mode triple-band bandpass filters using three pairs of degenerate modes in a ring resonator," *IEEE Transactions on Microwave Theory and Techniques*, vol. 59, no. 5, pp. 1222–1229, May 2011.

[39] S. Luo, L. Zhu, and S. Sun, "New studies on microstrip ring resonators for compact dual-mode dual- and triple-band bandpass filters," *Proceedings of IEEE Electrical Design of Advanced Packaging and Systems Symposium*, 2011, pp. 1–4.

[40] M. Makimoto and S. Yamashita, "Bandpass filters using parallel coupled stripline stepped impedance resonators," *IEEE Transactions on Microwave Theory and Techniques*, vol. 28, no. 12, pp. 1413–1417, Dec. 1980.

[41] S. Chang, Y. Jeng, and J. Chen, "Dual-band step-impedance bandpass filter for multimode wireless LANs," *IET Electronics Letters*, vol. 40, no. 1, pp. 38–39, Jan. 2004.

[42] S. Sun and L. Zhu, "Compact dual-band microstrip bandpass filter without external feed," *IEEE Microwave and Wireless Components Letters*, vol. 15, no. 10, pp. 644–646, Oct. 2005.

[43] S. Sun and L. Zhu, "Novel design of dual-band microstrip bandpass filters with good in-between isolation," *Proceedings of Asia-Pacific Microwave Conference*, 2005, pp. 4–7.

[44] M. Mokhtaari, J. Bornemann, and S. Amari, "New reduced-size step-impedance dual-band filters with enhanced bandwidth and stopband performance," *Proceedings of IEEE MTT-S International Microwave Symposium Digest*, 2006, pp. 1181–1184.

[45] C. Rao, T. Wong, and M. Ho, "A parallel doubly coupled dual-band bandpass filter," *Proceedings of IEEE MTT-S International Microwave Symposium Digest*, 2006, pp. 511–514.

[46] J. Kuo and H. Lin, "Dual-band bandpass filter with improved performance in extended upper rejection band," *IEEE Transactions on Microwave Theory and Techniques*, vol. 57, no. 4, pp. 824–829, April 2009.

[47] Lin-Chuan Tsai, "Design of dual-band bandpass filters using stepped-impedance resonators," *Proceedings of 2nd International Conference on Artificial Intelligence, Management Science and Electronic Commerce (AIMSEC)*, 2011, pp. 6602–6605.

[48] Y. Chang, C. Kao, M. Weng, and R. Yang, "Design of the compact dual-band bandpass filter with high isolation for GPS/WLAN applications," *IEEE Microwave and Wireless Components Letters*, vol. 19, no. 12, pp. 780–782, Dec. 2009.

[49] C. Ye, Y. Su, M. Weng, and R. Yang, "Design of a high selectivity dual-band bandpass filter with stepped impedance resonator," *Proceedings of 39th European Microwave Conference*, 2009, pp. 311–313.

[50] T. Lin, U. Lok, and J. Kuo, "New dual-mode dual-band bandpass filter with quasi-elliptic function passbands and controllable bandwidths," *Proceedings of IEEE MTT-S International Microwave Symposium Digest*, 2010, p. 1.

[51] J. Kuo and T. Lin, "Dual-mode dual-band ring resonator bandpass filter with transmission zeros," *Proceedings of Asia-Pacific Microwave Conference*, 2010, pp. 1865–1870.

[52] M. Weng, H. Wu, and Y. Su, "Compact and low loss dual-band bandpass filter using pseudo-interdigital stepped impedance resonators for WLANs," *IEEE Microwave and Wireless Components Letters*, vol. 17, no. 3, pp. 187–189, March 2007.

[53] Q. Chu and F. Chen, "A compact dual-band bandpass filter using meandering stepped impedance resonators," *IEEE Microwave and Wireless Components Letters*, vol. 18, no. 5, pp. 320–322, May 2008.

[54] D. Zayniyev and D. Budimir, "Compact microstrip dual-band pseudo-interdigital stepped impedance bandpass filters for wireless applications," *Proceedings of IEEE Antennas and Propagation Society International Symposium*, 2009, pp. 1–4.

[55] X. Zhang, J. Chen, J. Shi, and Q. Xue, "High-selectivity dual-band bandpass filter using asymmetric stepped-impedance resonators," *IET Electronics Letters*, vol. 45, no. 1, pp. 63–64, Jan. 2009.

[56] H. Yuan and Y. Fan, "Compact microstrip dual-band filter with stepped-impedance resonators," *IET Electronics Letters*, vol. 47, no. 24, p. 1328, Nov. 2011.

[57] M. Hayati, L. Noori, and A. Adinehvand, "Compact dual-band bandpass filter using open loop resonator for multimode WLANs," *IET Electronics Letters*, vol. 48, no. 10, pp. 573–574, May 2012.

[58] C. Tang and P. Wu, "Design of a planar dual-band bandpass filter," *IEEE Microwave and Wireless Components Letters*, vol. 21, no. 7, pp. 362–364, July 2011.

[59] X. Ming Lin and Q. Chu, "Design of triple-band bandpass filter using tri-section stepped-impedance resonators," *Proceedings of International Conference on Microwave and Millimeter Wave Technology*, 2007, pp. 1–3.

[60] C. Hsu, C. Lee, and Y. Hsieh, "Tri-band bandpass filter with sharp passband skirts designed using tri-section SIRs," *IEEE Microwave and Wireless Components Letters*, vol. 18, no. 1, pp. 19–21, Jan. 2008.

[61] X. Yang, L. Dai, and R. Zhou, "The tri-band filter design based on SIR," *Proceedings of International Conference on Audio, Language and Image Processing*, 2008, pp. 211–214.

[62] J. Hu, G. Li, H. Hu, and H. Zhang, "A new wideband triple-band filter using SIR," *Journal of Electromagnetic Waves and Applications*, vol. 25, no. 16, pp. 2287–2295, Oct. 2011.

[63] J. Hu, G. Li, H. Hu, and H. Zhang, "A new wideband triple-band filter using SIR," *Journal of Electromagnetic Waves and Applications*, vol. 25, no. 16, pp. 2287–2295, Oct. 2011.

[64] J. Xu, C. Miao, L. Cui, Y.-X. Ji, and W. Wu, "Compact high isolation quad-band bandpass filter using quad-mode resonator," *IET Electronics Letters*, vol. 48, no. 1, pp. 28–29, Jan. 2012.

[65] L. Zhu, H. Bu, and K. Wu, "Miniaturized multi-pole broadband microstrip bandpass filter: concept and verification," *Proceedings of 30th European Microwave Conference*, Paris, 2000, pp. 1–4.

[66] X. Zhang, J. Chen, Q. Xue, and S. Li, "Dual-band bandpass filters using stub-loaded resonators," *IEEE Microwave and Wireless Components Letters*, vol. 17, no. 8, pp. 583–585, Aug. 2007.

[67] P. Mondal and M. Mandal, "Design of dual-band bandpass filters using stub-loaded open-loop resonators," *IEEE Transactions on Microwave Theory and Techniques*, vol. 56, no. 1, pp. 150–155, Jan. 2008.

[68] B. Virdee, M. Farhat, and K. Ahmed, "Dual-band bandpass filter using open-loop resonators," *Proceedings of Asia-Pacific Microwave Conference*, 2011, pp. 1054–1057.

[69] E. Babu, M. Ramesh, and A. Kalghatgi, "Compact high isolation dual-band bandpass filter," *Proceedings of 41st European Microwave Conference*, 2011, pp. 748–750.

[70] W. He, Z. Ma, C. Chen, T. Anada, and Y. Kobayashi, "A novel dual-band bandpass filter using microstrip stub-loaded two-mode resonators with source and load coupling," *Proceedings of Asia-Pacific Microwave Conference*, 2008, pp. 1–4.

[71] L. Li, T. An, W. Wang, and W. Zhang, "A square ring dual-band filter with adjustable second-passband by the open stubs," *Proceedings of International Conference on Microwave and Millimeter Wave Technology*, 2008, pp. 291–293.

[72] X. Ma and H. Zheng, "Compact dual-band bandpass filter with adjustable passbands using open stubs," *Proceedings of 7th International Conference on Wireless Communications, Networking and Mobile Computing (WiCOM)*, 2011, pp. 23–25.

[73] X. Wu, Q. Chu, and L. Fan, "Compact dual-band bandpass filter with controllable bandwidths," *Proceedings of International Conference on Microwave and Millimeter Wave Technology (ICMMT)*, 2010, pp. 1305–1307.

[74] L. Wang, C. Zhao, C. Li, and X. Lin, "Dual-band bandpass filter using stub loaded resonators with multiple transmission zeros," *Proceedings of 9th International Symposium on Antennas Propagation and EM Theory (ISAPE)*, 2010, pp. 1208–1211.

[75] J. Lee and Y. Lim, "A dual-band bandpass filter using dual and triple-mode resonators," *Proceedings of IEEE Radio and Wireless Symposium (RWS)*, 2012, pp. 143–146.

[76] M. Doan, W. Che, and W. Feng, "Novel compact dual-band bandpass filter with multiple transmission zeros and good selectivity," *Proceedings of International Conference on Microwave and Millimeter Wave Technology (ICMMT)*, 2012, pp. 1–4.

[77] M. Zhou, X. Tang, and F. Xiao, "Compact dual band bandpass filter using novel E-type resonators with controllable bandwidths," *IEEE Microwave and Wireless Components Letters*, vol. 18, no. 12, pp. 779–781, Dec. 2008.

[78] M. Doan and W. Che, "Compact microstrip dual-band bandpass filter using short-stub loaded half-wavelength resonator," *Proceedings of International Conference on Advanced Technologies for Communications (ATC)*, 2011, pp. 203–206.

[79] F. Chen, Q. Chu, and Z. Tu, "Tri-band bandpass filter using stub loaded resonators," *IET Electronics Letters*, vol. 44, no. 12, pp. 747–749, June 2008.

[80] J. Chen, N. Wang, Y. He, and C. Liang, "Fourth-order tri-band bandpass filter using square ring loaded resonators," *IET Electronics Letters*, vol. 47, no. 15, pp. 858–859, July 2011.

[81] M. Doan, W. Che, and W. Feng, "Tri-band bandpass filter using square ring short stub loaded resonators," *IET Electronics Letters*, vol. 48, no. 2, pp. 106–107, Jan. 2012.

[82] J.-Y. Wu and W.-H. Tu, "Design of quad-band bandpass filter with multiple transmission zeros," *IET Electronics Letters*, vol. 47, no. 8, pp. 502–503, April 2011.

[83] X. Ma and H. Zheng, "A compact dual-band bandpass filter with adjustable second passband using S-shaped stepped impedance resonator embedded by open stubs," *Proceedings of Cross Strait Quad-Regional Radio Science and Wireless Technology Conference (CSQRWC)*, 2011, pp. 671–674.

[84] S. Gao, H. Hu, and Z. Xiao, "A novel compact dual-band bandpass filter using SIRs with open-stub line," *Proceedings of Microwave Conference China–Japan Joint*, 2008, pp. 464–466.

[85] M. del Castillo Velazquez-Ahumada, J. Martel, F. Medina, and F. Mesa, "Design of a dual band-pass filter using modified folded stepped-impedance resonators," *Proceedings of IEEE MTT-S International Microwave Symposium Digest*, 2008, pp. 857–860.

[86] J. Hong and W. Tang, "Dual-band filter based on non-degenerate dual-mode slow-wave open-loop resonator," *Proceedings of IEEE MTT-S International Microwave Symposium Digest*, 2009, pp. 861–864.

[87] F. Chen and Q. Chu, "Novel multistub loaded resonator and its application to high-order dual-band filters," *IEEE Transactions on Microwave Theory and Techniques*, vol. 58, no. 6, pp. 1551–1556, June 2010.

[88] H. Liu, Z. Zhang, S. Wang, L. Zhu, X. Guan, J. Lim, and D. Ahn, "Compact dual-band bandpass filter using defected microstrip structure for GPS and WLAN applications," *IET Electronics Letters*, vol. 46, no. 21, pp. 1444–1445, Oct. 2010.

[89] Q. Chu, Z. Li, F. Chen, and X. Wu, "Design of dual-band bandpass filter using asymmetrical stub-loaded resonators," *Proceedings of 41st European Microwave Conference (EuMC)*, 2011, pp. 767–769.

[90] W. Feng, W. Che, and Q. Xue, "Novel dual-band bandpass filter using multi-mode resonator," *Proceedings of 4th International High Speed Intelligent Communication Forum (HSIC)*, 2012, pp. 1–4.

[91] Z. Li and Q. Chu, "Compact dual-band bandpass filter using a novel dual-mode resonator," *Proceedings of International Conference on Microwave and Millimeter Wave Technology (ICMMT)*, 2012, pp. 1–4.

[92] W. Chen, M. Weng, and S. Chang, "A new tri-band bandpass filter based on stub-loaded step-impedance resonator," *IEEE Microwave and Wireless Components Letters*, vol. 22, no. 4, pp. 179–181, April 2012.

[93] Q. Yin, L. Wu, L. Zhou, and W. Yin, "A tri-band filter using tri-mode stub-loaded resonators (SLRs)," *Proceedings of IEEE Electrical Design of Advanced Packaging & Systems Symposium*, 2010, pp. 1–4.

[94] W. Liu and Y. Tung, "Tri-band filter with partly combined stepped-impedance resonators," *Proceedings of Asia-Pacific Microwave Conference*, 2011, pp. 975–978.

[95] A. Liu and F. Zheng, "A new compact tri-band bandpass filter using step impedance resonators with open stubs," *Journal of Electromagnetic Waves and Applications*, vol. 26, no. 1, pp. 130–139, Jan. 2012.

[96] J. Xiao and W. Zhu, "H-shaped SIR bandpass filter with dual and tri-band," *Microwave and Optical Technology Letters*, vol. 54, no. 7, pp. 1686–1688, July 2012.

[97] Z. Li, T. Su, and C. Zhao, "A novel compact triple-passband filter design based on stepped impedance stub-loaded resonators," *Microwave and Optical Technology Letters*, vol. 54, no. 9, pp. 2106–2108, Sept. 2012.

[98] F. Wei., Q.-L. Huang, W.-T. Li, and X.-W. Shi, "A compact quad-band band-pass filter using novel stub-loaded SIR structure," *Microwave and Optical Technology Letters*, vol. 56, no. 3, pp. 538–542, March 2014.

[99] C. Chen and C. Hsu, "A simple and effective method for microstrip dual-band filters design," *IEEE Microwave and Wireless Components Letters*, vol. 16, no. 5, pp. 246–248, May 2006.

[100] X. Zhang, J. Shi, J. Chen, and Q. Xue, "Dual-band bandpass filter design using a novel feed scheme," *IEEE Microwave and Wireless Components Letters*, vol. 19, no. 6, pp. 350–352, June 2009.

[101] Z. Zhang, Y. Jiao, X. Wang, and S. Cao, "Design of a compact dual-band bandpass filter using opposite hook-shaped resonator," *IEEE Microwave and Wireless Components Letters*, vol. 21, no. 7, pp. 359–361, July 2011.

[102] Q. Chu, L. Ye, and F. Chen, "Design of compact dual-band bandpass filter using λ/4 resonators," *Proceedings of Asia-Pacific Microwave Conference*, 2008, pp. 1–4.

[103] C. Tang, W. Liu, M. Chen, and Y. Lin, "Design of dual-band bandpass filters with short-end coupled lines," *Proceedings of European Microwave Conference*, 2009, pp. 1385–1388.

[104] H. Wang and Q. Chu, "A compact dual-band filter with adjustable transmission zeros," *Proceedings of European Microwave Conference*, 2009, pp. 117–120.

[105] M. Farhat and B. Virdee, "Dual-band bandpass filter using embedded short-circuited resonators in the open-loop c-shape resonator," *Proceedings of European Microwave Conference*, 2011, pp. 759–762.

[106] M. Doan and W. Che, "Dual-band bandpass filter using quarter-wavelength resonator with controllable bandwidths and good selectivity," *Proceedings of International Conference on Advanced Technologies for Communications*, 2011, pp. 199–202.

[107] G. Dai, Y. Guo, and M. Xia, "Dual-band bandpass filter using parallel short-ended feed scheme," *IEEE Microwave and Wireless Components Letters*, vol. 20, no. 6, pp. 325–327, June 2010.

[108] A. Gorur and C. Karpuz, "Compact dual-band bandpass filters using dual-mode resonators," *Proceedings of IEEE/MTT-S International Microwave Symposium*, 2007, pp. 905–908.

[109] C. Karpuz, A. Gorur, E. Gunturkun, and A. Gorur, "Asymmetric response dual-mode dual-band bandstop filters having simple and understandable topology," *Proceedings of Asia Pacific Microwave Conference*, 2009, pp. 925–928.

[110] C. Karpuz and A. Gorur, "Dual-mode dual-band microstrip filters," *Proceedings of European Microwave Conference*, 2009, pp. 105–108.

[111] A. Gorur and C. Karpuz, "A novel perturbation arrangement for dual-mode resonators and its dual-band bandpass filter applications," *Proceedings of European Microwave Conference*, 2011, pp. 468–471.

[112] X. Wu, Q. Chu, and F. Chen, "Dual-band bandpass filter with controllable bandwidth and good selectivity by using stub-loaded resonators," *Microwave and Optical Technology Letters*, vol. 54, no. 6, pp. 1525–1528, June 2012.

[113] Y. Cho, X. Wang, and S. Yun, "Design of dual-band interdigital bandpass filters using both series and shunt resonators," *IEEE Microwave and Wireless Components Letters*, vol. 22, no. 3, pp. 111–113, March 2012.

[114] J. Wang, L. Ge, K. Wang, and W. Wu, "Compact microstrip dual-mode dual-band bandpass filter with wide stopband," *IET Electronics Letters*, vol. 47, no. 4, pp. 263–265, Feb. 2011.

[115] J. Konpang, "A dual-band filter using coupled-feed with asymmetrical stepped-impedance resonator," *Proceedings of Second International Conference on Digital Information and Communication Technology and Its Applications*, 2012, pp. 333–337.

[116] J. Konpang, "A dual-band filter using stepped-impedance resonators with coupled-feed lines," *Proceedings of IEEE 13th Annual Wireless and Microwave Technology Conference*, 2012, pp. 1–4.

[117] X. Wei, Y. Shi, P. Wang, J. Liao, Z. Xu, and B. Yang, "Compact dual-band bandpass filter with improved stopband characteristics," *IET Electronics Letters*, vol. 48, no. 12, pp. 704–705, June 2012.

[118] J. Lee and Y. Lim, "Compact dual-band bandpass filter with good frequency selectivity," *IET Electronics Letters*, vol. 47, no. 25, pp. 1376–1377, Dec. 2011.

[119] P. Chen, H. Ho, and M. Ho, "Dual-band bandpass filter using the combined λ/2 and λ/4 SIRS for WLAN applications," *Proceedings of IEEE Region 10 Conference TENCON*, 2007, pp. 1–3.

[120] H. Deng, Y. Zhao, Y. Fu, X. Zhou, and Y. Liu, "Compact and high selectivity dual-mode dual-band microstrip BPF with QWR and SLR," *Microwave and Optical Technology Letters*, vol. 54, no. 12, pp. 2702–2705, Dec. 2012.

[121] P. Singh, S. Basu, and Y. Wang, "Miniature dual-band filter using quarter wavelength stepped impedance resonators," *IEEE Microwave and Wireless Components Letters*, vol. 18, no. 2, pp. 88–90, Feb. 2008.

[122] M. Zhou, X. Tang, and F. Xiao, "Compact dual band transversal bandpass filter with multiple transmission zeros and controllable bandwidths," *IEEE Microwave and Wireless Components Letters*, vol. 19, no. 6, pp. 347–349, June 2009.

[123] K. Kim, H. Pyo, J. An, and Y. Lim, "Dual-band filter using half wavelength resonators and dual-mode resonator," *Proceedings of 6th International Conference on Electrical Engineering/Electronics, Computer, Telecommunications and Information Technology*, 2009, pp. 928–931.

[124] C. Tseng and H. Shao, "A new dual-band microstrip bandpass filter using net-type resonators," *IEEE Microwave and Wireless Components Letters*, vol. 20, no. 4, pp. 196–198, April 2010.

[125] X. Zhang, C. Chan, Q. Xue, and B. Hu, "Dual-band bandpass filter with controllable bandwidths using two coupling paths," *IEEE Microwave and Wireless Components Letters*, vol. 20, no. 11, pp. 616–618, Nov. 2010.

[126] Y. Kuo, J. Lu, C. Liao, and C. Chang, "Dual-band filters with frequency-separated dual-path coupling scheme," *Proceedings of 40th European Microwave Conference*, 2010, pp. 553–336.

[127] Z. Khan, M. Mohd Salleh, and M. Abdul Latip, "Microstrip dual-band dual-path bandpass filter," *Proceedings of Asia-Pacific Microwave Conference*, 2011, pp. 916–918.

[128] X. Luo, S. Sun, and E. Li, "Miniaturized dual-band bandpass filter using λ/2 spiral-resonator and loaded open-stub," *Proceedings of IEEE MTT-S International Microwave Symposium Digest*, 2012, pp. 1–3.

[129] C. Hsu, C. Chen, and H. Chuang, "A miniaturized dual-band bandpass filter using embedded resonators," *IEEE Microwave and Wireless Components Letters*, vol. 21, no. 12, pp. 658–660, Dec. 2011.

[130] X. Luo, H. Qian, J. Ma, K. Ma, and K. Yeo, "Compact dual-band bandpass filters using novel embedded spiral resonator (ESR)," *IEEE Microwave and Wireless Components Letters*, vol. 20, no. 8, pp. 435–437, Aug. 2010.

[131] [131] X. Luo, J. Ma, and K. Yeo, "Dual-band bandpass filter using embedded spiral resonator and broadside-coupled meander slot-line," *IET Electronics Letters*, vol. 46, no. 16, pp. 1135–1137, Aug. 2010.

[132] C. Chen, C. Hsu, and H. Chuang, "Design of miniature planar dual-band filter using dual-feeding structures and embedded resonators," *IEEE Microwave and Wireless Components Letters*, vol. 16, no. 12, pp. 669–671, Dec. 2006.

[133] A. Djaiz, T. Denidni, and M. Nedil,, "Dual-band filter using multilayer structures and embedded resonators," *IET Electronics Letters*, vol. 43, no. 9, pp. 527–528, April 2007.

[134] F. Chen and Q. Chu, "Design of compact tri-band bandpass filters using assembled resonators," *IEEE Transactions on Microwave Theory and Techniques*, vol. 57, no. 1, pp. 165–171, Jan. 2009.

[135] X. Zhang, Q. Xue, and B. Hu, "Planar tri-band bandpass filter with compact size," *IEEE Microwave and Wireless Components Letters*, vol. 20, no. 5, pp. 262–264, May 2010.

[136] R. Geschke, B. Jokanovic, and P. Meyer, "Filter parameter extraction for triple-band composite split-ring resonators and filters," *IEEE Transactions on Microwave Theory and Techniques*, vol. 59, no. 6, pp. 1500–1508, June 2011.

[137] M. Doan, W. Che, K. Deng, and W. Feng, "Compact tri-band bandpass filter using stub-loaded resonator and quarter-wavelength resonator," *Proceedings of China-Japan Joint Microwave Conference Proceedings*, 2011, pp. 1–4.

[138] Q. Chu, X. Wu, and F. Chen, "Novel compact tri-band bandpass filter with controllable bandwidths," *IEEE Microwave and Wireless Components Letters*, vol. 21, no. 12, pp. 655–657, Dec. 2011.

[139] X. Lin, "Design of compact tri-band bandpass filter using $\lambda/4$ and stub loaded resonators," *Journal of Electromagnetic Waves and Applications*, vol. 24, no. 14, pp. 2029–2035, Oct. 2010.

[140] C.-F. Yang, Y.-C. Chen, C.-Y. Kung, J.-J. Lin, and T.-P. Sun, "Design and fabrication of a compact quad-band bandpass filter using two different parallel positioned resonators," *Progress in Electromagnetics Research*, vol. 115, pp. 159–172, 2011.

[141] J.-C. Liu, J.-W. Wang, B.-H. Zeng, and D.-C. Chang, "CPW-fed dual-mode double-square-ring resonators for quad-band filters," *IEEE Microwave and Wireless Components Letters*, vol. 20, no. 3, pp. 142–144, March 2010.

[142] H.-W. Wu and R.-Y. Yang, "A new quad-band bandpass filter using asymmetric stepped impedance resonators," *IEEE Microwave and Wireless Components Letters*, vol. 21, no. 4, pp. 203–205, April 2011.

[143] K.-W. Hsu and W.-H. Tu, "Sharp-rejection quad-band bandpass filter using meandering structure," *IET Electronics Letters*, vol. 48, no. 15, pp. 935–936, July 2012.

[144] J.-K. Xiao, Y.-F. Zhu, Y. Li, and X.-W. Li, "Miniature quad-band bandpass filter with passband individually controllable using folded SIR," *IET Electronics Letters*, vol. 50, no. 9, pp. 679–680, April 2014.

[145] F.-C. Chen and Q.-X. Chu, "Design of quad-band bandpass filter using assembled resonators," *Microwave and Optical Technology Letters*, vol. 53, no. 6, pp. 1305–1308, June 2011.

[146] H. Zhu, L. Gao, X. Y. Zhang, and B.-J. Hu, "Design of quad-band bandpass filter using open- and short-stub-loaded resonators," *Proceedings of Cross Strait Quad-Regional Radio Science and Wireless Technology Conference*, 2011, pp. 661–663.

[147] S. Sun, B. Wu, S. Yang, K. Deng, and C. Liang, "A novel quad-band filter using centrally shorted-stub loaded resonator and stepped impedance resonator," *Proceedings of IEEE MTT-S International Microwave Symposium Digest*, 2012, pp. 1–3.

[148] Z.-C. Zhang, Q.-X. Chu, and F.-C. Chen, "Novel quad-band filter with high frequency ration and controllable bandwidths using SLHSIRs and SLQSIRs," *Microwave and Optical Technology Letters*, vol. 56, no. 12, pp. 2845–2848, Dec. 2014.

[149] S.-C. Lin, "Microstrip dual/quad-band filters with coupled lines and quasi-lumped impedance inverters based on parallel-path transmission," *IEEE Transactions on Microwave Theory and Techniques*, vol. 59, no. 8, pp. 1937–1946, Aug. 2011.

[150] C.-F. Chen, "Design of a compact microstrip quint-band filter based on the tri-mode stub-loaded stepped-impedance resonators," *IEEE Microwave and Wireless Components Letters*, vol. 22, no. 7, pp. 357–359, July 2012.

[151] H. Wu, Y. Chen, and Y. Chen, "Multi-layered dual-band bandpass filter using stub-loaded stepped-impedance and uniform-impedance resonators," *IEEE Microwave and Wireless Components Letters*, vol. 22, no. 3, pp. 114–116, March 2012.

[152] P. Hsiao, R. Weng, and Y. Chang, "A miniaturized dual-band bandpass filter using open-loop and SIR-DGS resonators," *Proceedings of IEEE Topical Meeting on Silicon Monolithic Integrated Circuits in RF Systems*, 2008, pp. 191–194.

[153] B. Wu, C. Liang, Q. Li, and P. Qin, "Novel dual-band filter incorporating defected SIR and microstrip SIR," *IEEE Microwave and Wireless Components Letters*, vol. 18, no. 6, pp. 392–394, June 2008.

[154] E. Corrales, O. Menendez, and P. de Paco, "Compact dual-band filter based on microstrip-DGS resonator pairs," *Proceedings of 41st European Microwave Conference*, 2011, pp. 464–467.

[155] M. Weng, H. Wu, K. Shu, J. Chen, R. Yang, and Y. Su, "A novel triple-band bandpass filter using multilayer-based substrates for WiMAX," *Proceedings of European Microwave Conference*, 2007, pp. 325–328.

[156] L. Ren, "Tri-band bandpass filters based on dual-plane microstrip/DGS slot structure," *IEEE Microwave and Wireless Components Letters*, vol. 20, no. 8, pp. 429–431, Aug. 2010.

[157] X. Lai, C. Liang, H. Di, and B. Wu, "Design of tri-band filter based on stub loaded resonator and DGS resonator," *IEEE Microwave and Wireless Components Letters*, vol. 20, no. 5, pp. 265–267, May 2010.

[158] C.-M. Cheng and C.-F. Yang, "Develop quad-band (1.57/2.45/3.5/5.2 GHz) bandpass filters on the ceramic substrate," *IEEE Microwave and Wireless Components Letters*, vol. 20, no. 5, pp. 268–270, May 2010.

[159] B. Wu, C. Liang, P. Qin, and Q. Li, "Compact dual-band filter using defected stepped impedance resonator," *IEEE Microwave and Wireless Components Letters*, vol. 18, no. 10, pp. 674–676, Oct. 2008.

[160] C. Chio, S. Ting, E. Lim, and K. Tam, "Compact dual-band quasi-elliptic bandpass filter with transmission-zero control," *Proceedings of IEEE MTT-S International Microwave Symposium Digest*, 2012, pp. 1–3.

[161] H. Liu, L. Shen, Z. Zhang, J. Lim, and D. Ahn, "Dual-mode dual-band bandpass filter using defected ground waveguide," *IET Electronics Letters*, vol. 46, no. 13, pp. 895–897, June 2010.

[162] L. Ren and H. Huang, "Dual-band bandpass filter based on dual-plane microstrip/interdigital DGS slot structure," *IET Electronics Letters*, vol. 45, no. 21, pp. 1077–1079, Oct. 2009.

[163] J. Chen, T. Yum, J. Li, and Q. Xue, "Dual-mode dual-band bandpass filter using stacked-loop structure," *IEEE Microwave and Wireless Components Letters*, vol. 16, no. 9, pp. 502–504, Sept. 2006.

[164] V. Radonić and V. Crnojević-Bengin, "Multilayer dual-mode dual-band filter using square loop resonators," *Proceedings of 40th European Microwave Conference*, 2010, pp. 1245–1248.

[165] R.-Y. Yang, C. Y. Hung, and J.-S. Lin, "Design and fabrication of a quad-band bandpass filter using multi-layered SIR structure," *Progress In Electromagnetics Research*, vol. 114, pp. 457–468, 2011.

[166] K.-W. Hsu and W.-H. Tu, "Compact wide-stopband quad-band bandpass filter with tunable transmission zeros," *Proceedings of IEEE MTT-S International Microwave Symposium Digest*, 2012, pp. 1–3.

[167] K.-W. Hsu, W.-C. Hung, and W.-H. Tu, "Compact quint-band microstrip bandpass filter using double-layered substrate," *Proceedings of IEEE MTT-S International Microwave Symposium Digest*, 2013, pp. 1–4.

[168] C. Tang, S. You, and I. Liu, "Design of a dual-band bandpass filter with low-temperature co-fired ceramic technology," *IEEE Transactions on Microwave Theory and Techniques*, vol. 54, no. 8, pp. 3327–3332, Aug. 2006.

[169] K. Chin, J. Hung, C. Huang, J. Fu, B. Chen, and T. Chen, "LTCC dual-band stepped-impedance-stub filter constructed with vertically folded structure," *IET Electronics Letters*, vol. 46, no. 23, pp. 1554–1556, Nov. 2010.

[170] Y. Guo, L. Ong, M. Chia, and B. Luo, "Dual-band bandpass filter in LTCC," *Proceedings of IEEE MTT-S International Microwave Symposium Digest*, 2005, p. 4.

[171] K. Qian and X. Tang, "Compact dual-band semi-lumped bandpass filter with LTCC technology," *IET Electronics Letters*, vol. 47, no. 13, pp. 755–757, June 2011.

[172] C. Chen, Y. Lin, T. Horng, and S. Wu, "Highly flexible and miniaturized triple-band bandpass filter design using coupled stacked spiral resonators," *Proceedings of IEEE MTT-S International Microwave Symposium Digest*, 2012, pp. 1–3.

[173] K. Huang, T. Chiu, and H. Wu, "Compact LTCC tri-band filter design," *Proceedings of Asia-Pacific Microwave Conference*, 2007, pp. 1–4.

[174] L. Wang, M. Chang, S. Li, J. Liou, C. Shiau, C. Hsu, and C. Lee, "Quarter-wavelength stepped-impedance YBCO resonators for miniaturized dual-band high-Tc superconducting filters," *IEEE Transactions on Applied Superconductivity*, vol. 19, no. 3, pp. 895–898, June 2009.

[175] K. Yeo and M. Lancaster, "8 pole high temperature superconductor microstrip dual band bandpass filter design," *Proceedings of IEEE MTT-S International Microwave Symposium Digest*, 2011, pp. 1–4.

[176] A. Abu Hudrouss, A. Jayyousi, and M. Lancaster, "Dual-band HTS filter using modified dual-spiral resonators," *IET Electronics Letters*, vol. 46, no. 3, pp. 221–222, Feb. 2010.

[177] A. Abu-Hudrouss, A. Jayyousi, and M. Lancaster, "Triple-band HTS filter using dual spiral resonators with capacitive-loading," *IEEE Transactions on Applied Superconductivity*, vol. 18, no. 3, pp. 1728–1732, Sept. 2008.

[178] H. Joshi and W. Chappell, "Dual-band lumped-element bandpass filter," *IEEE Transactions on Microwave Theory and Techniques*, vol. 54, no. 12, pp. 4169–4177, Dec. 2006.

[179] Y. Cho, H. Baek, H. Lee, and S. Yun, "A dual-band combline bandpass filter loaded by lumped series resonators," *IEEE Microwave and Wireless Components Letters*, vol. 19, no. 10, pp. 626–628, Oct. 2009.

[180] K.-W. Hsu, J.-H. Lin, W.-C. Hung, and W.-H. Tu, "Design of compact quad-band bandpass filter using semi-lumped resonators," *Proceedings of IEEE MTT-S International Microwave Symposium Digest*, 2013, pp.1–4.

[181] C. Tseng and T. Itoh, "Dual-band bandpass and bandstop filters using composite right/left-handed metamaterial transmission lines," *Proceedings of IEEE MTT-S International Microwave Symposium Digest*, 2006, pp. 931–934.

[182] M. Studniberg and G. Eleftheriades, "A dual-band bandpass filter based on generalized negative-refractive-index transmission-lines," *IEEE Microwave and Wireless Components Letters*, vol. 19, no. 1, pp. 18–20, Jan. 2009.

[183] P. Kapitanova, D. Kholodnyak, S. Humbla, R. Perrone, J. Mueller, M. Hein, and I. Vendik, "Multi-band and tunable multi-band microwave resonators and filters based on cascaded left/right-handed transmission line sections," *Proceedings of IEEE EUROCON*, 2009, pp. 39–45.

[184] S. Kahng, G. Jang, and J. Anguera, "Metamaterial dual-band bandpass filters using CRLH zero-order-resonators and improving its intermediate stopband," *Proceedings of 4th European Conference on Antennas and Propagation*, 2010, pp. 1–3.

[185] V. Radonic, V. Crnojevic-Bengin, and B. Jokanovic, "Novel left-handed unit cell for multi-band filtering applications," *Proceedings of 38th European Microwave Conference*, 2008, pp. 694–697.

[186] V. Radonic, N. Jankovic, and V. Crnojevic-Bengin, "Left-handed unit cell based on Hilbert grounded patch for multi-band filtering applications," *Proceedings of 9th International Conference on Telecommunication in Modern Satellite, Cable, and Broadcasting Services*, 2009, pp. 147–150.

[187] M. Studniberg and G. V. Eleftheriades, "A quad-band bandpass filter using negative-refractive-index transmission-line (NRI-TL) metamaterials," *Proceedings of International Symposium of Antennas and Propagation Society*, 2007, pp. 4961–4964.

[188] R. J. Cameron, M. Yu, and Y. Wang, "Direct-coupled microwave filters with single and dual bandstops," *IEEE Transactions on Microwave Theory and Techniques*, vol. 53, no. 11, pp. 3288–3297, Nov. 2005.

[189] T. Faraji and A. R. Hazeri, "Microstrip dual-band bandstop filter to suppress DGS, DMB, GPS, UTMTS, WCDMA, WIBRO, and WLAN bands," *Microwave and Optical Technology Letters*, vol. 54, no. 4, pp. 854–856, April 2012.

[190] V. K. Velidi, A. Guntupalli, and S. Sanyal, "Compact tapped stepped impedance open stub dual-band bandstop filters with sharp rejection characteristics," *Microwave and Optical Technology Letters*, vol. 51, no. 10, pp. 2274–2277, Oct. 2009.

[191] K.-S. Chin, J.-H. Yeh, and S.-H. Chao, "A compact dual-band bandstop filters using stepped-impedance resonators," *IEEE Microwave and Wireless Components Letters*, vol. 17, no. 12, pp. 849–851, Dec. 2007.

[192] C.-K. Lung, K.-S. Chin, and J. S. Fu, "Tri-section stepped-impedance resonators for design of dual-band bandstop filter," *Proceedings of 39th European Microwave Conference*, 2009, pp.771–774.

[193] X. Y. Zhang, X. Huang, H.-L. Zhang, and B.-J. Hu, "Novel dual-band bandstop filter with controllable stopband frequencies," *Microwave and Optical Technology Letters*, vol. 54, no. 5, pp. 1203–1206, May 2012.

[194] L. Gao, S. W. Cai, X.-Y. Zhang, and Q. Xue, "Dual-band bandstop filter using open and short stub-loaded resonators," *Proceedings of Microwave and Millimeter Wave Technology International Conference*, vol. 4, pp.1–3, 2012.

[195] F.-C. Chen, J.-M. Qiu, and Q.-X. Chu, "Dual-band bandstop filter using stub-loaded resonators with sharp rejection characteristic," *IET Electronics Letters*, vol. 49, no. 5, pp. 351–352, Feb. 2013.

[196] S. Vegesna and M. Saed, "Microstrip dual-band bandpass and bandstop filters," *Microwave and Optical Technology Letters*, vol. 54, no. 1, pp. 168–171, Jan. 2012.

[197] C. Karpuz, A. Gorur, E. Gunturkun, and A. K. Gorur, "Asymmetric response dual-mode dual-band bandstop filters having simple and understandable topology," *Proceedings of Asia Pacific Microwave Conference*, 2009, pp. 925–928.

[198] K. W. Wong, L. Chiu, and Q. Xue, "A compact dual-band filter," *Microwave and Optical Technology Letters*, vol. 51, no. 12, pp. 2952–2954, Dec. 2009.

[199] M. A. Kakhki and M. H. Neshati "Experimental investigation of a dual band-reject filter using C-shaped DGS with improved Q-factor," *Proceedings of 5th International Symposium on Telecommunications*, 2010, pp. 332–335.

[200] X. Hu, Q. Zhang, and S. He, "Dual-band rejection filter based on split ring resonator (SRR) and complimentary SRR," *Microwave and Optical Technology Letters*, vol. 51, no. 10, pp. 2519–2522, Oct. 2009.

[201] J. Wang, H. Ning, L. Mao, and M. Li, "Miniaturized dual-band bandstop filter using defected microstrip structure and defected ground structure," *Proceedings of IEEE MTT-S International Microwave Symposium Digest*, 2012, pp. 1–3.

[202] K. Singh, K. Ngachenchaiah, D. Bhatnagar, and S. Pal, "Wideband, compact microstrip band stop filter for triband operations," *Proceedings of International Conference on Recent Advances in Microwave Theory and Applications*, 2008, pp. 96–98.

[203] Z. Wang, F. Nasri, and C.-W. Park, "Compact tri-band notched UWB bandpass filter based on interdigital hairpin finger structure," *Proceedings of IEEE 12th Annual Wireless and Microwave Technology Conference*, 2011, pp. 1–4.

[204] J.-K. Xiao and W.-J. Zhu, "New defected microstrip structure bandstop filter," *Proceedings of Progress In Electromagnetics Research Symposium*, 2011, pp. 1471–1474.

[205] S. H. Han, X. L. Wang, and Y. Fan, "Analysis and design of multiple-band bandstop filters," *Progress In Electromagnetics Research*, vol. 70, pp. 297–306, 2007.

[206] S. Awasthi, A. Biswas, and M. J. Akhtar, "Direct coupled quad-band bandstop filter synthesis using frequency transformation," *Proceedings of Asia-Pacific Microwave Conference*, 2013, pp. 893–895.

3 Techniques for the synthesis of multi-band transfer functions

Tobias Gerhardus Brand, Riana Helena Geschke, and Petrie Meyer

3.1 Introduction

The previous chapters have given a broad overview of the present state-of-the-art of multi-band filters. It is clear that no single standard technique exists to design these filters, and that a number of authors have proposed different approaches to this problem. To summarize, most of these techniques can be classified as belonging to one of three main design methodologies, namely the interconnection of single-band filters [1–2], the usage of multiple resonant resonators [3–7], and the construction and subsequent synthesis of multi-band approximation functions [8–24]. The interconnection of single-band filters is by far the simplest option, but invokes large overheads in terms of space and weight due to the combining and dividing circuitry, and suffers from the frequency responses of these additional circuits. In contrast, filters using resonators with multiple resonating frequencies offer a much more compact solution, but are more difficult to design, and significantly more difficult to implement for practical filters of high orders. They do offer the advantage that the passbands of the filter can be positioned at frequencies which are quite apart from each other.

Of the three approaches, construction and synthesis of multi-band transfer functions offer the most general solutions. In practice, the transfer function itself can often be bypassed by simply replacing elements in a filter with frequency transformation sub-circuits [8–12], yet this inherently limits the possible topologies, and the full classical network synthesis approach remains the most general. Popular techniques to find such a transfer function include the direct optimization of pole- and zero-positions in the multi-band frequency domain [13–19], and the mapping of the poles and zeros of a single-band function onto the multi-band frequency plane using mathematical functions [20–24].

In 2007, Tamiazzo *et al.* proposed the use of quadratic polynomials to obtain dual-band transfer functions, which were then used to synthesize single-structure multi-band filters [20–21]. Lee and Sarabandi [22, 23] introduced mapping functions for filters with up to three passbands, but with limited isolation between bands. This is a general issue with mapping functions, as available functions often do not map the full single-band stopband to all the stopbands of the multi-band filter. Mohan proposed a method for N-band filters using successive applications of the classical low-pass to bandpass transformation [24], followed by the collection of all these poles and zeros into one function, and subsequent optimization. In a recent article by the authors, a very general rigorous technique that requires no optimization was applied to coupled-resonator filters, [25]. This technique constructs a mapping function from classical reactance functions and is completely unlimited in the number of bands that can be created. In addition, the centre frequency and bandwidth of each band can be chosen completely arbitrarily, and the result is a multi-band filter function of which each band is an exact bandwidth-scaled and frequency-translated copy of the original low-pass function. The approach leads to circuit topologies that are very well suited to coupled-resonator implementations, as each resonator is simply expanded into a separate coupled-resonator chain.

This chapter gives an overview, supported by examples, of two techniques for creating multi-band transfer functions: one technique based on optimization of poles and zeros, and another based on reactance transformations. To put this into perspective, a brief overview of classical and existing techniques is first given, followed by detailed expositions.

3.2 Basic transfer function theory

3.2.1 Basic low-pass transfer functions

To set up the theoretical discussion of mapping functions, it is first necessary to briefly review the basic theory of filter transfer functions, specifically low-pass transfer functions. Most modern filter realizations are based on the insertion loss method, which in this chapter forms the underlying basis for all filter realizations. The design of filters using the insertion loss method includes three stages: approximation, synthesis, and realization. The approximation stage involves finding a mathematical transfer function to approximate the ideal filter response. For our purposes, the transfer function is a rational function in s and is typically a power or voltage ratio.

An ideal filter amplitude-frequency transfer function is constant in the passband, has infinite selectivity at the passband edges and a linear phase response in the passband. However, this type of response is not physically realizable. A realistic transfer function therefore has to approximate the desired response, bearing in mind that the linear phase and high selectivity are competing goals.

The concept of a filter prototype is often used to simplify the procedure. A prototype is a filter that exists in a domain normalized with respect to frequency and impedance. The design specifications are usually mapped from the actual frequency domain to the prototype frequency domain using an appropriate normalization function. There are numerous well-known normalization functions available in the literature [26], e.g. low-pass-to-high-pass transformation, low-pass-to-bandpass transformation, and low-pass-to-band-stop transformation. Filter designs typically employ up to three frequency domains: a low-pass prototype domain Ω_s, sometimes an intermediate prototype domain Ω_m, and the final multi-band frequency domain ω. Frequency transformation is detailed in Section 3.2.2.

The transmission zeros or attenuation poles of the prototype are the frequencies where perfect attenuation occurs, while reflection zeros are frequencies where there is no reflection from the filter. Transmission zeros are positioned outside the passband and the reflection zeros inside the passband. The exact positions of the reflection and transmission zeros depend on the type of filter response, e.g. Chebychev, Butterworth, Elliptic, Quasi-Elliptic, or Linear Phase.

For filters referred to as "all-pole" filters, all the filter transmission zeros are located at infinity. This type of response can be accommodated in synthesis by a LC-ladder network. It is possible to move some of the transmission zeros at infinity to finite frequencies, referred to as finite (frequency) transmission zeros. Nevertheless, the total number of transmission zeros always remains the same as the order of the filter.

Approximation is the procedure of finding network functions that approximate the ideal requirements. Filters are passive and reciprocal two-port circuits, which are therefore described by three network functions – two driving point functions and one transfer function. In classical synthesis, it is common practice to define all of the network functions in terms of the characteristic polynomials, $P(s)$, $F(s)$, and $E(s)$. This is normally done in the (low-pass) prototype domain, implying normalization. The roots of $F(s)$ form the reflection zeros of the transfer function and the roots of $P(s)$ the transmission zeros. Permissible zero locations in the s-plane are detailed in [27]. If $P(s) = 1$ then all of the prototype's transmission zeros are located at infinity.

The network functions and the polynomials all derive their forms from the unitary properties of the scattering matrix of a lossless passive two-port network [27]. These properties are essentially a restatement of the conservation of the energy principle. The transmission and reflection coefficients are related to the characteristic polynomials by

$$S_{21}(s) = \frac{P(s)}{\epsilon E(s)}$$
$$S_{11}(s) = \frac{F(s)}{\epsilon_r E(s)}. \tag{3.1}$$

The magnitude of the transmission coefficient S_{21} should approximate unity inside the passband and zero in the stopband. The roots of S_{11} are the reflection zeros in the prototype domain. The roots of S_{21} are the transmission zeros in the prototype domain. These are normalized frequencies associated with the points of perfect transmission and perfect rejection, respectively.

Characteristic polynomials are in turn related to the filtering function. The filtering function, also known as the filter characteristic function, is defined as

$$C(\Omega_s) = \frac{F(\Omega_s)}{P(\Omega_s)}. \tag{3.2}$$

The filtering function defines the shape of a filter response. This is normally a real-valued function with small values in the passband frequencies and large values outside the passbands. The poles of $C(\Omega_s)$ are the filter transmission zeros and the zeros are the reflection zeros of the prototype filter. For an all-pole filter, the characteristic function for a Chebychev response with an equi-ripple response in the passband is

$$C_N(\Omega_s) = \cosh[N\cosh^{-1}(\Omega_s)]. \tag{3.3}$$

To represent an equal ripple response in the passband for filters with finite transmission zeros, a general Chebychev response is described by

$$C_N(\Omega_s) = \cosh\left[\sum_{n=1}^{N} \cosh^{-1}(x_n(\Omega_s))\right], \tag{3.4}$$

where N is the order of the filter, Ω_{sn} is the transmission zero, either at infinity or finite frequency, and

$$x_n(\Omega_s) = \frac{\Omega_s - 1/\Omega_{sn}}{1 - \dfrac{\Omega_s}{\Omega_{sn}}}. \tag{3.5}$$

The approximation procedure typically starts by constructing $C(\Omega_s)$, after which the characteristic polynomials $F(s)$ and $P(s)$ are calculated, and from them $E(s)$ [27], using the Feldtkeller equation

$$\frac{F(s)F(s)^*}{\epsilon_r^2} + \frac{P(s)P(s)^*}{\epsilon^2} = E(s)E(s)^*. \tag{3.6}$$

The relationship between the transmission coefficient and the filtering function can be expressed as

$$|S_{21}(\Omega_s)|^2 = \frac{1}{1 + \dfrac{\epsilon^2}{\epsilon_r^2}\left|\dfrac{F(\Omega_s)}{P(\Omega_s)}\right|^2}. \tag{3.7}$$

The polynomials $F(s)$ and $P(s)$ are normalized to monic format using the real constants ϵ_r and ϵ, respectively. The characteristic polynomials $P(s)$ and $F(s)$ are found from the filtering function using

$$s = j\Omega_s. \tag{3.8}$$

The construction of a filtering function that meets the design requirements in the prototype domain, is therefore a fundamental part of the filter design process. Once the characteristic polynomials are known, a circuit can be synthesized from the three network functions associated with the filter.

In this chapter, the design of a normalized (low-pass) single band filter is the basis of each of the techniques presented. We focus on coupled-resonator filters and therefore employ the coupling matrix synthesis methods available in literature [27]. The coupling matrices that result from these synthesis procedures then describe prototype circuits. In order to obtain a final design, one must de-normalize the circuit using the initial normalization function. At this point in the design procedure, the filter is described by an LC network. Apart from specialized multi-layer techniques, it is not possible to construct the LC networks at microwave frequencies directly. The insertion loss method for microwave filters therefore includes a final stage, referred to as implementation. During this stage, one finds a microwave structure corresponding to the circuit model found during the synthesis procedure.

3.2.2 Transfer function synthesis of multi-band filters through transformation

The classical low-pass-to-bandpass transformation forms the basis for most single-band bandpass filters. In addition, it is also used extensively in the design of multi-band filters, where multi-band prototype functions are transformed as a final step of the synthesis. We therefore start this section with a brief overview of this well-known transformation, given in (3.9). Note that, for single-band filters, $\Omega = \Omega_s$, whereas $\Omega = \Omega_m$ in the case of multi-band filters

$$\begin{aligned} \Omega &= \frac{1}{\Delta}\left(\frac{\omega}{\omega_0} - \frac{\omega_0}{\omega}\right) \\ \omega_0 &= \sqrt{\omega_2 \omega_1} \\ \Delta &= \frac{\omega_2 - \omega_1}{\omega_0} \end{aligned} . \tag{3.9}$$

Normally, ω_2 and ω_1 are the upper and lower passband frequencies of the transformed filter, but in the case of multi-band filters, ω_2 is the highest frequency of the highest passband and ω_1 is the lowest frequency of the lowest passband. The centre frequency is defined as ω_0 and the fractional bandwidth is defined as Δ. This transformation is non-linear. For single-band filters, this has no impact on the realized bandwidth of a filter after performing the transformation, as illustrated in Figure 3.1, since, as expected, the two passband edges simply map to

$$\Omega = \pm 1. \tag{3.10}$$

However, the effect is of importance when transforming a multi-band prototype filter. The bandwidth ratios are not preserved after the transformation due to the non-linearity, as shown in Figure 3.2 for a bandpass filter used as prototype, and transformed using (3.9). This is especially visible for filters with multiple passbands with larger bandwidths or larger spacing between passbands, since the degree of non-linearity is a function of fractional bandwidth. In this case, again the mapping of (3.9) holds for the outer passband edges. The inner passband limits are influenced by the non-linear mapping. The result is that a prototype filter that is symmetric in amplitude with respect to $\Omega = 0$ does not map to an amplitude response with equal bandwidths centered around ω_0. The solution is to compensate for this by using an asymmetric prototype. The implication is a transfer function with complex coefficients, and asynchronously tuned resonators in the coupled-resonator circuit.

One of the first papers that presented such a transformation [28] used a modified low-pass transformation, which resulted in the concept of appending resonators to a prototype circuit to create more passbands

$$\Omega = b_1\left(\frac{\omega}{\omega_{01}} - \frac{\omega_{01}}{\omega}\right) - \frac{1}{b_2\left(\frac{\omega}{\omega_{02}} - \frac{\omega_{02}}{\omega}\right)}. \tag{3.11}$$

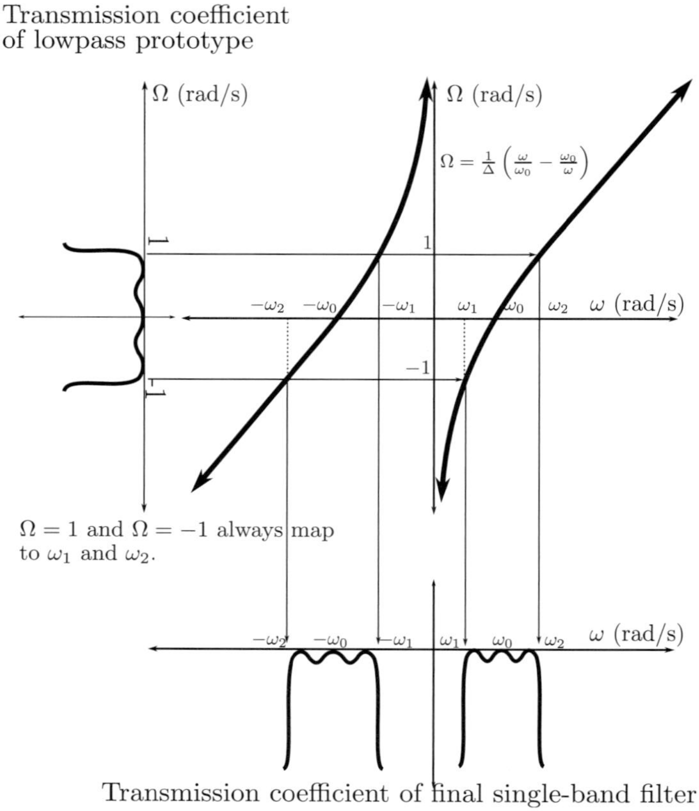

Figure 3.1 Basic low-pass to bandpass transformation for a single-band filter.

The variables ω_{01} and ω_{02} represent the self-resonant frequencies of the prototype resonators and the appended resonators, respectively while b_1 and b_2 are the susceptance slope parameters of each resonator, respectively. This transformation effectively replaces each single-band resonator in the prototype by a dual-band resonator, which is a parallel resonator in parallel with a series resonator, or equivalently two parallel resonators coupled by an inverter.

This specific implementation was for dual-band transformations that created two passbands separated by finite transmission zeros, all placed at the same frequency. The appended resonators are referred to as rejection resonators as these create transmission zeros to split the passbands. Various authors continued with similar transformations and demonstrated multi-band responses with up to three passbands [23, 29, 30]. None of these methods can be extended to more than three passbands.

In a more general paper, García-Lampérez *et al.* presented frequency transformation for general multi-band responses realized by appending resonators to a single-band filter [31]. The technique replaces each resonator with a Foster-type circuit, which effects a frequency transformation. While not suffering from many of the drawbacks of other techniques, this approach is still not completely general, and the computation of the mapping function is not straightforward.

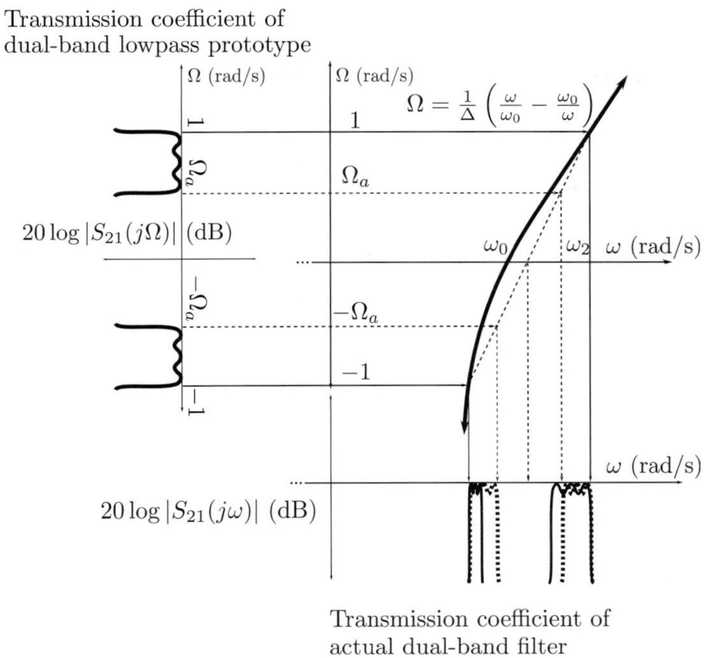

Figure 3.2 Low-pass to bandpass transformation for a multi-band prototype.

The technique proposed by the authors [25], referred to in the introduction, presents the next step in the evolution of mapping functions. It is quite general, and in one of its forms resembles the technique in [31]. A detailed exposition is given in Section 3.3.

3.2.3 Transfer function synthesis of multi-band filters through optimization

It is often the case that the transformation methods presented in the literature do not provide the exact filter response required. In many cases, bandwidths or reflection loss levels may need adjustment. These require an optimization step to fine-tune the reflection and transmission zero positions of the filter. In these cases, the zeros are approximately placed by the transformation method and are then optimized further to provide the desired passband widths and equi-ripple responses.

An alternative presented by [14] is to rely completely on optimization to provide a desired multi-band response, where quad-band filters of up to sixteenth order are presented based on pure optimization, i.e. both approximation and synthesis steps are based on optimization. First, in the approximation step, the filter polynomials for multi-band filters are obtained using an iterative procedure, although of course a frequency transformation method could also have been used in this step. In synthesis, the coupling matrix representation for a desired coupling topology, i.e. with entries only in specified positions, is optimized to match the required transfer function response. The optimization cost function is the sum of the squares of the difference in frequency position of each of the filter reflection and transmission zeros [14].

The benefit of this approach is that convenient coupling topologies in a given technology may be described by a coupling matrix and its allowed non-zero entries obtained directly from the approximation function. In this method, the coupling topology selected has to be compatible with the type of response required.

In Section 3.3, this chapter presents an alternative optimization approach for the design of multi-band filters. The method constructs multi-band filtering functions directly in the multi-band frequency domain, by random placement of the transmission zeros in the stopbands and reflection zeros in the passbands. This method avoids using single-band prototypes, frequency transformations, or an intermediate prototype domain. The zero positions are then optimized to meet the design specification.

3.3 General synthesis of transfer functions using optimization of poles and zeros

In this section, an optimization-based method to design multi-band filters, utlizing the technique by Zhang *et al.* [32] is presented. The technique finds expressions for multi-band filtering fuctions using an iterative approximation algorithm, followed by classical synthesis. The section starts with a qualitative description of the approximation algorithm, followed by a detailed mathematical exposition, and finally a general design method and design examples to validate the theory.

3.3.1 Qualitative description of approximation algorithm

From the discussion in the previous section, we can conclude that the filtering function is a rational function in the real frequency variable and therefore completely defined by its zeros and poles, i.e. the points of perfect transmission and perfect attenuation, respectively, in a transfer function based on this filtering function. It follows that such a filtering function can be constructed in full from its poles and zeros. The algorithm presented here starts by constructing an initial filtering function by arbitrarily placing poles and zeros in appropriate starting positions on the frequency axis. The chosen set of poles and zeros are then iteratively improved, in a converging manner, until an optimal filtering function is found.

Appropriate starting positions for poles and zeros can be identified by noting that points of perfect transmission (filtering function zeros) are always confined to the passbands of a filter. Likewise points of zero transmission (filtering function poles) must be confined to the stopbands of a filter. The number of zeros correspond to the order of a filter response and therefore also to the order of the filtering function. The total number of poles is equal to the total number of zeros. Some of the poles can be located at finite frequencies and the rest at infinity.

Typical responses for the magnitude transfer function and filtering function of a dual-band filter are shown in Figures 3.3 and 3.4. Figure 3.3 presents a plot of the magnitudes of the transmission and reflection coefficients of a ninth-order prototype dual-band filter. The first passband contains four reflection zeros and the second passband five

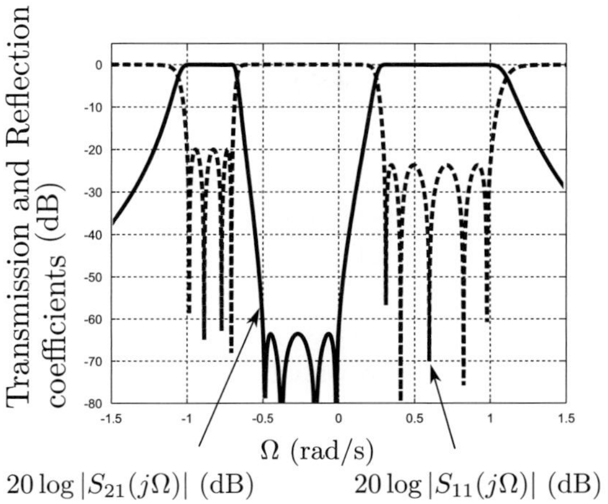

Figure 3.3 Magnitude transfer function response of a dual-band prototype filter. Note that in the general case, different passband ripple levels can occur.

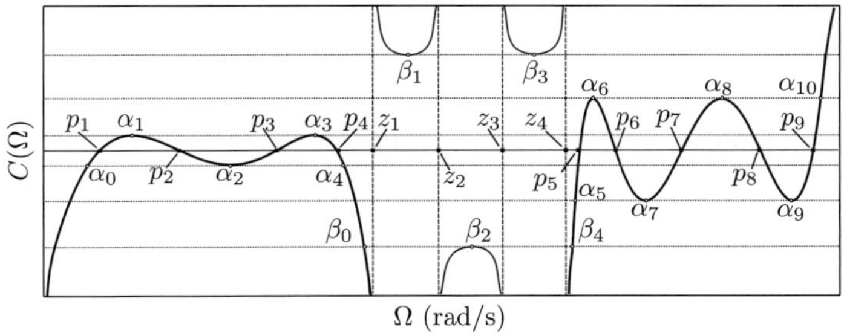

Figure 3.4 The filtering function $C(\Omega)$ for the dual-band filter prototype in Figure 3.3.

reflection zeros. The inner stopband contains four transmission zeros. The remaining transmission zeros are located at infinity. Figure 3.4 shows the filtering function associated with the transmission and reflection coefficients plotted in Figure 3.3, where $\Omega = \Omega_m$. The filtering function is of ninth order, with the reflection zeros denoted by $p_1, \ldots, p_9$ and the transmission zeros by $Z_1, \ldots, Z_4$

The filtering function in Figure 3.4 has a constant, not necessarily equal ripple in both its passbands and inside the inner stopband. Filters that have equi-ripple stopbands and passbands are very desirable as they have constant minimum return loss levels in their passbands and constant minimum attenuation levels in their stopbands. Filtering functions with equi-ripple passbands and stopbands represent a class of filters called elliptic filters. The filter in Figures 3.3 and 3.4 exhibits this property in three of its bands and is therefore termed *quasi-elliptic*. The behavior of the outer stopbands of the filter depend on the number of finite frequency transmission zeros. If the number of finite frequency transmission zeros is equal to the order of the filter, the outer stopbands will have finite minimum attenuation levels. If the number of finite frequency transmission zeros is less than the order of the filter, the outer stopbands will increase monotonically to infinity.

To achieve such a response, it should be clear that the zeros of the filtering function must be confined to the passbands, the poles of the function must be confined to the stopbands, and both the poles and zeros must be positioned in such a manner as to force an equi-ripple condition in each passband and stopband. The bounds of the passbands and stopbands can easily be obtained by mapping the passband and stopband requirements to the prototype domain. Denoting the passband edges α_0, α_5, and α_{10}, and the stopband edges β_0 and β_4, an initial non-optimal filtering function can be constructed by arbitrarily placing the desired number of poles and zeros in the appropriate bands. Next, the variation in the respective amplitudes, the passbands, and the stopbands is characterized by the turning points of the functions within the bands, denoted by α_n for the passbands and by β_n for the stopbands. These points are referred to as critical points in the following sections.

The minimum return loss level inside any passband is directly related to the maximum ripple value of the filtering function evaluated at the passband edge. If $|C_N(\Omega)|$ assumes the same value at the passband edges and at all the turning points inside a passband, the minimum return loss level in that passband will be constant. Note that the different passbands in Figure 3.3 have different ripple levels and therefore different minimum return loss figures. For the inner stopband, which extends from β_0 to β_4, the turning points of $C_N(\Omega)$ inside the stopband are marked as $\beta_1,\ldots,\beta_3$. If $|C_N(\Omega)|$ assumes the same value at the edges of a stopband and at all the turning points inside that stopband, the minimum attenuation level in that stopband will also be constant.

The arbitrary placement of reflection zeros and transmission zeros in the multi-band prototype domain will not result in a filtering function such as that depicted in Figure 3.4. A realistic result would be a filtering function where the ripple level is not constant in any of the passbands or the stopbands, as shown in Figure 3.5 as "iteration 1." The figure shows the absolute value of the filtering function for different iterations of an approximation algorithm, with iteration 1 showing an initial filtering function that was obtained by arbitrarily placing transmission zeros and reflection zeros in their appropriate bands. For a band to display an equi-ripple characteristic (referred to as an equi-ripple state henceforth), it can be shown that all the critical points associated with that band must be collinear. Thus, the dotted lines in Figure 3.5 must be straight and parallel to the Ω-axis. An equi-ripple state can therefore be obtained in each passband by moving all the zeros in that band in such a way that all the critical points become collinear, with each band being adjusted individually. This is achieved in a systematic way by choosing pairs of critical points iteratively. It is simple to show that the function $|C_N(\Omega)|$, evaluated at the critical point between the two zeros, becomes larger if the distance between the two zeros is increased and vice versa. This property can be exploited to predict how reflection zeros (or transmission zeros) must be adjusted to obtain equi-ripple bands. Note that the first and last critical points in each band must remain stationary for the bandwidth to remain constant. Figure 3.5 shows the execution of the algorithm for the first couple of iterations.

In the approximation algorithm, when two critical points are not collinear, the transmission zero or reflection zero between them is moved in such a manner as to force them to become collinear. This process is performed for each pair of critical points

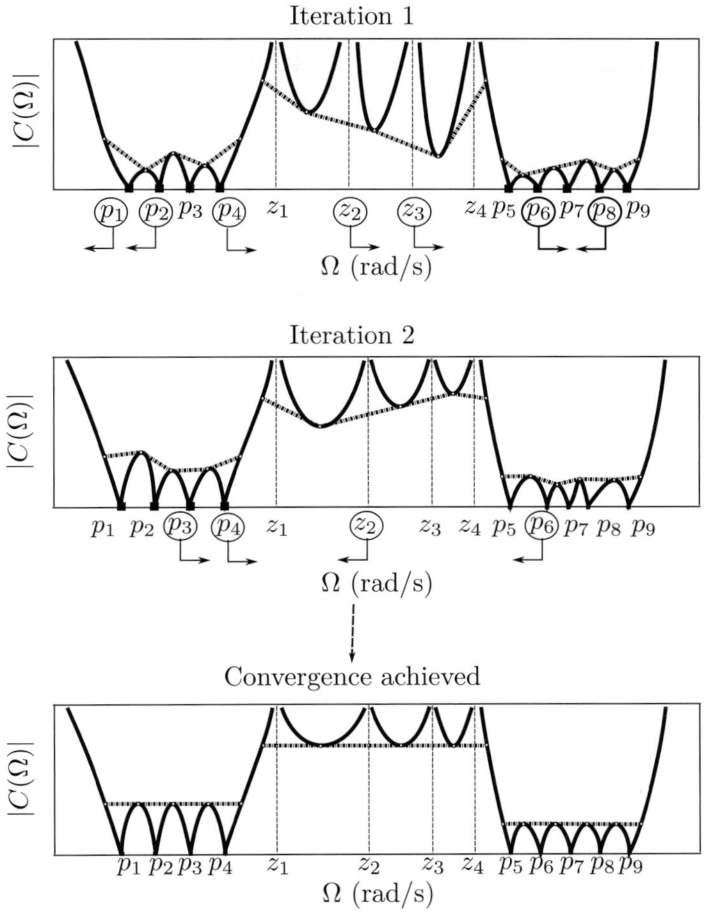

Figure 3.5 Execution of the approximation algorithm.

in each of the bands, with one cycle adjusting all the transmission and reflection zeros referred to as one approximation iteration. During a single iteration, each zero is adjusted assuming that it is not influenced by the adjustment of the other zeros, and that different passbands and stopbands do not influence each other. In reality, these assumptions are evidently not true, and the algorithm requires multiple iterations to converge to a valid equi-ripple filtering function. In practice, the algorithm always converges if one forces it not to adjust zeros to locations outside their associated bands.

3.3.2 Identification of critical points

As shown in the previous section, the critical points associated with a passband/or a stopband are the edge frequencies of the band and all the turning points located inside that band. The edge frequencies of a band are usually defined in the actual frequency domain. These frequencies must be mapped to the prototype domain using an inverse low-pass-to-bandpass transformation. The turning points of a filtering function are the roots of the gradient of said function. Mathematically, this can be expressed as

$$\begin{gathered}\frac{\mathrm{d}C_N(\Omega)}{\mathrm{d}\Omega}=\frac{\left(\frac{\mathrm{d}F(\Omega)}{\mathrm{d}\Omega}\right)P(\Omega)-F(\Omega)\left(\frac{\mathrm{d}P(\Omega)}{\mathrm{d}\Omega}\right)}{[P(\Omega)]^2}=0\\ \Leftrightarrow\left(\frac{\mathrm{d}F(\Omega)}{\mathrm{d}\Omega}\right)P(\Omega)-F(\Omega)\left(\frac{\mathrm{d}P(\Omega)}{\mathrm{d}\Omega}\right)=0\end{gathered}. \tag{3.12}$$

The roots of (3.12) that are located inside the edges of a given band are therefore the turning points of that specific band. These critical points can be used to derive equations to adjust the pole and zero positions.

3.3.3 Pole and zero adjustment formulas

At the core of the adjustment algorithm is a procedure to update reflection and transmission zeros in order to improve the transfer function in a systematic manner. In this section, $C_{N,k}(\Omega)$ refers to the filtering function after the kth zero has been updated, where N is the number of reflection zeros.

3.3.3.1 Update of reflection zeros

For equi-ripple passbands, the filtering function should evaluate to equal but opposite signed values at consecutive critical points. Mathematically this can be expressed as

$$C_{N,k}(\alpha_{i-1}) = -C_{N,k}(\alpha_i) \text{ for } i\in[1,N]. \tag{3.13}$$

With an initial filtering function $C_{N,0}(\Omega)$, the first reflection zero is updated to obtain $C_{N,1}(\Omega)$ as

$$C_{N,1}(\Omega)=\left(\frac{\Omega-\hat{p}_1}{\Omega-p_1}\right)C_{N,0}(\Omega). \tag{3.14}$$

Here p_1 is the initial reflection zero and $\hat{p}_1$ the update. Following this technique for a single approximation iteration results in the modification of N reflection zeros. Using the notation above, we can rewrite (3.14) as

$$\left(\frac{\alpha_{i-1}-\hat{p}_i}{\alpha_{i-1}-p_i}\right)C_{N,k-1}(\alpha_{i-1})=-\left(\frac{\alpha_i-\hat{p}_i}{\alpha_i-p_i}\right)C_{N,k-1}(\alpha_i). \tag{3.15}$$

Solving (3.15) for $\hat{p}_i$ results in

$$\hat{p}_i=\frac{p_i[\alpha_{i-1}C_{N,k-1}(\alpha_{i-1})+\alpha_iC_{N,k-1}(\alpha_i)]-\alpha_{i-1}\alpha_i[C_{N,k-1}(\alpha_{i-1})+C_{N,k-1}(\alpha_i)]}{p_i[C_{N,k-1}(\alpha_{i-1})+C_{N,k-1}(\alpha_i)]-[\alpha_iC_{N,k-1}(\alpha_{i-1})+\alpha_{i-1}C_{N,k-1}(\alpha_i)]}. \tag{3.16}$$

Equation (3.16) thus calculates the new position of p_i so that the adjacent critical points ($\alpha_{\mathrm{i}-1}$ and α_{i}) become collinear.

3.3.3.2 Update transmission zeros

A similar approach is followed for the update of transmission zeros. For equi-ripple stopbands, the filtering function should also evaluate to equal but opposite signed values at consecutive critical points. Mathematically, this can be expressed as

$$C_{N,k}(\beta_{i-1}) = -C_{N,k}(\beta_i) \text{ for } i \in [1, n_{fz}]. \tag{3.17}$$

Here n_{fz} is the number of finite frequency transmission zeros. Using an initial filtering function $C_{N,0}(\Omega)$, the first transmission zero is updated to obtain $C_{N,1}(\Omega)$ as

$$C_{N,1}(\Omega) = \left(\frac{\Omega - z_1}{\Omega - \hat{z_1}}\right) C_{N,0}(\Omega). \tag{3.18}$$

Here z_1 is the initial transmission zero and $\hat{z_1}$ is the update. Following this technique for a single approximation iteration results in the modification of n_{fz} reflection zeros. Using the notation above, we can rewrite (3.17) as

$$\left(\frac{\beta_{i-1} - z_i}{\beta_{i-1} - \hat{z_i}}\right) C_{N,k-1}(\beta_{i-1}) = -\left(\frac{\beta_i - z_i}{\beta_i - \hat{z_i}}\right) C_{N,k-1}(\beta_i). \tag{3.19}$$

Solving (3.19) for $\hat{z_i}$ results in

$$\hat{z_i} = \frac{z_i[\beta_i C_{N,k-1}(\beta_{i-1}) + \beta_{i-1} C_{N,k-1}(\beta_i)] - \beta_{i-1}\beta_i[C_{N,k-1}(\beta_{i-1}) + C_{N,k-1}(\beta_i)]}{z_i[C_{N,k-1}(\beta_{i-1}) + C_{N,k-1}(\Omega_i)] - [\beta_{i-1} C_{N,k-1}(\beta_{i-1}) + \beta_i C_{N,k-1}(\beta_i)]}. \tag{3.20}$$

Equation (3.20) calculates the new position of z_i so that the adjacent critical points (β_{i-1} and β_i) become collinear.

3.3.3.3 Convergence condition

Using the update algorithms in the previous sections, a systematic improvement of zero locations can be achieved. To complete the optimization algorithm, only a termination convergence condition is therefore required.

Suppose Φ_k is the kth set of reflection and transmission zeros. We can then define an objective function

$$U(\Phi_k) = \sum_{w=1}^{W} |C_{N,k(N+n_{fz})}(\Omega_w) - C_{N,(k-1)(N+n_{fz})}(\Omega_w)|^2. \tag{3.21}$$

where Ω_1, Ω_2, …, Ω_W are frequency points spanning the frequency range of interest. Note that from the previous sections, the filter has N reflection zeros and n_{fz} finite frequency transmission zeros, and given an initial filtering function of $C_{N,0}$, the filtering function after one approximation iteration is $C_{N,N+n_{fz}}$. The objective function, $U(\Phi_k)$, is therefore a scalar function that measures the degree of change in C_N from one approximation iteration to the next. The objective function approximates zero when the algorithm converges to an optimum solution, i.e. when further updating of the zeros does not improve filter performance. The goal of the approximation algorithm is to reduce $U(\Phi_k)$ to a value below some pre-determined value, or to achieve the condition in (3.22)

$$U(\Phi_k) \leq \epsilon_{\text{limit}}. \quad (3.22)$$

Here ϵ_{limit} is a small numerical value and is used to define the accuracy of the approximated response. A typical value for ϵ_{limit} is 10^{-3}.

3.3.4 General algorithm

To summarize the procedure discussed in Section 3.3.3, a flowchart of the algorithm is given in Figure 3.6. The technique is as follows:

1. Define the passband and stopband edge frequencies in the multi-band prototype domain. This is done by mapping the passband and stopband edges from the actual frequency domain to the multi-band prototype domain using the low-pass-to-bandpass transformation (3.9).
2. Choose the number of reflection zeros in each passband and transmission zeros in each stopband. Note that the number of reflection zeros is equal to the order of the filter, and that the number of transmission zeros must equal the number of reflection zeros. This number in turn is equal to the minimum number of reactive elements in the prototype filter.
3. Synthesize an initial filtering function by uniformly distributing the reflection zeros in the passbands and the transmission zeros in the stopbands.
4. Find all the critical points associated with the different bands, i.e. find the coordinates indicated with α_n and β_n in Figure 3.4.
5. Find a new filtering function that is a better approximation of the design specifications. This is done by finding new positions for the reflection zeros and transmission zeros of the filter, using (3.16) for reflection zero positions and (3.20) for the transmission zero positions. The new zeros are used to find a new filtering function. Note that it is not always necessary to update both the reflection zeros and the transmission zeros. If equi-ripple stopbands are not required, the transmission zeros can be predetermined/predecided.
6. Apply the convergence condition of (3.22). If the convergence condition is not satisfied, the current filtering function is inadequate and a new filtering function must be synthesized using steps 4, 5, and 6 again. If the convergence condition is satisfied, we can construct the characteristic polynomials from the filtering function. A multi-band filter circuit can then be synthesized using classic theory [27].

3.3.5 Example 1

To illustrate the technique presented in the previous sections, this subsection contains two examples illustrating the design of multi-band filters using optimization. The first example is a dual-band stripline filter with the following half-power bandwidth specification:

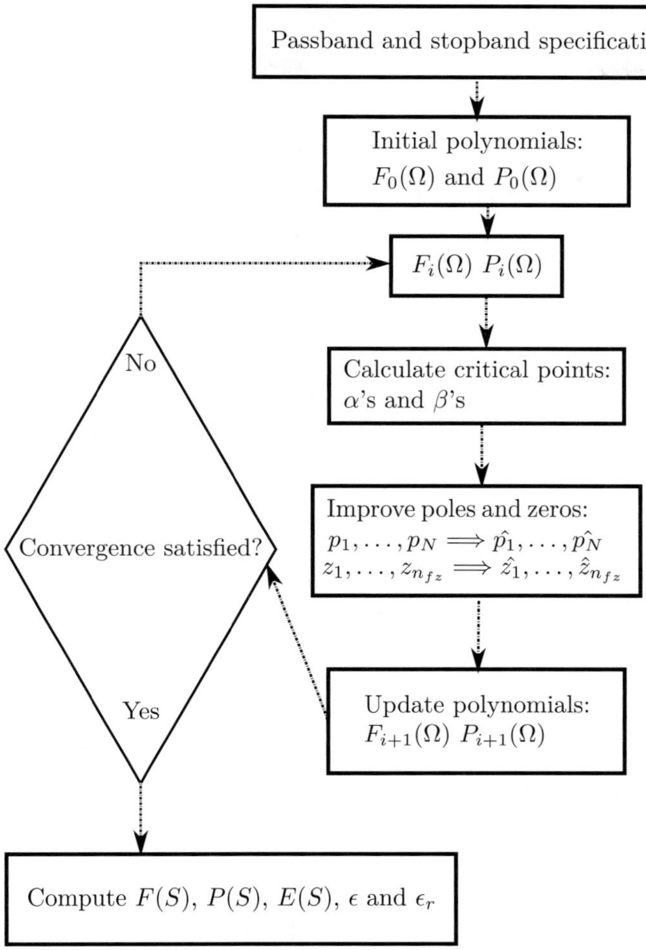

Figure 3.6 Flowchart of the iterative approximation technique.

- Half-power passband 1: 2.4 GHz to 2.5 GHz.
- Half-power passband 2: 2.688GHz to 2.8 GHz.
- Minimum return loss in first passband: 20dB.
- Order of filter: 8.

Approximation

The approximation algorithm presented in the previous sections defines the passband edges using equi-ripple frequencies. As there is no simple relationship between the equi-ripple bandwidth and the half-power bandwidth of a multi-band filter, the approximation problem must incorporate the equi-ripple passband edge frequencies in the actual frequency domain as arguments. For this example, the following equi-ripple bandwidths correspond to the desired half-power bandwidths:

- Pass-band 1: 2408.35 MHz to 2483.5 MHz.
- Pass-band 2: 2705.90 MHz to 2790.35 MHz.

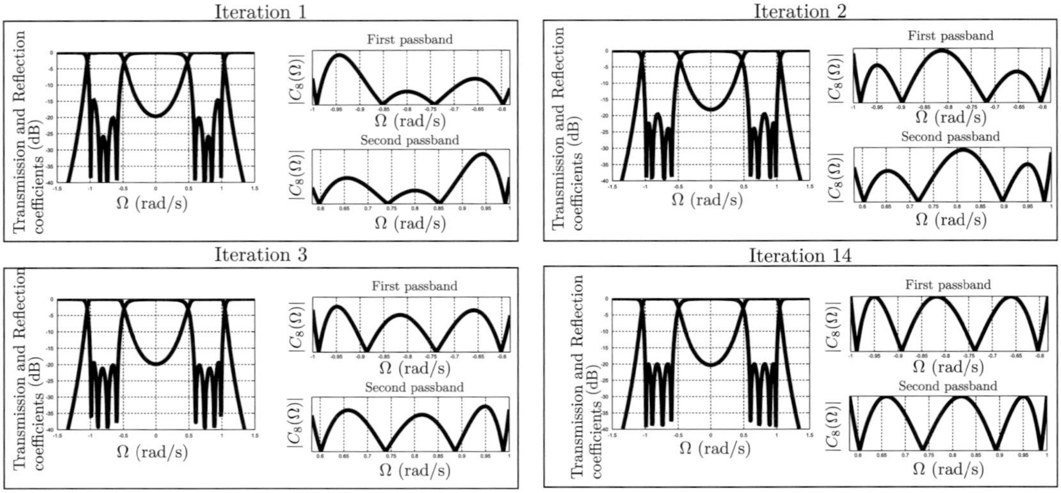

Figure 3.7 Graphical illustration of the execution of the approximation algorithm for example 1.

These equi-ripple passbands can be mapped to the prototype frequency domain through the inverse low-pass-to-bandpass transformation, resulting in a dual-band filter prototype with the following specifications:

- Pass-band 1: $\Omega \in [-1;-0.5822]$.
- Pass-band 2: $\Omega \in [0.5822;1]$.
- Minimum return loss in passband 1: 20 dB.
- Order of dual-band filter: 8.
- All the transmission zeros must be located at infinity.

This prototype specification is symmetric around $\Omega = 0$, which implies that the characteristic polynomials will all have real coefficients. Choosing four reflection zeros in each passband, and all transmission zeros at infinity, the iterative approximation algorithm is used to generate a valid equi-ripple filtering function. This process is illustrated in Figure 3.7, with the amplitude-frequency response shown on the left and the filtering function on the right. One iteration constitutes the modification of all of the reflection zeros. With a convergence limit of $\epsilon_{\text{limit}} = 10^{-3}$, the algorithm converges after 14 iterations. The resulting filtering function is

$$C_8(\Omega) = \Omega^8 - 2.6779\Omega^6 + 2.58\Omega^4 - 1.054\Omega^2 + 0.1534. \tag{3.23}$$

With $s = j\Omega$, the roots of (3.23) result in the characteristic polynomials

$$\begin{aligned} F(s) &= s^8 + 2.678s^6 + 2.5802s^4 + 1.0541s^2 + 0.1534 \\ jP(s) &= j. \end{aligned} \tag{3.24}$$

Using the conservation of energy principle $|S_{21}|^2 + |S_{11}|^2 = 1$, the normalization factor ϵ can be found as

$$\epsilon = \frac{1}{\sqrt{10^{\mathrm{RL}/10} - 1}} \left| \frac{P(j\Omega'_c)}{F(j\Omega'_c)} \right|. \tag{3.25}$$

Here Ω_c is the passband edge frequency. For this example, $\epsilon = 67.3703$. If the order of the filter is N and the number of finite frequency transmission zeros n_{fz}, ϵ_r can be found as [27]

$$\epsilon_r = \begin{cases} 1 & n_{fz} < N \\ \dfrac{\epsilon}{\sqrt{\epsilon^2 - 1}} & n_{fz} = N \end{cases}. \tag{3.26}$$

For this example, $\epsilon_r = 1$. Finally, $E(s)$ can be found using either the Feldtkeller equation (3.6) or the alternating pole method [27]

$$\begin{aligned} E(s) = s^8 + 0.9088s^7 + 3.0909s^6 + 1.9878s^5 + 3.1953s^4 \\ +1.324s^3 + 1.2655s^2 + 0.2582s + 0.1541 \end{aligned}. \tag{3.27}$$

At this stage, all the dual-band characteristic polynomials are known, and the approximation problem is solved. The next step in the design procedure is to synthesize a filter from the characteristic polynomials. For this example, we synthesize a coupling matrix from the characteristic polynomials and then realize a coupled stripline filter from the synthesized matrix.

Synthesis

A coupling matrix can be synthesized from the calculated characteristic polynomials using the methods in [27], giving the N+2 x N+2 matrix as

$$M = \begin{bmatrix} 0 & 0.674 & 0 & 0 & 0 & 0 & 0 & 0 & 0 & 0 \\ 0.674 & 0 & 0.835 & 0 & 0 & 0 & 0 & 0 & 0 & 0 \\ 0 & 0.835 & 0 & 0.367 & 0 & 0 & 0 & 0 & 0 & 0 \\ 0 & 0 & 0.367 & 0 & 0.75 & 0 & 0 & 0 & 0 & 0 \\ 0 & 0 & 0 & 0.75 & 0 & 0.309 & 0 & 0 & 0 & 0 \\ 0 & 0 & 0 & 0 & 0.309 & 0 & 0.75 & 0 & 0 & 0 \\ 0 & 0 & 0 & 0 & 0 & 0.75 & 0 & 0.367 & 0 & 0 \\ 0 & 0 & 0 & 0 & 0 & 0 & 0.367 & 0 & 0.835 & 0 \\ 0 & 0 & 0 & 0 & 0 & 0 & 0 & 0.385 & 0 & 0.674 \\ 0 & 0 & 0 & 0 & 0 & 0 & 0 & 0 & 0.674 & 0 \end{bmatrix}. \tag{3.28}$$

From the coupling matrix, we can obtain a lumped element coupled-resonator circuit, containing lumped element resonators and ideal admittance inverters. This circuit is shown in Figure 3.8, together with simulated circuit responses. It is clear that the first half-power passband extends from 2.4 GHz to 2.5 GHz, and the second from 2.688 GHz to 2.8 GHz. The first ripple passband extends from 2.408 GHz to 2.483 GHz, and the second from 2.705 GHz to 2.790 GHz. For this example, the circuit is realized using half-wavelength stepped-impedance resonators [33] to yield the coupled-line stepped-impedance filter shown in Figure 3.9, using NI AWR Microwave Office (MWO) models. Figure 3.10 shows the simulation results for the filter for both the linear MWO analysis, and a full-wave electromagnetic analysis using Sonnet. From Figure 3.10(a), it is clear that the initial Sonnet simulation predicts a deviation from the circuit model. Following a space-mapping electromagnetic optimization of the circuit [34], the final circuit is obtained. In space-mapping, we optimize a highly accurate but computationally

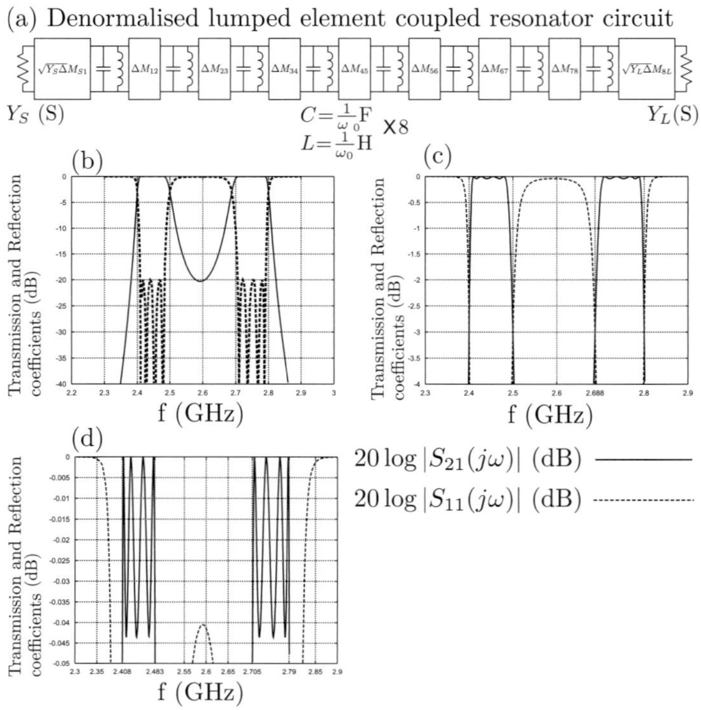

Figure 3.8 (a) The lumped element coupled-resonator circuit for example 1. (b)-(d) Simulated amplitude-frequency responses.

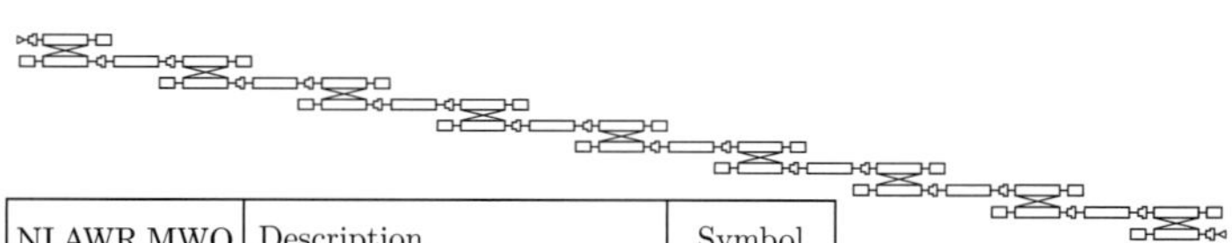

NI AWR MWO	Description	Symbol
GMCLIN	Multiple coupled lines on general multi-layer substrate	
GM1LIN	Single line on general multi-layer substrate	
PORT	Port	▷
SSTEP0	Stripline step in conductor width	
SLEF	Open stripline with end effect	

Figure 3.9 NI AWR MWO model of the filter in example 1.

expensive model by predicting its response using a fairly accurate but computationally inexpensive model in an optimization loop. The computationally inexpensive model is referred to as a surrogate model. The surrogate model for this design consists of electromagnetic simulations of small separate sections, interconnected in Microwave

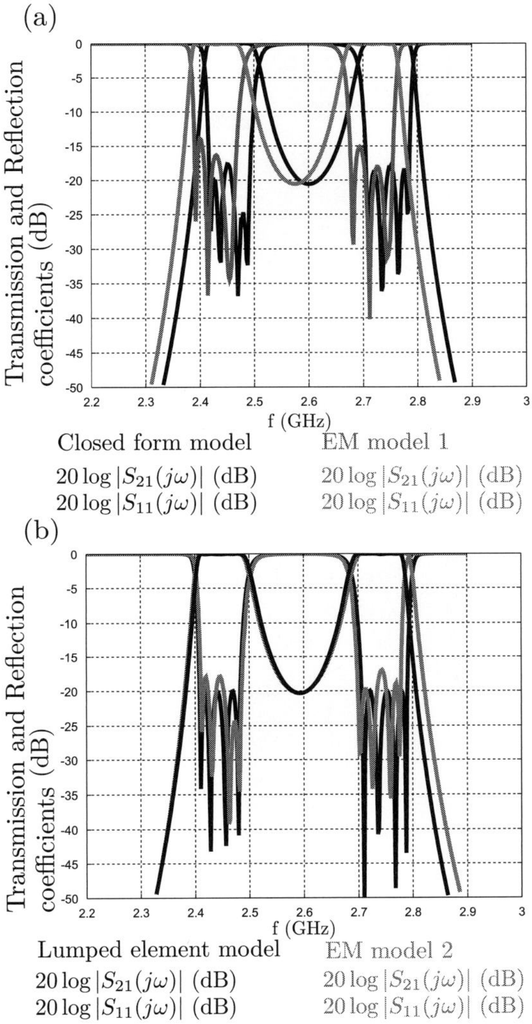

Figure 3.10 (a) Initial simulated performance of the filter. (b) Simulated performance after Space-Mapping optimization.

Office using closed form models, as shown in Figure 3.11. The closed form models can be tuned to obtain the final dimensions of the physical electromagnetic structure. The simulation results of the final electromagnetic model are shown in Figure 3.10(b).

The filter was fabricated using a photo-etching procedure on a 1.524 mm Rogers 4003C substrate, using a co-axial to stripline transition optimized in CST Microwave Studio as shown in Figure 3.12. The ground planes of the substrates were connected to the metal housing using a conductive adhesive, and the two stripline layers bonded using the Arlon CuClad 6250 bonding film. The fabricated filter is shown in Figure 3.13. The measured frequency response, compared to a lossy electromagnetic model in Sonnet, is shown in Figure 3.14.

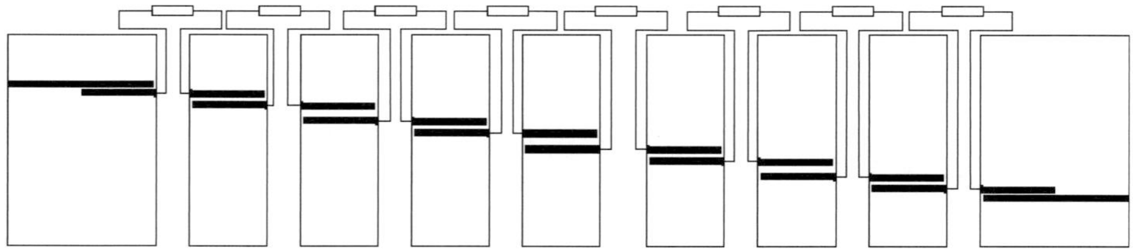

NI AWR MWO	Description	Symbol
GM1LIN	Single line on general multi-layer substrate	

Figure 3.11 A surrogate model for the cross-coupled stepped-impedance resonator.

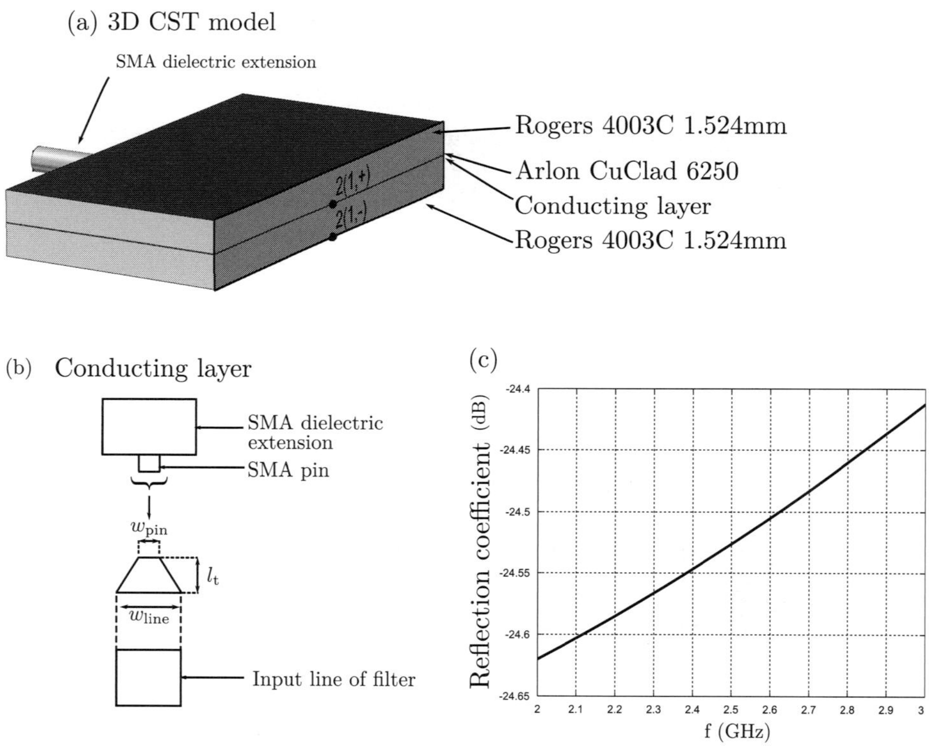

Figure 3.12 CST analysis of the co-axial to stripline transition.

3.3.6 Example 2

In this example, we design a cascaded triplet filter in stripline that has the following half-power bandwidth specification:

- Half-power passband 1: 5.15 GHz to 5.35 GHz.
- Half-power passband 2: 5.75 GHz to 5.95 GHz.

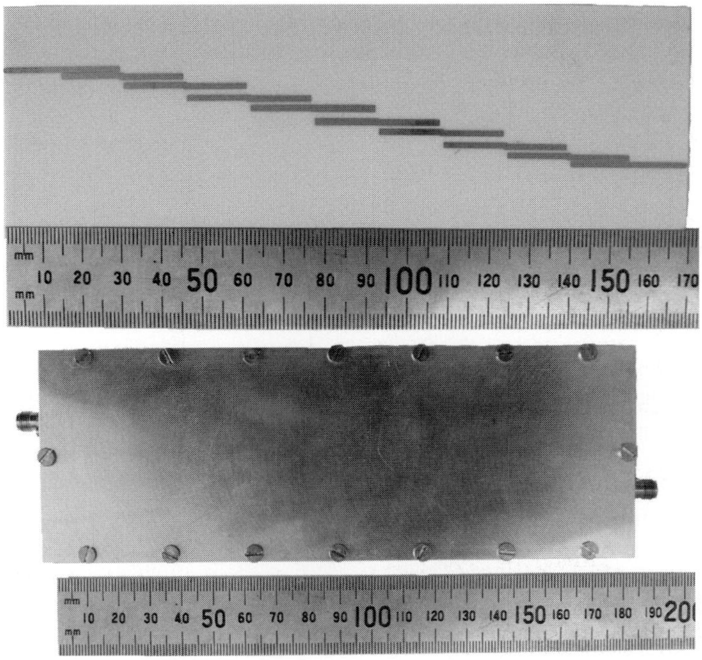

Figure 3.13 The filter in example 1.

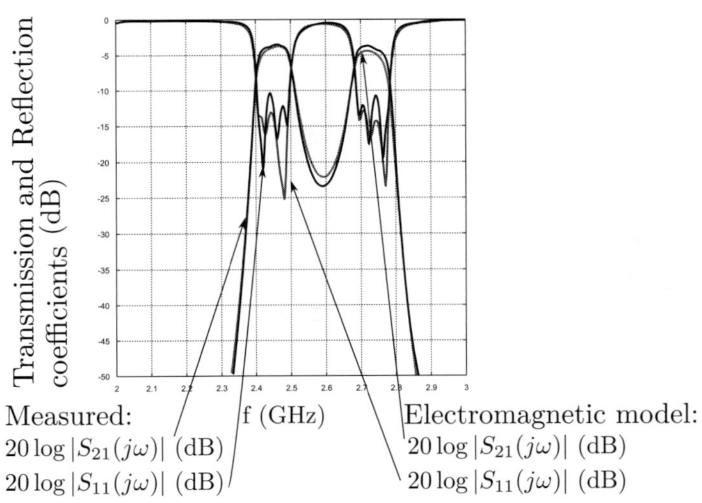

Figure 3.14 Measured and predicted responses of the filter.

- Minimum return loss in first passband: 20dB.
- Order of filter: 4.
- Two finite frequency transmission zeros are located between the passbands for isolation.

Approximation

As in the previous example, the bandwidth specification above is given in terms of half-power bandwidths. Using the same approach, the following equi-ripple bandwidths are associated with the desired half-power bandwidths:

- Stop-band 1: 0 GHz to 5.1 GHz.
- Pass-band 1: 5.2559 GHz to 5.3872 GHz.
- Stop-band 2: 5.59 GHz to 5.6748 GHz.
- Pass-band 2: 5.785 GHz to 5.862 GHz.
- Stop-band 3: 6.03 GHz to ∞.

The equi-ripple passband frequencies above can be mapped to the prototype frequency domain through the inverse low-pass-to-bandpass transformation, resulting in a dual-band filter prototype with the following specifications:

- Pass-band 1: $\Omega \in [-1;-0.5822]$.
- Pass-band 2: $\Omega \in [0.7575;1]$.
- Minimum return loss in passband 1: 20 dB.

The synthesis of the filtering function is illustrated in Figure 3.15, with the frequency responses in each case shown on the left, and the filtering function on the right. As before, one iteration constitutes modification of all of the reflection zeros. With the convergence limit $\epsilon_{\text{limit}} = 10^{-3}$, the algorithm converges after seven iterations, yielding the filtering function

$$C_4(\Omega) = \frac{\Omega^4 - 0.2122\Omega^3 - 1.3531\Omega^2 + 0.1749\Omega + 0.4195}{\Omega^2 - 0.5477\Omega + 0.0653}. \quad (3.29)$$

Here the finite frequency transmission zeros have been positioned at $\Omega = 0.1752$ and $\Omega = 0.3725$. This prototype specification is asymmetric around $\Omega = 0$, therefore the

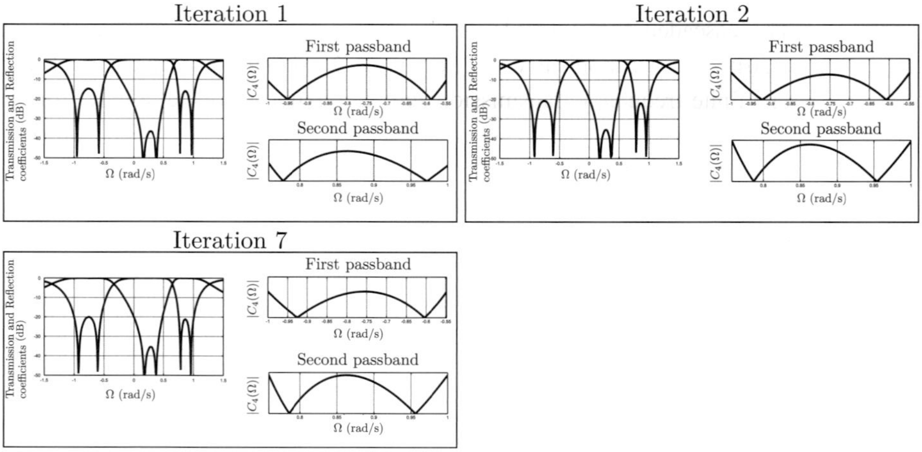

Figure 3.15 Graphical illustration of the execution of the approximation algorithm for example 2.

characteristic polynomials will have complex coefficients. As before, the roots of (3.29) yield the characteristic polynomials

$$\begin{aligned} F(s) &= s^4 - 0.2122js^3 + 1.3531s^2 - 0.1749js + 0.4195 \\ jP(s) &= js^2 + 0.5477s - 0.0653j \end{aligned} \tag{3.30}$$

and

$$\begin{aligned} \epsilon &= 1.5615 \\ \epsilon_r &= 1 \end{aligned}. \tag{3.31}$$

Finally $E(s)$ is calculated

$$\begin{aligned} E(s) = s^4 &+ (1.0701 - 0.2122j)s^3 + (1.9257 - 0.4080j)s^2 \\ &+ (0.6604 - 0.4899j)s + 0.3762 - 0.1903j \end{aligned}. \tag{3.32}$$

At this stage, the dual-band characteristic polynomials are known, and the approximation problem solved. For this example, we synthesize a coupling matrix from the characteristic polynomials and then realize a cascaded triplet filter using stepped-impedance resonators in stripline.

Synthesis

A coupling matrix is synthesized from the calculated characteristic polynomials from [27], and transformed using the methods presented in [39] to obtain a N+2 x N+2 coupling matrix in the cascaded triplet topology

$$M = \begin{bmatrix} 0 & 0.0923 & 0.7256 & 0 & 0 & 0 \\ 0.0923 & -0.0733 & 0.8017 & 0 & 0 & 0 \\ 0.7256 & 0.8017 & -0.031 & 0.6078 & 0 & 0 \\ 0 & 0 & 0.6708 & 0.3504 & 0.6929 & -0.7259 \\ 0 & 0 & 0 & 0.6929 & -0.4584 & 0.09 \\ 0 & 0 & 0 & -0.7259 & 0.09 & 0 \end{bmatrix}. \tag{3.33}$$

The cascaded triplet topology is obtained by first transforming the initial coupling matrix into a canonical wheel topology. Next multiple triplets, each associated with one finite frequency transmission zero, are extracted one at a time from the wheel and moved to their final positions in the filter [39] to give the lumped element circuit shown in Figure 3.16. Phase-reversing transformers are included in the model to ensure that the lumped element resonators have phase shifts corresponding to that of half-wavelength distributed resonators. Accurate representation of the phase at the start and end of each resonator is required as the cross-couplings are to be realized using delay lines. A delay line between the source/load and one of the internal resonators introduces a finite frequency transmission zero if the phase shift along the two paths differs by 180°. As before, the method in [33] is used to design the main coupling path using stepped-impedance resonators, yielding the ideal circuit with its simulated results in Figure 3.16. Figure 3.17 shows the closed-form Microwave Office model, with simulated results in Figure 3.18.

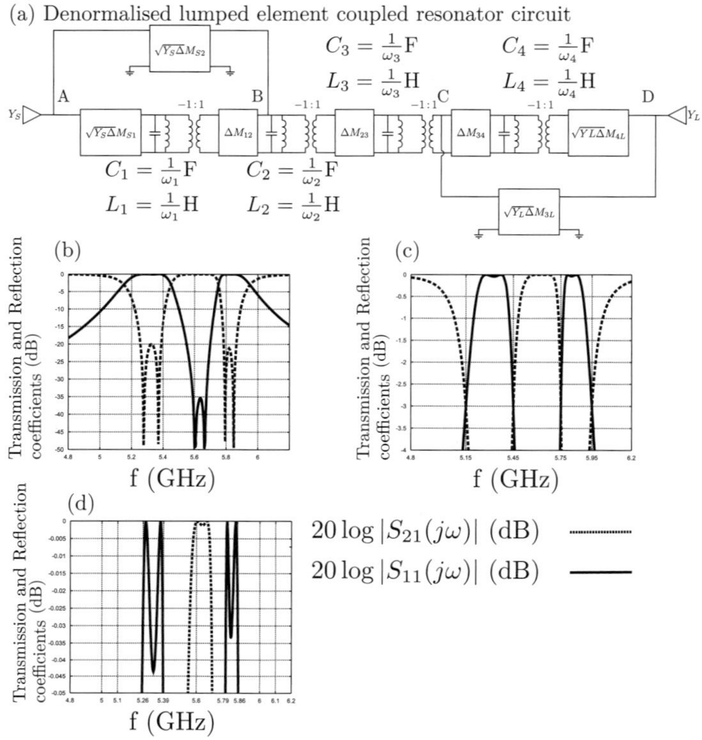

Figure 3.16 (a) Filter circuit for example 2. (b)–(d) Simulation results.

For the realization, the main path is first realized, followed by the systematic realization of each triplet in turn using delay lines. The lengths of the delay lines, their characteristic impedances, and the coupling gaps at the internal resonators can all be used to construct the required triplets. A full-wave electromagnetic analysis using Sonnet is shown together with the closed-form circuit analysis in Figure 3.18, as well as the circuit after optimization using space-mapping. The space-mapping surrogate model is constructed in the same way as for example 1, and shown in Figure 3.19.

The filter was fabricated using a photo-etching procedure on a 1.524 mm Rogers 4003C substrate. The ground planes of the substrates were connected to the metal housing using a conductive adhesive, and the two stripline layers were bonded using the Arlon CuClad 6250 bonding film. The fabricated filter is shown in Figure 3.20, with measured results compared to a lossy electromagnetic simulation in Figure 3.21.

3.4 Synthesis of transfer functions using reactance-based mapping functions

3.4.1 Transfer functions for coupled-resonator filters

In contrast to optimization-based approaches, this section presents a rigorous frequency mapping approach to derive multi-band transfer functions from low-pass transfer functions. While in principle quite general, this approach is particularly well suited to

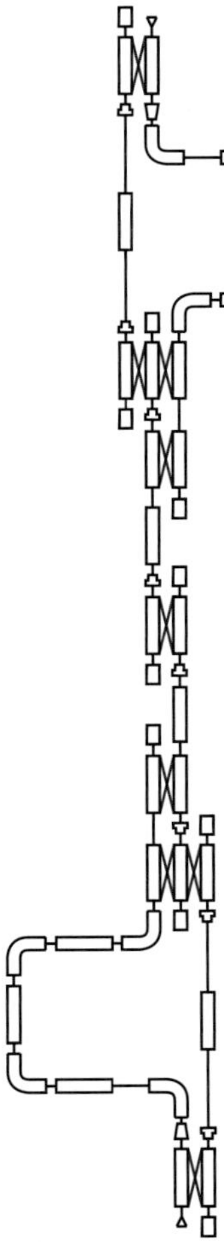

NI AWR MWO	Description	Symbol
GMCLIN	Multiple coupled lines on general multi-layer substrate	
GMCLIN	Multiple coupled lines on general multi-layer substrate	
GM1LIN	Single line on general multi-layer substrate	
MTAPER	A linear microstrip taper	
MCURVE	Microstrip radiused corner	
PORT	Port	
SSTEP0	Stripline step in conductor width	
SLEF	Open stripline with end effect	

Figure 3.17 A closed form MWO model of the cross-coupled stepped-impedance resonator filter.

coupled-resonator filters; therefore, the discussion will centre around this specific application.

Transfer functions derived by transformation-based procedures typically start with a normalized low-pass filtering function which needs to be transformed to a multi-band function. To construct such a multi-band filtering function, a mathematical relationship is required between a normalized single-band frequency variable Ω_s and a multi-band frequency variable, ω. By simple substitution, a multi-band filtering function $H(\omega)$ can then be obtained from a single-band filtering function $H_s(\Omega_s)$ through the mapping function $\mathrm{H}(\omega) = H_s\Big(\omega(\Omega_s)\Big)$.

As discussed in the previous sections, modern multi-band designs often first generate (typically by optimization) a transfer function that is not symmetrical with respect to the zero-frequency axis, but which contains the correct number of bands. Such a function can only be realized by non-physical components. This function is then

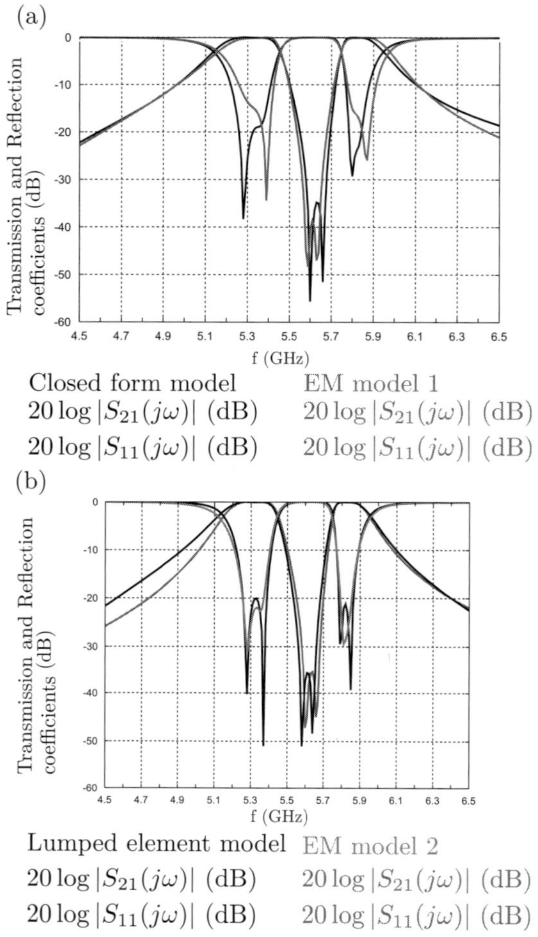

Figure 3.18 (a) Initial simulated performance of the filter. (b) Simulated performance after Space-Mapping optimization.

transformed through the classical low-pass-to-bandpass transformation to obtain the final, realizable function. This process is illustrated in Figure 3.22, with Ω_m denoting the intermediate frequency axis, and $\mathrm{H_m}(\Omega_\mathrm{m}) = \mathrm{H_s}\left(\Omega_\mathrm{m}(\Omega_\mathrm{s})\right)$ the intermediate filtering function.

3.4.2 Low-pass to intermediate multi-band transformation

We can illustrate the required mapping regions graphically, as shown in Figure 3.23, where allowed regions (where a mapping may exist) and disallowed regions are, respectively, designated by unshaded and shaded areas. As this step is an intermediate step, the function can be positioned arbitrarily with respect to the zero-frequency axis, in this case with all the passbands fitting between –1 and +1. The requirements of this mapping function, with reference to Figure 3.23, are as follows:

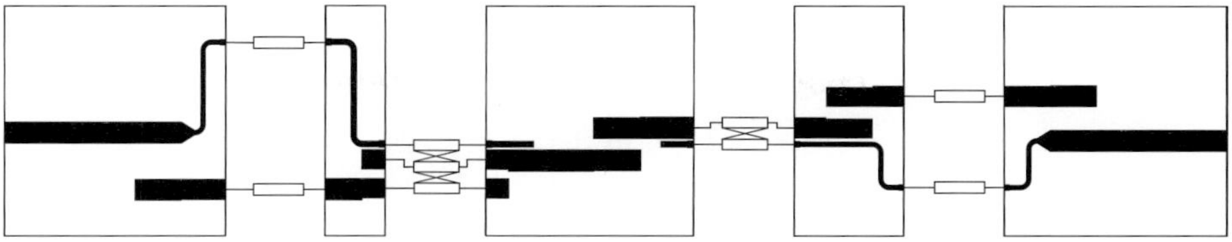

NI AWR MWO	Description	Symbol
GM1LIN	Single line on general multi-layer substrate	
GMCLIN	Multiple coupled lines on general multi-layer substrate	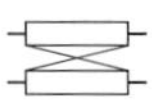

Figure 3.19 A surrogate model for the cross-coupled stepped-impedance resonator filter.

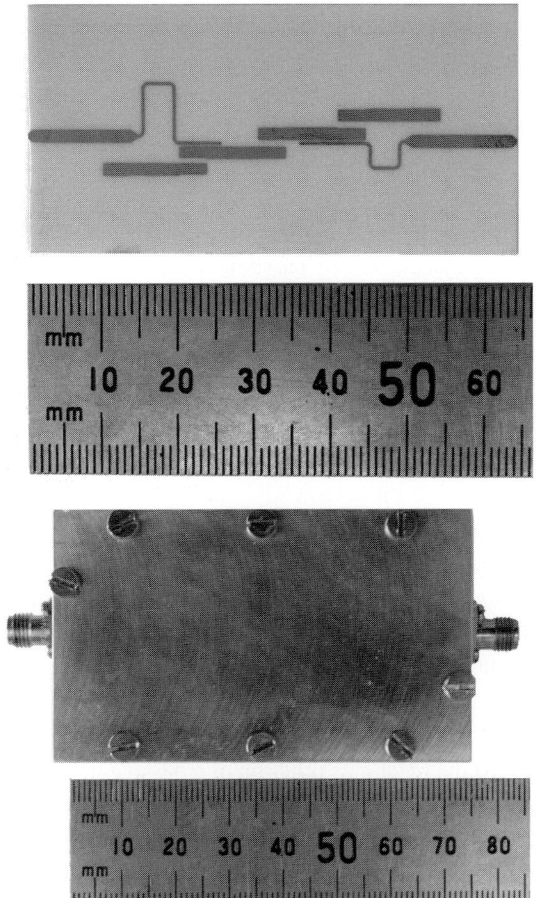

Figure 3.20 The filter in example 2.

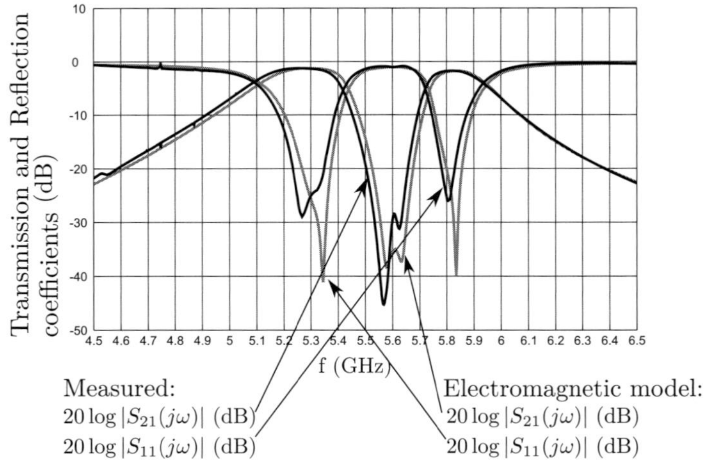

Figure 3.21 Measured and predicted responses of the cascaded triplet filter.

- $|\Omega_s| \leq 1$ should map to passbands. This section of the mapping function should be monotonically rising, and preferably linear to prevent severe phase and group delay distortion in the final function.
- $|\Omega_s| > 1$ should not map to passbands. This is guaranteed if this section of the mapping function also rises monotonically.
- $|\Omega_s| = 0$ should map to the centre frequency of every band, here denoted by z_n.
- $|\Omega_s| = \pm\infty$ should preferably map to frequencies between every passband (here denoted by p_n) and to $\pm\infty$ as such a mapping will ensure a point of infinite isolation between each adjacent set of passbands for a low-pass function with attenuation poles at $\pm\infty$.

A stringent requirement to avoid the shaded regions, while seemingly obvious, does place quite severe restrictions on possible mapping functions, especially if poles are required between adjacent passbands. A classical function which does satisfy all these requirements is a rational function with single poles and zeros interspersed on the Ω_s axis. Such a function is known to rise monotonically, and when plotted on the same axis system in Figure 3.23, can be seen to yield the required mapping function. From classical network synthesis [35], it is also well-known that the reactance functions of passive LC circuits are monotonically increasing rational functions with single poles and zeros interspersed on the frequency axis, with either poles or zeros at zero and infinite frequencies. When such a classical reactance function is transformed using a standard bandpass-to-low-pass transformation, the intermediate filtering function in Figure 3.22 is found exactly.

In [25], the authors proposed the reactance function of a passive, lossless LC network, transformed through a bandpass to low-pass transformation, as a mapping function

$$\Omega_s(\Omega_m) = \frac{\alpha_N \Omega_m^N + \alpha_{N-1}\Omega_m^{N-1} + \cdots + \alpha_1 \Omega_m + \alpha_0}{\beta_{N-1}\Omega_m^{N-1} + \cdots + \beta_1 \Omega_m + 1} = \frac{P(\Omega_m)}{Q(\Omega_m)}. \tag{3.34}$$

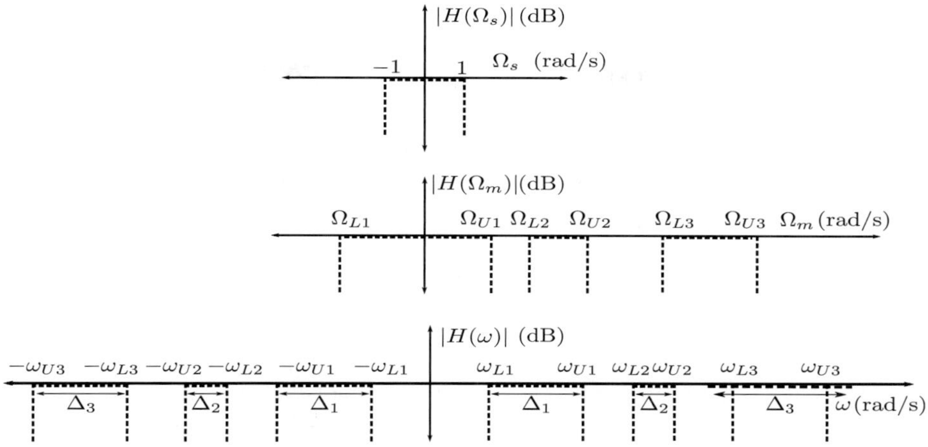

Figure 3.22 Low-pass to multi-band response transformation using an unsymmetrical intermediate response. © 2015 Wiley Periodicals, Inc. Int J RF and Microwave CAE, 2015.

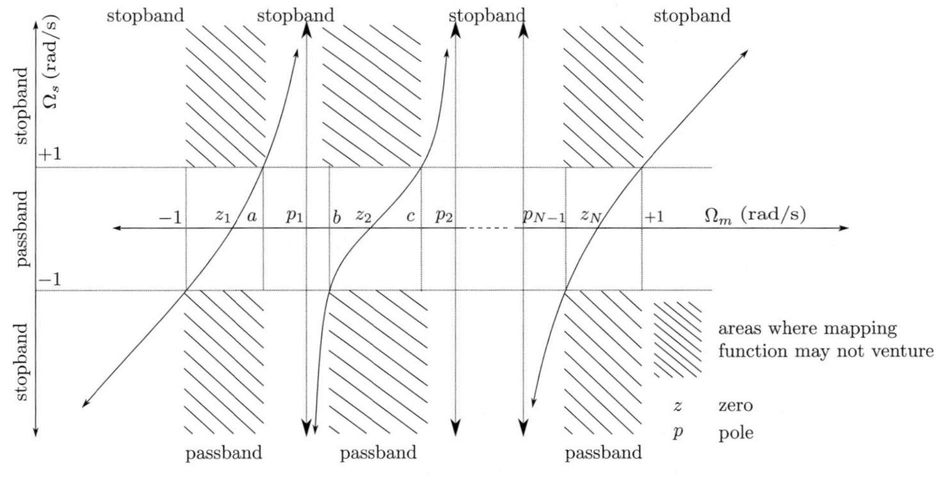

Figure 3.23 Graphical representation of the ideal frequency mapping function. © 2015 Wiley Periodicals, Inc. Int J RF and Microwave CAE, 2015.

The rational function in (3.34) has N zeros $z_1 \ldots z_N$ and N poles $p_1 \ldots p_{N-1}, \pm\infty$ with $z_1 < p_1 < z_2 < p_2 < \ldots < z_{N-1} < p_{N-1} < z_N$. The low-pass filtering function evaluated at its centre frequency maps to the N zeros of the mapping function in (3.34), while frequencies at $\pm\infty$ maps to the poles of the function. As a consequence, each $p_i \rightarrow z_i \rightarrow p_{i+1}$ region on the multi-band frequency axis is a bandwidth-scaled replica of the low-pass filtering function. In addition to satisfying the mapping requirements, this choice of mapping function has the following important advantages:

- There is no theoretical limit on the number of passbands which can be obtained.
- Each passband has the exact same reflection and transmission responses as the original low-pass transfer function in terms of amplitude and phase.
- The centre frequency and bandwidth for each band can be chosen completely arbitrarily.
- By simply choosing the correct poles and zeros, or equivalently the passband edges for each band, a valid mapping function can be obtained using a single matrix equation. No optimization is required.
- Given a low-pass filter function which can be realized with a circuit containing only positive inductors and capacitors, the application of the mapping function, followed by a low-pass to bandpass transform, will result in a classical reactance function transformation, and will therefore guarantee a multi-band filter function which can also be realized with positive-valued inductors and capacitors.
- The multi-band circuit can simply be realized by replacing each reactive element of any low-pass network with a network obtained by synthesizing the mapping function as a scaled impedance or admittance one-port. The process places no restrictions on the low-pass function, therefore any low-pass network can be used.
- No complicated cross-couplings are incurred by the low-pass to multi-band mapping, as each reactive element is realized in isolation.
- The final multi-band function need never be realized directly, only the low-pass function and an impedance or admittance in the form of the mapping function.

The coefficients of the rational function in (3.34) are calculated directly from the multi-band filter specification. Starting with the band-edge frequencies of the final filter (ω_m) a bandpass-to-low-pass transformation is performed to yield a set of normalized multi-band frequencies (Ω_m), with the low band-edge of the lowest passband at $\Omega_m = -1$, the high band-edge of the highest passband at $\Omega_m = +1$ and the remaining band edges at frequencies between these two edges. This results in a set of normalized multi-band frequencies $\Omega_\mathrm{m} \in [-1, a] \bigcap [b; \ldots 1]$. The mapping function is now required to map each of these frequency points to either of the band edges of a normalized low-pass filtering function with a passband defined by $\Omega_s \in [-1;1]$, leading to a set of $2N$ coordinate pairs

$$\{\ \Omega_{mi}, \Omega_{si}\} = \{(-1,\ -1); (a, 1); (b,\ -1); (c, 1) \ldots (1, 1)\} \qquad i \in [1, 2N]. \quad (3.35)$$

By applying (3.34) to each of these pairs, a set of $2N$ linear equations in the unknown coefficients are obtained

$$P(\Omega_{mi}) - \Omega_{si} Q(\Omega_{mi}) = 0\ ; \qquad i \in [1, 2N]. \quad (3.36)$$

This can be expressed in matrix notation as AX-B or in detailed format

$$\begin{bmatrix} \Omega_{m1}^{N} & \Omega_{m1}^{N-1} & \cdots & \Omega_{m1}^{1} & 1 & \Omega_{s1}\Omega_{m1}^{N-1} & \Omega_{s1}\Omega_{m1}^{N-3} & \cdots & \Omega_{s1}\Omega_{m1}^{1} \\ \Omega_{m2}^{N} & \Omega_{m2}^{N-1} & \cdots & \Omega_{m2}^{1} & 1 & \Omega_{s2}\Omega_{m2}^{N-1} & \Omega_{s2}\Omega_{m2}^{N-3} & \cdots & \Omega_{s2}\Omega_{m2}^{1} \\ \vdots & \vdots & \vdots & \vdots & \vdots & \vdots & \vdots & \vdots & \vdots \end{bmatrix}$$
$$\begin{bmatrix} \alpha_N \\ \alpha_{N-1} \\ \vdots \\ \alpha_1 \\ \alpha_0 \\ -\beta_{N-1} \\ -\beta_{N-3} \\ \vdots \\ -\beta_1 \end{bmatrix} = \begin{bmatrix} \Omega_{s1} \\ \Omega_{s2} \\ \vdots \end{bmatrix}. \tag{3.37}$$

Naturally $A \in \mathbb{R}^{2N \times 2N}$, $X \in \mathbb{R}^{2N \times 1}$, and $B \in \mathbb{R}^{2N \times 1}$ with the necessary and sufficient condition for a solution to the matrix equation that the rank of A and that of the augmented matrix $A|B$ must be $2N$. This condition is always true as both A and $A|B$ always have $2N$ linearly independent row vectors. We can confirm this by constructing the following vector sum

$$c_1 A_{\,11:2N} + c_2 A_{\,21:2N} + \ \ldots\ c_{2N} A_{\,2N1:2N} = 0. \tag{3.38}$$

Equation (3.38) only holds if $\Omega_{mi} \neq \Omega_{mj}$ for $i \neq j$ and $c_1 = 0, c_2 = 0, \ldots c_{2N} = 0$. Hence A is always of rank $2N$. A similar argument holds for the augmented matrix $A|B$. We can therefore conclude that a solution to the coefficients of (3.34) always exists.

3.4.3 Intermediate to final multi-band transformation

Once a normalized, unsymmetrical multi-band response is obtained, the final frequency transformation is the standard low-pass to bandpass transformation

$$\Omega_m = \frac{1}{\Delta}\left(\frac{\omega}{\omega_0} - \frac{\omega_0}{\omega}\right). \tag{3.39}$$

where $\Delta = (\omega_{NU} - \omega_{IL})/\omega_o$ and $\omega_o = \sqrt{\omega_{NU}\omega_{IL}}$, if we denote the high band-edge of the highest passband as ω_{NU} and the low band-edge of the lowest passband as ω_{IL}. Due to the nature of the mapping function in (3.34), the combination of the two transformations (3.34) and (3.39), when applied to a valid low-pass network function, will again result in a valid network function showing N bands with identical amplitude and phase responses, but arbitrary centre frequencies and bandwidths.

3.4.4 Implementation as coupled-resonator structure

Once a transformation function is found, a multi-band coupled-resonator filter structure can simply be implemented by designing a low-pass coupled-reactance filter using

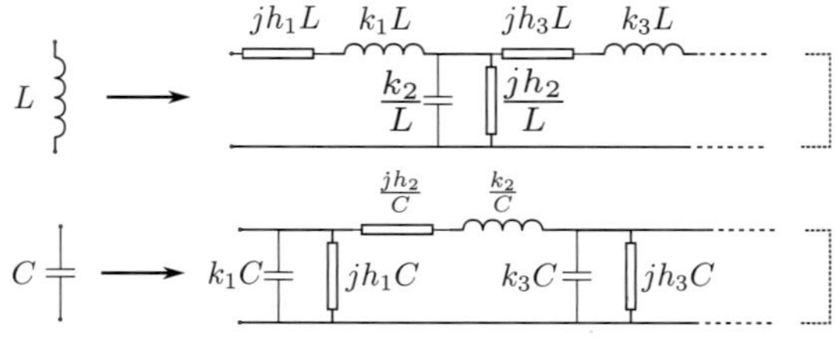

Figure 3.24 Cauer I forms of the low-pass to multi-band mapping function. © 2015 Wiley Periodicals, Inc. Int J RF and Microwave CAE, 2015.

standard techniques, followed by a step in which each reactive element is replaced by a reactive circuit. This is done in a four-step procedure.

Step 1. As a first step, each reactive element in the low-pass filter is replaced with a circuit, which is the direct network expansion (as admittance or impedance) of the low-pass to multi-band mapping function. There are numerous classical techniques available in the literature to synthesize the required transformation sub-circuit from the reactance (susceptance) function, with two of the simplest forms being the Cauer I & II topologies [35]. For a network containing frequency independent reactances, the Cauer I expansion is shown in (3.40), with the circuit form in Figure 3.24

$$j\Omega_s(\Omega_m) = j\Omega_m k_1 + jh_1 + \cfrac{1}{j\Omega_m k_2 + jh_2 + \cfrac{1}{\ddots}}. \tag{3.40}$$

Step 2. The Cauer I expansion in Figure 3.24 is transformed to a coupled-reactance low-pass structure with equal shunt capacitance or series inductance values coupled by J or K inverters, respectively. The frequency-independent reactances/susceptances are also transformed using the same inverter constants.

Step 3. Each frequency dependent element in the resulting coupled-reactance structure is replaced by the standard resonant circuit obtained from the transform in (3.39). This results in a classical coupled-resonator structure with the addition of a frequency-independent reactance or susceptance in each arm. Note that, as all the frequency dependent elements in a low-pass coupled-reactance circuit are normally chosen to be equal, all of the expanded networks will be identical.

Step 4. An equivalent resonator is constructed for each arm by the combination of the resonant circuit obtained by application of (3.39) with the frequency-independent reactance in that arm, as shown in Figure 3.25. For the case of a shunt capacitance for instance, the admittance of a branch in the low-pass coupled-resonator circuit will be given by

$$Y(j\Omega_m) = j\Omega_m C + jB. \tag{3.41}$$

Using the transform in (3.39), we obtain the admittance of an arm in the final multi-band filter as

$$Y'(j\omega_i) = j\frac{C}{\Delta}\left(\frac{\omega_i}{\omega_0} - \frac{\omega_0}{\omega_i}\right) + jB = j\omega_i C' + \frac{1}{j\omega_i L'} \tag{3.42}$$

Table 3.1 Bandwidth specifications for example 1

Band	Lower edge	Upper edge
1	2.4 GHz	2.5 GHz
2	2.7 GHz	2.8 GHz

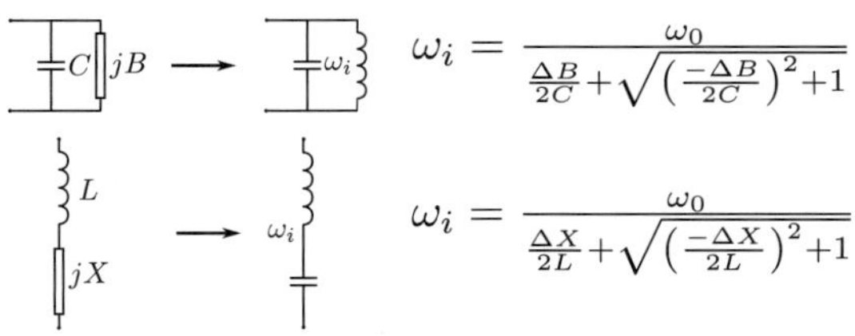

Figure 3.25 Equivalent Resonators. © 2015 Wiley Periodicals, Inc. Int J RF and Microwave CAE, 2015.

where C' and L' are the element values for the equivalent resonator, and ω_i the new resonant frequency. To calculate these values, we note that at resonance the input susceptance of the resonator should equal zero, thus $Y'(j\omega_i) = 0$. The frequency at which this is valid is

$$\omega_i = \frac{\omega_0}{\frac{\Delta B}{2C} + \sqrt{\left(\frac{-\Delta B}{2C}\right)^2 + 1}}. \quad (3.43)$$

The circuit transformations and appropriate expressions describing this approximation are provided in Figure 3.25. Note that the equivalent resonator has the same susceptance slope at the new resonant frequency as that of the resonator at ω_o in the case where $B = 0$, resulting in an equivalence over a fairly wide bandwidth. For typical coupled-resonator filters with bandwidths below 10% (which in this case should include the lowest edge of the lowest band and the highest edge of the highest band), the equivalence is sufficiently accurate to cause no deviation from ideal behavior in the final filter.

This procedure results in a coupled-resonator implementation of the complete mapping function. It should be clear that, given a coupled-reactance low-pass prototype, the replacement of each of its elements with a branch described in this section will lead to a full multi-band coupled-resonator filter structure. We now illustrate the application of the presented theory using two examples.

3.4.5 Example 1

As a first example, a dual-band filter with 20 dB return loss in each passband is designed for the set of passbands defined in Table 3.1. Note that the choice of the number of passbands is not limited in any way by the procedure, and that of the actual frequencies only by the normal coupled-resonator narrow-band condition.

Table 3.2 Prototype passband edge frequencies

Ω_{mi}	Ω_{si}
–1	–1
–0.47	1
0.527	–1
1	1

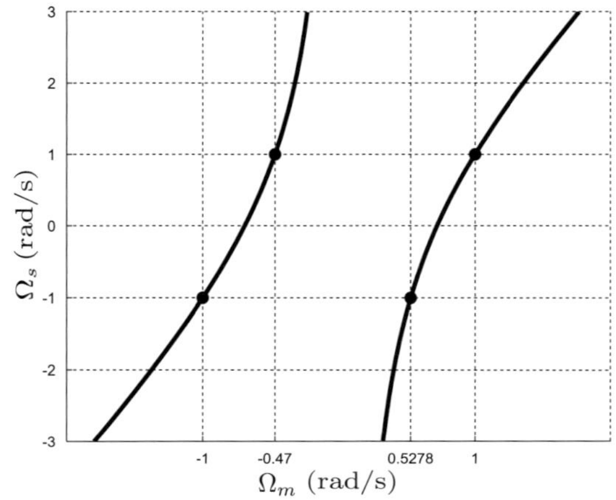

Figure 3.26 The reactance transformation function.

To transform to the intermediate frequency domain, the inverse low-pass-to-bandpass transformation is applied to each frequency of interest, with

$$f_0 = \sqrt{2.4 \times 2.8} = 2.592 \text{ GHz and } \Delta = 0.1543. \tag{3.44}$$

The resulting coordinate set $\{\Omega_{mi}, \Omega_{s1}\}$ that must be interpolated by the mapping function is listed in Table 3.2.

The matrices required for the calculation of the coefficients of the reactance transformation can be constructed from Table 3.2 as

$$A = \begin{bmatrix} 1 & -1 & 1 & -1 \\ 0.2209 & -0.47 & 1 & 0.47 \\ 0.2785 & 0.5278 & 1 & 0.5278 \\ 1 & 1 & 1 & -1 \end{bmatrix}, \quad B = \begin{bmatrix} -1 \\ +1 \\ -1 \\ +1 \end{bmatrix} \quad \text{and} \quad X = \begin{bmatrix} -34.6154 \\ 1 \\ 17.2692 \\ -17.3462 \end{bmatrix} \tag{3.45}$$

The desired reactance transformation is therefore

$$\Omega_s(\Omega_m) = \frac{-34.6153846\Omega_m^2 + \Omega_m + 17.2692}{-17.3451\Omega_m + 1} \tag{3.46}$$

A plot of the reactance transformation is shown in Figure 3.26. It is clear that the function performs the mapping as designed.

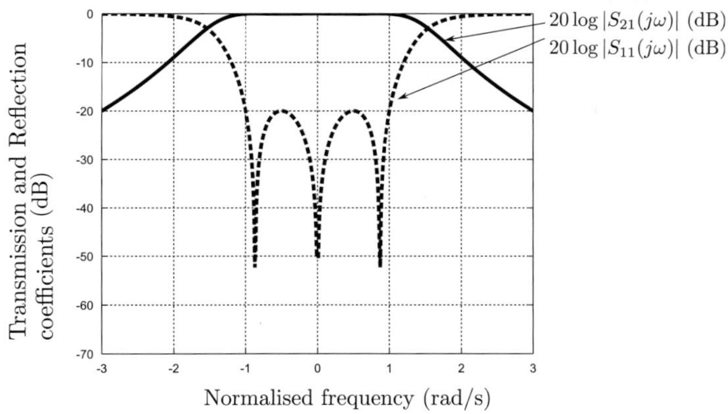

Figure 3.27 The low-pass frequency response.

For the low-pass filtering function, a generalized third-order Chebyshev polynomial is used. Cameron's method [36] is used to find the low-pass filtering function

$$C_3(\Omega_s) = \Omega_s^3 - 0.75\Omega_s. \tag{3.47}$$

Here C_3 is a third-order Chebyshev filtering function. The characteristic polynomials are now constructed and a ripple factor is calculated to obtain a minimum passband return loss level of 20 dB [36]. A plot of the low-pass frequency response is provided in Figure 3.27.

The first step in the synthesis procedure is to determine the low-pass circuit for the filtering function in (3.4.14). Using the methods by Cameron [36, 37], the coupling matrix obtained is

$$M = \begin{bmatrix} 0 & 1.08246 & 0 & 0 & 0 \\ 1.08246 & 0 & 1.0303 & 0 & 0 \\ 0 & 1.0303 & 0 & 1.0303 & 0 \\ 0 & 0 & 1.0303 & 0 & 1.08246 \\ 0 & 0 & 0 & 1.08246 & 0 \end{bmatrix}. \tag{3.48}$$

In order to simplify the synthesis discussion, a set of symbols representing sub-circuits is shown in Figure 3.28.

The single-band prototype circuit associated with the coupling matrix in (3.48) is shown in Figure 3.29, with $jb = 0$ in each case.

The next step is to implement the reactance transformation circuit. Equation (3.46) is expanded in the form of (3.40) to find the h and k factors. Alternatively, a series of pole and constant reactance extractions can be performed on (3.46) using classical extraction techniques [35]. The results are listed in Table 3.3.

The calculated h and k factors now describe circuits of the form depicted in Figure 3.24. Immittance inverters can be inserted into these circuits to transform them into the topology illustrated in Figure 3.30(b).

Table 3.3 Reactance transformation coefficients

i	k_i	h_i
1	1.995565410199558	0.057394014778687
2	1.007803762818456	−0.058099551736764

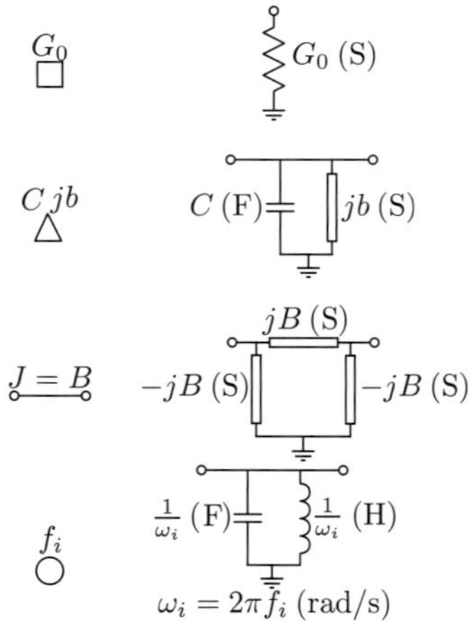

Figure 3.28 Symbols and their circuit representations. © 2015 Wiley Periodicals, Inc. Int J RF and Microwave CAE, 2015.

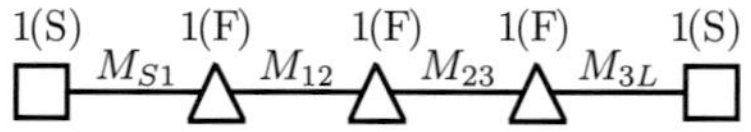

Figure 3.29 The low-pass prototype circuit.

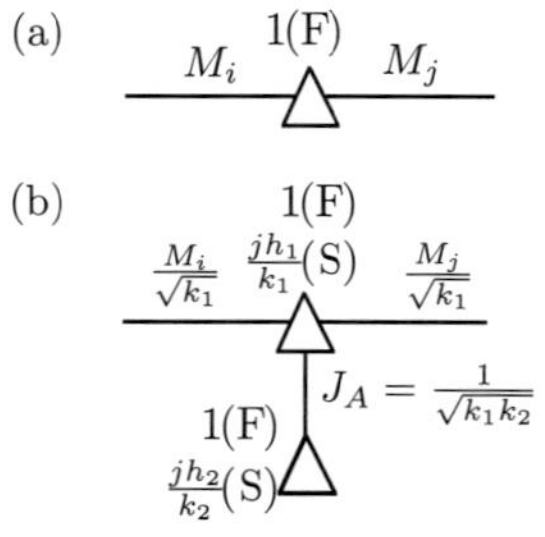

Figure 3.30 (a) Low-pass element, (b) Multi-band expansion circuit. © 2015 Wiley Periodicals, Inc. Int J RF and Microwave CAE, 2015.

The circuit in Figure 3.29 is now transformed into a multi-band prototype by substituting sections corresponding to Figure 3.30(a) with the sub-circuit in 30(b). The final circuit follows using the procedure of steps 3 and 4, and is shown in Figure 3.31. As with all coupled-resonator filters, each resonator can be chosen

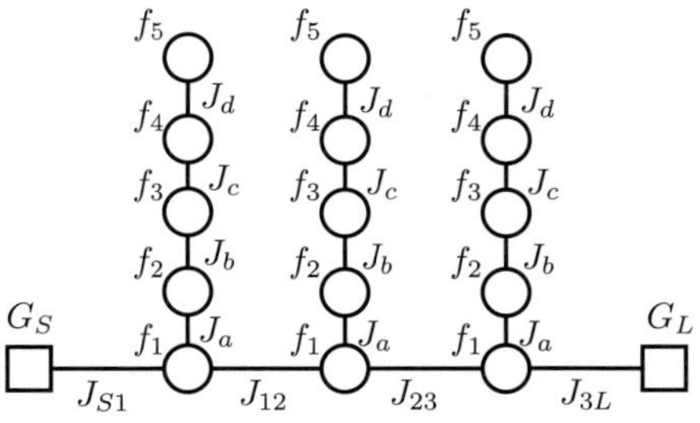

Figure 3.31 Dual-band filter circuit.

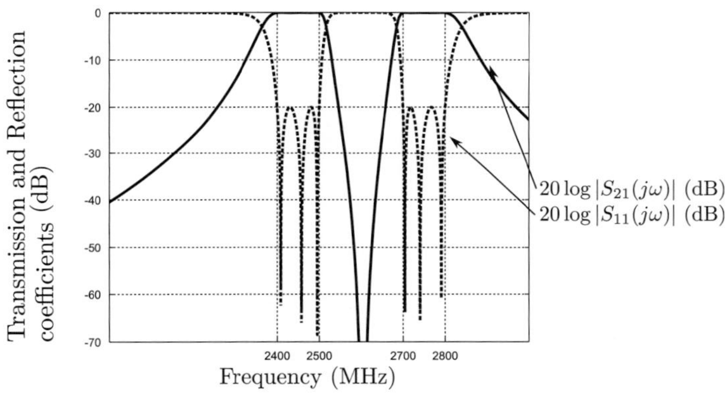

Figure 3.32 Circuit simulation of the dual-band filter.

arbitrarily, as all are coupled by inverters. The couplings of the final circuit in Figure 3.31 are defined as follows:

$$\begin{aligned} \text{Main coupling path} &: J_{ij} = \frac{\Delta M_{ij}\sqrt{b_i b_j}}{k_1} \quad i \neq j;\ i,j \in [1;3] \\ \text{Input} &: J_{S1} = \sqrt{\frac{\Delta b_1 G_s}{k_1}} M_{s1} \\ \text{Output} &: J_{L4} = \sqrt{\frac{\Delta b_4 G_L}{\mathrm{k}_1}} M_{L4} \\ \text{Branch} &: J_x = J_X \Delta\sqrt{b_x b_{x+1}}\ x \in \{a,b\} \end{aligned} \tag{3.49}$$

In each case, b_i is the susceptance slope of resonator i [38]. G_S and G_L represent the source and load conductances, respectively, Δ is the fractional bandwidth of the entire frequency response, k_1 is a scalar defined in (3.40), and M_{ij} is an appropriate matrix entry from (3.48). The centre frequencies of the different nodes are found using the frequency invariant susceptances shown in Figure 3.30 and the formulas provided in Figure 3.25. The circuit simulation is shown in Figure 3.32. Note that the response meets the design specifications exactly.

Table 3.4 Bandwidth specifications for five-band filter

Band	Lower edge	Upper edge
1	4.1 GHz	4.2 GHz
2	4.35 GHz	4.5 GHz
3	4.55 GHz	4.6 GHz
4	4.65 GHz	4.7 GHz
5	4.95 GHz	5 GHz

Table 3.5 Prototype passband edge frequencies

Ω_{mi}	Ω_{si}
−1	−1
−0.756613	1
−0.402937	−1
−0.061728	1
0.0494505	−1
0.1594202	1
0.2682198	−1
0.3758865	1
0.8984287	−1
1	1

3.4.6 Example 2

A second example is that of a five-band filter with 20 dB return loss in each passband, designed for the set of passbands defined in Table 3.4.

To transform to the intermediate frequency domain, the inverse low-pass-to-bandpass transformation is again applied to each frequency of interest, with

$$f_0 = \sqrt{4.1 \times 5} = 4.5277 \quad \text{GHz} \quad \text{and} \quad \Delta = 0.198776. \tag{3.50}$$

The resulting coordinate set $\{\Omega_{mi}, \Omega_{s1}\}$ that must be interpolated by the mapping function is listed in Table 3.5.

The desired reactance transformation is therefore

$$\Omega_s(\Omega_m) \quad = \quad \frac{\alpha_N \Omega_m^N + \alpha_{N-1}\Omega_m^{N-1} + \cdots + \alpha_1 \Omega_m + \alpha_0}{\beta_{N-1}\Omega_m^{N-1} + \cdots + \beta_1 \Omega_m + 1}. \tag{3.51}$$

with the coefficients of the mapping function results in Table 3.6.

A plot of the resulting reactance transformation is shown in Figure 3.33. It is clear that the function performs the mapping as designed.

For this design example, we shall use the same single-band base filter used in the previous example. The next step is therefore to implement the reactance transformation circuit. The coefficients in Table 3.6 are used to find the h and k factors in (3.40). These are listed in Table 3.7.

Table 3.6 Coefficients of reactance transformation

Coefficient	Value
α_5	−998.5775311
α_4	264.68613428
α_3	808.336874
α_2	−136.084478
α_1	−41.1354804
α_0	3.794739672
β_4	−451.2586688
β_3	194.73996428
β_2	218.88253163
β_1	−62.343568376
β_0	1

Table 3.7 Reactance transformation coefficients

i	k_i	h_i
1	2.212871685626753	0.368410471031648
2	1.789050044954395	−0.213396706921520
3	1.266047423094517	−0.065762109265814
4	8.256013882770498	−0.970967729071748
5	7.847132435188047	−1.119934126892491

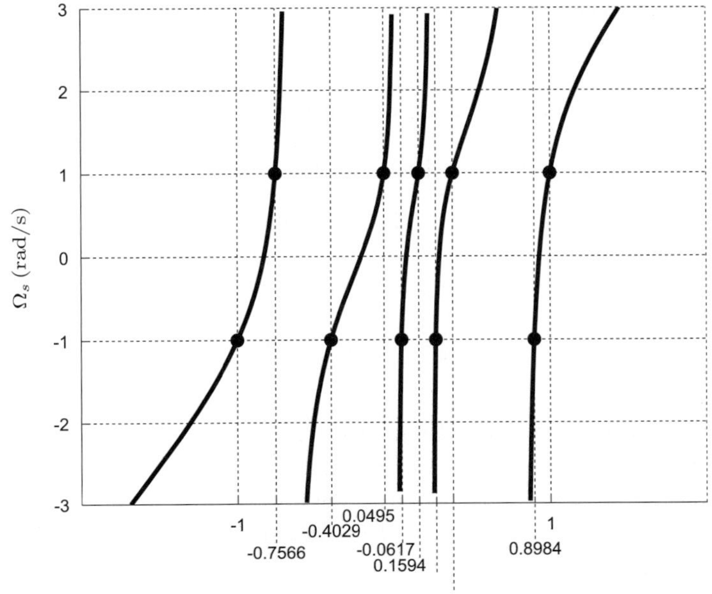

Figure 3.33 The reactance transformation function.

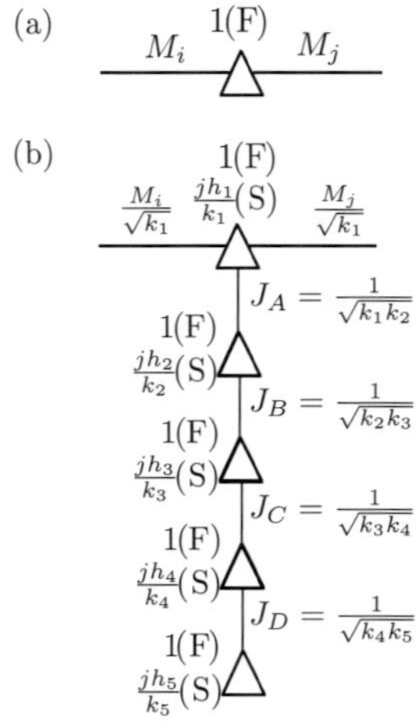

Figure 3.34 (a) Low-pass element. (b) Multi-band expansion circuit.

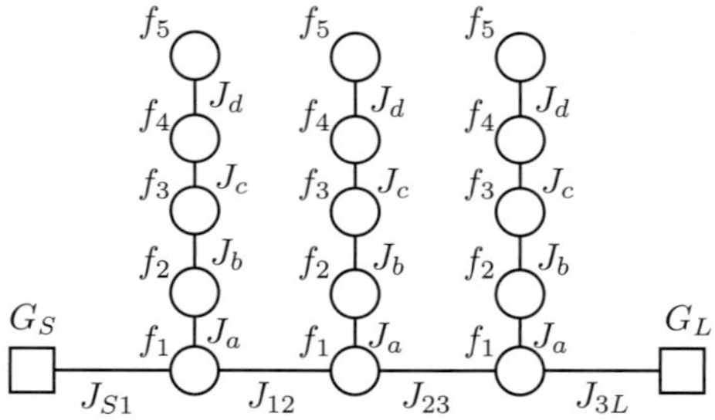

Figure 3.35 Circuit diagram of final five-band filter.

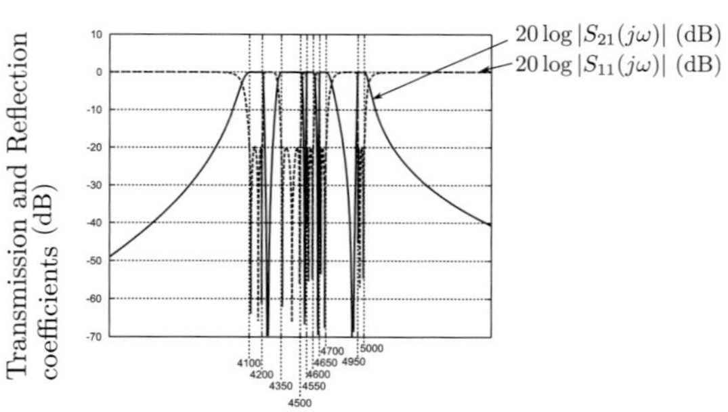

Figure 3.36 Circuit simulation of the 5-band filter.

The calculated h and k factors now describe circuits of the form depicted in Figure 3.24. Immittance inverters can be inserted into these circuits to transform them into the topology illustrated in Figure 3.34.

The circuit in Figure 3.29 is transformed into a multi-band prototype by substituting sections corresponding to Figure 3.30(a) with the sub-circuit in Figure 3.30(b). The final circuit follows using the procedure of steps 3 and 4 outlined in the previous sub-section, and is shown in Figure 3.35. Here all the circuit nodes have been replaced by resonators. The couplings of the final filter use the same equations as those in Example 1. The resonant frequencies are also calculated in the same manner. All couplings are calculated using appropriate susceptance slopes, and the simulated circuit response is shown in Figure 3.36.

3.5 Conclusions

In this chapter, recent developments in algorithms for the construction of multi-band transfer functions are described. Two approaches are presented in detail: one an optimized approach where poles and zeros of the filter function are systematically positioned and fine-tuned to achieve equi-ripple passbands and/or stopbands, and the other a rigorous approach proposed by the authors where the reactance function of a passive LC one-port is used as a frequency mapping function. In both cases, the mathematical theory is discussed, and examples presented. The optimization approach is very general with respect to the passband characteristics and allows for a wide range of asymmetrical passbands of different bandwidths and reflection coefficients, but is limited to an equi-ripple filtering response. It also suffers from the limitations of polynomial mapping functions, specifically stopband mapping and, in some cases, limitations on the level of reflection coefficients. On the other hand, the reactance-based mapping technique is rigorous, without any optimization, and allows for an arbitrary number of bands, bandwidths, and centre frequencies, but produces passbands with exactly the same amplitude and phase responses. Both techniques however have proved to produce excellent results. The construction of frequency mapping functions however still remains a challenging field of research, both in terms of optimization and rigorous mathematical approaches.

References

[1] X. Yu, X. Tang, F. Xiao, and F. Xu, "A new class of multi-band waveguide filters," *International Conference on Microwave and Millimeter Wave Technology (ICMMT)*, vol. 4, pp. 1–4, 2012.

[2] L.-C. Tsai and C.-W. Hsue, "Dual-band bandpass filters using equal-length coupled-serial-shunted lines and z-transform technique," *IEEE Transactions on Microwave Theory and Techniques*, vol. 52, no. 4, pp. 1111–1117, April 2004.

[3] M. T. Doan, W. Che, and W. Feng, "Novel compact dual-band bandpass filter with multiple transmission zeros and good selectivity," *IEEE International Conference on Microwave and Millimeter Wave Technology (ICMMT), 2012*, vol. 5, pp. 1–4., 2012.

[4] Y. Dong, C.-T. Wu, and T. Itoh, "Miniaturised multi-band substrate integrated waveguide filters using complementary split-ring resonators," *IET Microwaves, Antennas and Propagation*, vol. 6, no. 6, pp. 611–620, 2012.

[5] J.-K. Xiao and W.-J. Zhu, "Compact split ring sir bandpass filters with dual and tri-band," *Progress In Electromagnetics Research C*, vol. 25, pp. 93–105, 2012.

[6] W. Wang and X. Lin, "A dual-band bandpass filter using open-loop resonator," *IEEE 5th Global Symposium on Millimeter Waves (GSMM), 2012*, pp. 575–578, 2012.

[7] A. Genc, R. Baktur, and R. J. Jost, "Dual-bandpass filters with individually controllable passbands. IEEE Transactions on Components," *Packaging and Manufacturing Technology*, vol. 3, no. 1, pp. 105–112, Jan. 2013.

[8] X. Guan, Z. Ma, P. Cai, G. Li, Y. Kobayashi, T. Anada, and G. A. Hagiwara, "Dual-band bandpass filter synthesized by using frequency transformation and circuit conversion technique," *Asia-Pacific Microwave Conference Proceedings, 2005*, vol. 4, p. 4, Dec. 2005.

[9] M. Mokhtaari, J. Bornemann, K. Rambabu, and S. Amari, "Coupling-matrix design of dual and triple bandpass filters," *IEEE Transactions on Microwave Theory and Techniques*, vol. 54, no. 11, pp. 3940–3946, 2006.

[10] X. Guan, Z. Ma, P. Cai, Y. Kobayashi, T. Anada, and G. Hagiwara, "Synthesis of dual-band bandpass filters using successive frequency transformations and circuit conversions," *IEEE Microwave and Wireless Components Letters*, vol. 16, no. 3, pp. 110–112, 2006.

[11] A. Garcia-Lamperez and M. Salazar-Palma, "Dual band filter with split-ring resonators," *Proceedings of IEEE MTT-S International Microwave Symposium Digest, 2006*, pp. 519–522, 2006.

[12] A. Garcia Lamperez, "Analytical synthesis algorithm of dual-band filters with asymmetric passbands and generalized topology," *Proceedings of IEEE MTT-S International Microwave Symposium Digest, 2007*, pp. 909–912, 2007.

[13] H.-M. Lee and C.-M. Tsai, "Dual-band filter design with flexible passband frequency and bandwidth selections," *IEEE Transactions on Microwave Theory and Techniques*, vol. 55, no. 5, pp. 1002–1009, 2007.

[14] X. Shang, Y. Wang, G. Nicholson, and M. Lancaster, "Design of multiple-passband filters using coupling matrix optimization," *IET Microwaves, Antennas and Propagation*, vol. 6, no. 1, pp. 24–30, 2012.

[15] D. Deslandes and F. Boone, "An iterative design procedure for the synthesis of generalized dual-bandpass filters," *International Journal of RF and Microwave Computer-Aided Engineering*, vol. 19, no. 5, pp. 607–614, 2009.

[16] D. Deslandes and F. Boone, "Iterative design techniques for all-pole dual-bandpass filters," *IEEE Microwave and Wireless Components Letters*, vol. 17, no. 11, pp. 775–777, Nov. 2007.

[17] F. Seyfert and S. Bila, 'General synthesis techniques for coupled resonator networks," *IEEE Microwave Magazine*, vol. 8, no. 5, pp. 98–104, 2007.

[18] S. Bila, R. J. Cameron, P. Lenoir, V. Lunot, and F. Seyfert, "Chebyshev synthesis for multi-band microwave filters," *Proceedings of IEEE/MTT-S International Microwave Symposium Digest*, 2006, pp. 1221–1224.

[19] R. Cameron, "General coupling matrix synthesis methods for Chebyshev filtering functions," *IEEE Transactions on Microwave Theory and Techniques*, vol. 47, no. 4, pp. 433–442, April 1999.

[20] G. Macchiarella and S. Tamiazzo, "Dual-band filters for base station multi-band combiners," *Proceedings of IEEE/MTT-S International Microwave Symposium Digest, 2007*, pp. 1289–1292, 2007.

[21] G. Macchiarella and S. Tamiazzo, "Design techniques for dual-passband filters", *IEEE Transactions on Microwave Theory and Techniques*, vol. 53, no. 11, pp. 3265–3271, Nov. 2005.

[22] J. Lee and K. A Sarabandi, "A synthesis method for dual-passband microwave filters," *IEEE Transactions on Microwave Theory and Techniques*, vol. 55, no. 6, pp. 1163–1170, June 2007.

[23] J. Lee and K. Sarabandi, "Design of triple-passband microwave filters using frequency transformations," *IEEE Transactions on Microwave Theory and Techniques*, vol. 56, no. 1, pp. 187–193, Jan 2008.

[24] A. Mohan, S. Singh, and A. Biswas, "Generalized synthesis and design of symmetrical multiple passband filters," *Progress In Electromagnetics Research B*, vol. 42, pp. 115–139, 2012.

[25] T. G. Brand, P. Meyer, and R. Geschke, "Designing multiband coupled-resonator filters using reactance transformations," *International Journal of RF and Microwave Computer-Aided Engineering*, vol. 25, pp. 81–92, 2015.

[26] D. M. Pozar, *Microwave Engineering*, John Wiley, New Jersey, 2004.

[27] R. J. Cameron, R. Mansour, and C. M. Kudsia, *Microwave Filters for Communication Systems: Fundamentals, Design and Applications*, Wiley-Interscience, New Jersey, 2007.

[28] G. Macchiarella and S. Tamiazzo, "Design techniques for dual-passband filters," *IEEE Transactions on Microwave Theory and Techniques*, vol. 53, no. 11, pp. 3265–3271, Nov. 2005.

[29] An-Shyi Liu; Ting-Yi Huang, and Ruey-Beei Wu, "A dual wideband filter design using frequency mapping and stepped-impedance resonators," *IEEE Transactions on Microwave Theory and Techniques*, vol. 56, no.12, pp. 2921,2929, Dec. 2008.

[30] Hao Di, Wu Bian, Xin Lai, and Chang-Hong Liang, "Synthesis of cross-coupled triple-passband filters based on frequency transformation," *IEEE Microwave and Wireless Components Letters*, vol. 20, no. 8, pp. 432,434, Aug. 2010.

[31] A. Garcia-Lamperez and M. Salazar-Palma, "Single-band to multiband frequency transformation for multiband filters," *IEEE Transactions on Microwave Theory and Techniques*, vol. 59, no. 12, pp. 3048, 3058, Dec. 2011.

[32] Y. Zhang, K. Zaki, J. Ruiz-Cruz, and A. Atia, "Analytical synthesis of generalized multi-band microwave filters," *Proceedings of IEEE/MTT-S International Microwave Symposium Digest, 2007*, pp. 1273 –1276, June 2007.

[33] M. Makimoto and S. Yamashita, "Bandpass filters using parallel coupled stripline stepped-impedance resonators," *IEEE Transactions on Microwave Theory and Techniques*, vol. 28, no. 12, pp. 1413–1417, Dec. 1980.

[34] J. Meng, S. Koziel, J. Bandler, M. Bakr, and Q. Cheng, "Tuning space mapping: a novel technique for engineering design optimization," *Proceedings of IEEE MTT-S International Microwave Symposium Digest, 2008*, pp. 991–994, June 2008.

[35] G. C. Temes and J. W. LaPatra, *Introduction to Circuit Synthesis and Design*, McGraw-Hill International Student Edition, Kogakusha, 1977, pp. 51–55.

[36] R. Cameron, "General coupling matrix synthesis methods for Chebyshev filtering functions," *IEEE Transactions on Microwave Theory and Techniques*, vol. 47, no. 4, pp. 433–442, April 1999.

[37] R. Cameron, "Advanced coupling matrix synthesis techniques for microwave filters," *IEEE Transactions on Microwave Theory and Techniques*, vol. 51, no. 1, pp. 1–10, Jan. 2003.

[38] G. Matthaei, L. Young, and E. Jones, *Microwave Filters, Impedance-Matching Networks, and Coupling Structures*, McGraw-Hill, 1964.

[39] S. Tamiazzo and G. Macchiarella, "An analytical technique for the synthesis of cascaded n-tuplets cross-coupled-resonator microwave filters using matrix rotations," *IEEE Transactions on Microwave Theory and Techniques*, vol. 53, no. 5, pp. 1693–1698, May 2005.

4 Compact microwave multi-band bandpass filters based on quarter-wavelength resonators

Nikolina Janković and Vesna Crnojević-Bengin

4.1 Introduction

Design of multi-band filters represents a challenging task for a number of reasons. Not only is it required to achieve good performance in several passbands in terms of insertion loss and selectivity, but also multi-band filters are required to be as compact as possible and to offer the possibility of the independent control of passbands. The two latter properties represent the greatest challenge since they are in contradiction with the requirement for good performance. Namely, it is a difficult task to pack in a small area two or more filter elements that independently provide passbands. At the same time, a small area means strong coupling between the elements, which ultimately leads to performance degradation.

A traditional way for miniaturization is based on folding a straight-line resonator in order to fit it into a smaller area. Advanced approaches include fractal curves employment as well as multi-layer architecture and these will be discussed in the following chapters.

In Chapter 2, various design methods of multi-band filters were discussed. It was shown that compact filters can be realized with conventional $\lambda/4$ resonators which are twice as short as conventional $\lambda/2$ resonators. Using traditional techniques of miniaturization, i.e. folding of a resonator, overall dimensions of a $\lambda/4$ resonator can be additionally reduced.

As for the independent control of passbands, it was shown in Chapter 2 that the only method that allows complete independent control of passbands is the employment of two or more single-band filters.

In this chapter, we present dual-band and tri-band filters in which compactness and independent control of passbands are achieved by using single-band filters based on $\lambda/4$ resonators [1–2].

The proposed dual-band filter consists of two single-band filters, each of which consists of a pair of $\lambda/4$ resonators. In that manner, a very compact dual-band filter with excellent performance and the possibility of independent control of passbands is realized.

Introducing additional conductive segments into the structure of dual-band filter, a pair of ring-shaped $\lambda/2$ resonators is formed and it provides an additional passband. This tri-band filter is also characterized by small overall dimensions, as well as by excellent performance and the possibility of independent control of passbands.

In the first part of the chapter, the main properties of a quarter-wavelength resonator will be presented and discussed, and will be compared to those of a half-wavelength resonator.

In Chapter 2, the method that allows full independent control of passbands was discussed. In the third section of this chapter, this method will be discussed in more detail and special attention will be given to the filters based on quarter-wavelength resonators, their properties, and design procedures.

In the fourth section, the proposed dual-band filter of the second order operating at 2.4 and 3.5 GHz, will be analyzed in detail. In order to demonstrate the applicability of the structure, a dual-band filter of the fourth order and a dual-band filter operating at 3.5 and 5.2 GHz will be presented. In addition, filter performance will be compared to that of other published dual-band filters.

The fifth section is devoted to the proposed tri-band filter and analysis of its behavior. Its performance will be compared to that of other published tri-band filters.

Simulation tools from [3] and [4] have been used for electromagnetic simulations. Both filters have been fabricated in standard printed circuit board technology. The fabricated circuits were measured and comparison of the measured and simulation results will also be presented.

4.2 Quarter-wavelength resonator and its properties

A quarter-wavelength resonator is a structure with one end open and the other short-circuited, which exhibits the fundamental resonance at a frequency at which its electrical length is equal to $\pi/2$. Together with a conventional half-wavelength resonator, it is the most widely used resonator in filter design. In this section, its main properties will be presented and discussed, and they will be compared to those of the half-wavelength resonator.

Analysis of the behavior of a quarter-wavelength resonator can be observed through the input impedance of a transmission line; namely, the input impedance of a lossless transmission line loaded with the impedance Z_L can be expressed as

$$Z_{in} = Z_0 \frac{Z_L + jZ_0 \tan \beta l}{Z_0 + jZ_L \tan \beta l}, \tag{4.1}$$

where Z_0 is the characteristic impedance of the transmission line, β is the phase constant, and l is its physical length [5].

When the transmission line is short-circuited at one end, i.e. when it has the configuration of a quarter-wavelength resonator, the load impedance is equal to zero and thus the expression (4.1) takes the following form

$$Z_{in} = jZ_0 \tan \beta l. \tag{4.2}$$

In terms of Z parameters, the resonance condition corresponds to the frequency at which the imaginary part of the input impedance tends to infinity. In this case, the resonance condition, i.e. $Z_{in} \to \infty$, is met when the function $\tan\beta l$ tends to infinity, i.e. when the electrical length of the transmission line is equal to $\pi/2 + k\pi$, where k is an integer number. It can be seen that fundamental resonance occurs at a frequency at which the transmission line is $\lambda/4$ long. In addition, it should be noted that higher resonances occur at odd products of the fundamental resonance. However, in the case of half-wavelength resonators, higher resonances occur at integer products of the fundamental one. This means that a quarter-wavelength resonator is superior to conventional half-wavelength resonator in terms of minimizing interferences of higher resonances.

The input impedance of a lossy transmission line loaded with the impedance Z_L can be expressed as

$$Z_{in} = Z_0 \frac{Z_L + Z_0 \tanh \gamma l}{Z_0 + Z_L \tanh \gamma l}, \tag{4.3}$$

where Z_0 is the characteristic impedance of the transmission line, γ is the propagation constant, and l is its physical length.

In the case of the quarter-wavelength resonator, the load impedance is equal to zero, and thus the expression (4.3) obtains the following form

$$Z_{in} = Z_0 \tanh \gamma l, \tag{4.4}$$

i.e.

$$Z_{in} = Z_0 \tanh(\alpha + j\beta)l, \tag{4.5}$$

where α is the attenuation constant and β is the phase constant.

The previous expression can be rewritten as

$$Z_{in} = Z_0 \frac{1 - j \tanh \alpha l \cot \beta l}{\tanh \alpha l - j \cot \beta l}. \tag{4.6}$$

Let us assume that $l = \lambda/4$ at a frequency of $\omega = \omega_0$ and let $\omega = \omega_0 + \Delta\omega$. Then, it can be shown that the input impedance takes the following expression

$$Z_{in} = Z_0 \frac{1 + j \alpha l \pi \Delta\omega / 2\omega_0}{\alpha l + j \pi \Delta\omega / 2\omega_0} \cong \frac{Z_0}{\alpha l + j \pi \Delta\omega / 2\omega_0}, \tag{4.7}$$

since $\alpha l \pi \Delta\omega / 2\omega_0 << 1$. The previous expression has the same form as the impedance of parallel RLC circuit

$$Z_{in} = \frac{1}{(1/R) + 2j\Delta\omega C}. \tag{4.8}$$

In other words, the quarter-wavelength resonator behaves as a parallel resonant circuit and its elements R and C can be expressed as

$$R = \frac{Z_0}{\alpha l}, \tag{4.9}$$

$$C = \frac{\pi}{4\omega_0 Z_0}, \tag{4.10}$$

whilst inductance can be found as

$$L = \frac{1}{\omega_0^2 C}. \tag{4.11}$$

Using the previous expressions, the quality factor of quarter-wavelength resonator can be also determined

$$Q = \omega_0 RC = \frac{\pi}{4\alpha l} = \frac{\beta}{2\alpha}, \tag{4.12}$$

since $l = \pi/2\beta$ at the resonant frequency.

In practice, a quarter-wavelength resonator is realized by short-circuiting one end of a transmission line. In waveguide architecture, this is achieved by "closing" one end of a waveguide resonator. In the case of a microstrip resonator, a quarter-wavelength resonator is formed by grounding of its one end by a via. Nevertheless, the via cannot provide a perfect short circuit since it has a finite inductance. Thus, analysis of the behavior of the resonator is somewhat different than that of the ideal case.

Namely, the equivalent circuit of a quarter-wavelength resonator with a realistic via is shown in Figure 4.1 where the via is modeled as a lumped inductance L. The input impedance of the circuit is expressed as

$$Z_{in} = Z_0 \frac{j\omega L + jZ_0 \tan \beta l}{Z_0 - \omega L \tan \beta l}, \tag{4.13}$$

Since the resonance occurs when the previous expression tends to infinity, the resonance condition has the following form

$$Z_0 - \omega L \tan \beta l = 0, \tag{4.14}$$

i.e.

$$\tan \beta l = \frac{Z_0}{\omega L}. \tag{4.15}$$

Since the term on the right side is always positive, it implies that the fundamental resonance occurs at the frequency at which βl is lower than $\pi/2$. However, in the ideal case, the fundamental resonance occurs at the frequency at which βl is exactly equal to $\pi/2$. It can be concluded that the finite impedance of the via shifts the fundamental resonance to the lower frequencies. In other words, the quarter-wavelength transmission line is shorter in a realistic case than that in the ideal one.

From the previous analysis, it can be concluded that quarter-wavelength resonators have several advantages in comparison to conventional half-wavelength resonators.

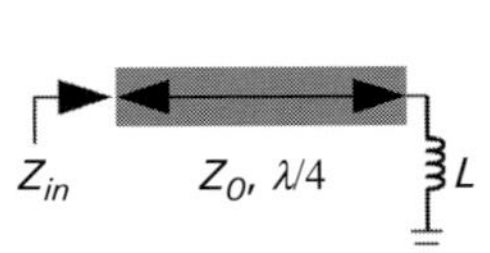

Figure 4.1 Configuration of the quarter-wavelength resonator with a realistic via.

Namely, they occupy an area twice as small as conventional resonators and they have better potential for minimizing interferences at higher resonances.

In [6], single-band filter designs using half-wavelength and quarter-wavelength resonators were compared. Namely, quarter-wavelength resonators have an inverting effect so that they behave like the series resonators at one end, and at their other ends like the shunt resonators. It can be shown that filters can be formed using quarter-wavelength resonators if they are coupled alternately by *K*- and *J*-inverters. Therefore, except for the use of two different inverters, the design procedure with quarter-wavelength resonators is very similar to the design of filters with half-wavelength ones.

Besides smaller size and a better harmonic response, quarter-wavelength resonators have the additional advantage in that their reactance or susceptance slope parameters are half as large as for the corresponding half-wavelength resonators. Because of this, for a given bandwidth and passband shape the couplings are considerably looser for the quarter-wavelength than for half-wavelength resonator types of filters. This implies larger capacitive gaps so that tolerances are less problematic, which leads to significantly higher maximum attenuation in the stopband above the passband.

However, quarter-wavelength resonators show some disadvantages in comparison to conventional resonators. The price that has to be paid is the fact that the fabrication process of quarter-wavelength resonators is more demanding since it requires fabrication of a grounding via. Apart from the fact that the resonant frequency is very sensitive to the via inductance and thus to the fabrication process, there is another important aspect related to vias. Namely, they act as inductances, and in filters with several resonators may be mutually coupled and affect each other's resonances, which makes the fabrication procedure more complex.

Nevertheless, in this chapter and Chapter 6 filters that employ vias will be presented. The filters have been fabricated in standard printed circuit board technology and it will be shown that seemingly demanding fabrication of vias does not have to be a challenging task.

4.3 Dual-band and tri-band filters based on quarter-wavelength resonators

In Chapter 2, it was shown that the only method that allows full independent control of passbands in multi-band filters is the one based on the employment of two or more single-band filters.

Although configuration that consists of several structures may imply large dimensions of filters, some of the most compact multi-band filter circuits with excellent performance have been designed using this method.

As was previously stated, the design of multi-band filters includes separate design of the comprising elements – single-band and dual-band filters which are afterwards put together to form a multi-band filter. Although the comprising filters are independent, they influence each other's performance and thus the final step in the design is optimization of the geometrical parameters of all filters in order to obtain the desired filter response.

Resonators that are used in the design of multi-band filters include conventional $\lambda/2$ and $\lambda/4$ resonators as well as more complex structures, such as dual-mode resonators with perturbation, SIR, and SLR structures. As for dual-band filters, one type of such filters are those in which resonators of the same type but with different resonant frequencies are employed. Dual-band filters with quarter-wavelength resonators were proposed in [7–11].

The first dual-band filter with quarter-wavelength resonators was proposed in [7]. The filter consists of two pairs of quarter-wavelength resonators that are capacitively coupled to the feeding lines. The two pairs provide two passbands at 2.4 and 5.2 GHz. However, the configuration is characterized by large overall size and extremely poor selectivity.

A very similar filter with the same configuration of the pairs of the resonators was proposed in [8]. In this filter, selectivity is significantly improved since source-load impedance was introduced. Furthermore, a new structure of the coupled-line is introduced to produce five transmission zeros at the adjacent two passbands that operate at 2.4 and 3.5 GHz. Thus, the bandwidths of this filter are controllable and they have good selectivity.

In [9], several dual-band filters were proposed, and each filter is a composite of two individual filtering sections which operate at different frequencies and share a common input and output. The double-diplexing T-junctions are used to feed the filter so that proper external coupling at each band can be satisfied in a relatively independent manner. For each of the filtering sections, two identical quarter-wavelength resonant lines are coupled through two separate paths, and a common magnetic coupled path which is constructed by one or two short-circuited stubs. Since the filtering sections share common vias, the proposed filters exhibit compact dimensions. Although each proposed dual-band filter exhibits several transmission zeros, none of them provides good selectivity of both passbands.

The dual-band filter proposed in [10] comprises the main open-loop resonators in which are embedded four quarter-wavelength resonators. Pairs of embedded resonators, which are designed to operate at different frequencies, generate two passbands at 2.8 and 4.2 GHz without increasing the size of the open-loop structure. The proposed dual-band filter generates four transmission zeros located between the passbands to exhibit high skirt selectivity, sharp frequency cut-off, and high isolation. In addition, it has good in-band characteristics. However, since the proposed filter includes an open-loop feeding scheme, it also has large overall dimensions.

In [11], a dual-band bandpass filter using a parallel shortened feed scheme was presented. Two sets of quarter-wavelength resonators operating at 0.85 and 1.45 GHz are employed to obtain dual-band responses as well as to minimize the overall size. A special feed scheme is proposed, which consists of two parallel short-ended strip branches. Each of them acts as a feed line at one passband and a loading element at the other one. This scheme provides two independent transmission paths. In addition, source-load coupling is realized to create four transmission zeros near the passbands' edges, resulting in high skirt selectivity. Although the filter enables independent control of the passbands, good in-band and out-of-band properties, it also has large overall dimensions – $0.1\lambda_g$ x $0.25\lambda_g$.

The previous analysis shows that although quarter-wavelength resonators offer good potential to achieve high-performance and compact dual-band filters, none of the proposed filters shows simultaneously good in-band and out-of-band properties, selectivity, and compactness.

As for tri-band filters based on combining dual-band and single-band filters, it was stated in Chapter 2 that, with the exception of the tri-band filter proposed in [12] that consists of three single-band filters, all configurations [12–17] are comprised of one single-band and one dual-band filter. However, only the structures in [15] and [17] employ a single-band filter based on a quarter-wavelength structure.

The tri-band filter in [15] consists of a pair of stub-loaded resonators inductively coupled to the feeding lines and of a pair of quarter-wavelength resonators capacitively coupled to the feeding lines. The stub-loaded resonators generate the first and third passbands, whilst the quarter-wavelength resonators are used to generate the second passband. Nevertheless, the filter exhibits large size and poor performance.

As for the filter proposed in [17], it is similar to the previous one with the difference that it employs a pair of stub-loaded resonators whose stubs are grounded by vias. Although it has somewhat better characteristics in terms of selectivity and in-band properties, this filter also has relatively large overall dimensions.

In the following sections, dual-band and tri-band filters will be presented, based on quarter-wavelength resonators that outperform not only filters based on $\lambda/4$ resonators, but also all other proposed structures.

4.4 Dual-band filter with $\lambda/4$ resonators

In this section, a dual-band filter based on $\lambda/4$ resonators, operating at 2.4 and 3.5 GHz will be presented. After detailed analysis of the filter configuration and behavior, the applicability of the proposed design procedure will be demonstrated in examples of a fourth-order dual-band filter and dual-band filter operating at 3.5 and 5.2 GHz. In addition, the proposed filter will be compared to other state-of-the-art dual-band filters.

4.4.1 Configuration and behavior of dual-band filter

The proposed dual-band filter is comprised of two single-band filters which are based on $\lambda/4$ resonators. As was stated in the Section 4.2, a quarter-wavelength resonator represents microstrip architecture, which is grounded by a via at one end and its name implies that the fundamental resonance occurs at the frequency at which its electrical length is equal to 90°. In practice, the via has a finite inductance and the electrical length deviates from 90° at the resonant frequency and deviation depends on the via impedance. In other words, the finite impedance of the via shifts the fundamental resonance of the quarter-wavelength resonator to the lower frequencies.

It was also stressed that higher-order harmonics in the λ/4 structure occur at odd products of the fundamental resonance. Therefore, not only does the λ/4 resonator have smaller dimensions, but it also has harmonics that are positioned farther than is the case in conventional λ/2 resonators.

The configuration of the proposed dual-band filter with the geometrical parameters indicated is shown in Figure 4.2(a). The parameters denoted by L_k represent the lengths of the conductive segments, whilst the parameter w is the width of the conductive lines, d is the via dimension, and g_1 and g_2 are the widths of the gaps between the quarter-wavelength resonators.

As was previously stated, the structure consists of two single-band filters, each of which represents a filter of the second order and comprises a pair of λ/4 resonators. Their passbands are positioned at two different frequencies – the lower part of the dual-band filter exhibits a passband at the frequency f_1, whilst the upper part exhibits a passband at the frequency f_2, Figure 4.2(b) and 4.2(c).

It can be noted that a quarter-wavelength resonator from the upper filter and a quarter-wavelength resonator from the lower filter share a common via which reduces the number of vias and eases the realization of the structure. Also, a structure comprised of two λ/4 resonators each of a different length can be considered as an asymmetrically grounded structure and its role will be explained and discussed later in this section.

It should be also noted that the two filters are inductively coupled to the feeding lines at the point of connection of the upper and lower resonators, which decreases insertion

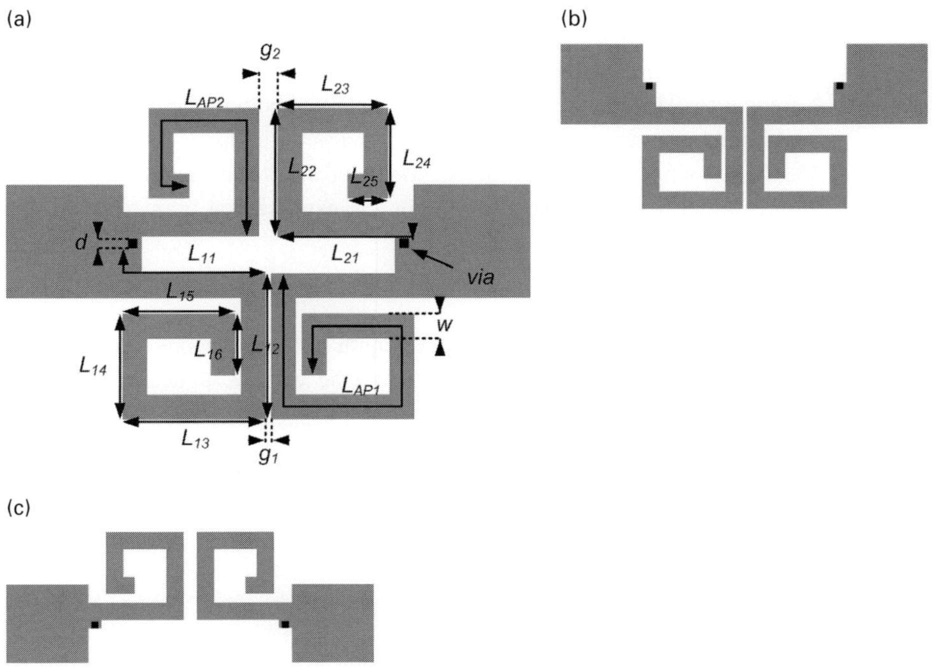

Figure 4.2 (a) Configuration of the proposed dual-band filter. (b) Single-band filter with a passband at f_1. (c) Single-band filter with a passband at f_2.

loss in the passbands. Another important aspect of the inductive coupling is the fact that it improves the quality factors of the passbands and allows their more flexible control. In addition, such a feeding scheme has a role in the formation of transmission zeros in the filter response, which will be discussed below.

The operating principle of the two single-band filters is the same. The first filter to be analyzed is the lower part of the dual-band filter, i.e. the single-band filter with the central frequency f_1. It is comprised of two $\lambda/4$ resonators that are capacitively coupled through the gap g_1. The length of the resonators is equal to $L_{11} + L_{AP1}$ where L_{AP1} is the length of the folded part of the resonator. The geometrical parameters L_{11} and L_{AP1} are chosen in such a manner as to provide resonance at the frequency f_1.

In the filter configuration, it can be seen that the resonators are folded, and the main purpose of the folding is reduction of the area that the resonators occupy. However, the folding of the resonators has another important role – to provide a transmission zero.

In [18], it was shown that the specific coupling configuration of transmission lines can provide transmission zeros. Figure 4.3 shows two types of the coupling between transmission lines – parallel and antiparallel.

In the case of parallel coupling, transmission becomes zero at a frequency where the electrical length of the coupled line becomes 180°. Therefore, to generate an attenuation pole at p x f_0, where f_0 is the central frequency, the electrical length of the coupled line at the central frequency θ_p, is chosen as $180°/p$. Since the length of a half-wavelength resonator is 180°, p will exceed 1, which implies that an attenuation pole can only be obtained in the upper side of passband.

As for antiparallel coupling structure, the open-ends of the parallel lines are positioned side by side. With this structure, a pole is generated at p x f_0 under the conditions of θ_{ap}, = $90°/p$, where θ_{ap} is the electrical length of the coupled line at the central frequency. The attenuation pole can be obtained both in the lower and upper side of the passband.

In this single-band filter, the segments of the $\lambda/4$ resonators with length L_{AP1} act as antiparallel structure, thus resulting in a transmission at frequency

$$f_{AP1} = f_1 \frac{\lambda_g}{4L_{AP1}}, \tag{4.16}$$

where λ_g represents the guided wavelength at the frequency f_1. Namely, the ratio of a quarter wavelength at the frequency f_1 and the length L_{AP1} determines the position of the

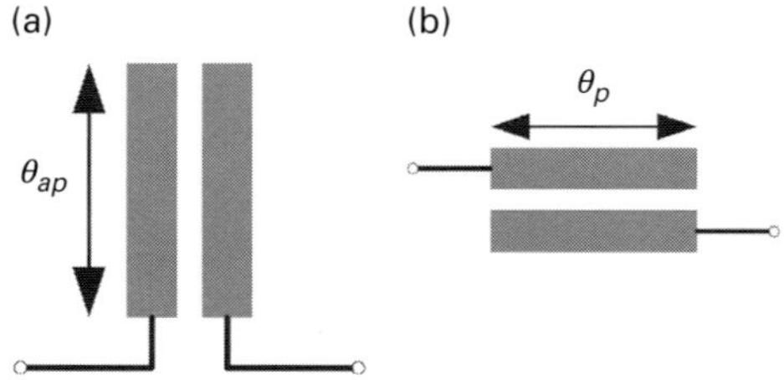

Figure 4.3 (a) Parallel coupling. (b) Antiparallel coupling.

transmission zero. The closer the ratio is to 1, the closer the transmission zero is to the passband.

It should be also noted that due to the resonator configuration, i.e. due to the fact that L_{AP1} is shorter than the overall length of the resonator, the transmission zero always occurs at the higher side of the passband. In addition, the position of the transmission zero depends on the coupling between the resonators, although it is predominantly determined by the length of L_{AP1}.

The upper part of the dual-band filter is a single-band filter of the second order with the central frequency positioned at f_2 and it operates by the same principle as the filter with the central frequency f_1. The length of the $\lambda/4$ resonators is equal to $L_{21} + L_{AP2}$ where L_{AP2} is the length of the folded part of the resonator. The geometrical parameters L_{21} and L_{AP2} are chosen in a manner to provide the resonance at the frequency f_2.

Analogous to the segments L_{AP1} in the single-band filter operating at f_1, the segments L_{AP2} long comprise an antiparallel structure that provides a transmission zero at a frequency that can be expressed in the following manner

$$f_{AP2} = f_2 \frac{\lambda_g}{4L_{AP2}}, \tag{4.17}$$

where λ_g represents the guided wavelength at the frequency f_2. Analogous to the lower single-band filter, L_{AP2} is shorter than the overall length of the resonator, and thus the transmission zero always occurs at higher side of the passband. The closer the ratio $\lambda_g/4L_{AP2}$ is to 1, the closer the transmission zero is to the passband.

The design procedure of the proposed filter consists of three major steps:

- design of the single-band filter that provides the lower passband,
- design of the single-band filter that provides the upper passband,
- merging of the two filters to form a dual-band filter and fine tuning of the geometrical parameters to obtain the desired response.

The filter is designed on the substrate Rogers RT/Duroid 5880 whose thickness is equal to 1.575 mm, dielectric constant $\varepsilon_r = 2.2$, and loss tangent $\tan\delta = 0.0009$.

In the following, the design procedure of a dual-band filter that operates at 2.4 and 3.5 GHz will be presented. In the first step, the single-band filter operating at 2.4 GHz is designed. The length of the $\lambda/4$ resonators at 2.4 GHz on a given substrate has been calculated using *TXLine* tool in *Microwave Office* design software. The initial length of the resonators is equal to 22.1 mm.

In order to obtain better selectivity of the passband, the transmission zero should be positioned as close as possible to the passband. This implies that the ratio of $(L_{11}+L_{AP1})$ and L_{AP1} should be approximately equal to 1, i.e. that L_{11} should be very short. On the other hand, the transmission zero should not be positioned too close to the passband since it can degrade the properties of the passband. Therefore, L_{11} is judiciously chosen to be approximately as long as one quarter of the $\lambda/4$ resonator. In that way, both good passband characteristics and the sharp roll-off skirt are preserved.

Once the length L_{11} is determined, the remaining parts of the resonators are folded in order to reduce the area they occupy, i.e. to miniaturize the final filter. However,

Table 4.1 Geometrical parameters of the single-band and dual-band structures

Initial structure	L_{11}	L_{12}	L_{13}	L_{14}	L_{15}	L_{16}	g_1	w
	7.4	7.6	6.2	4.5	4.9	1.2	0.6	1
	L_{21}	L_{22}	L_{23}	L_{24}	L_{25}	g_2	d	
	7	5.8	4.7	3.8	1.5	0.9	0.4	
Final structure	L_{11}	L_{12}	L_{13}	L_{14}	L_{15}	L_{16}	g_1	w
	7.4	6.2	6.2	4.5	4.9	2.6	0.3	1
	L_{21}	L_{22}	L_{23}	L_{24}	L_{25}	g_2	d	
	7	5.5	4.7	3.8	1.7	0.9	0.4	

Note: All dimensions are given in mm.

excessive folding may cause significant mutual inductance between adjacent straight segments and reduce the overall inductance of the resonator, leading to an increase of the resonant frequency and ultimately to higher losses. Hence, the miniaturization of the resonator and reduction of the insertion losses are contradictory goals, and a trade-off is needed.

By using these directions and by fine tuning of the parameters L_{12}, L_{16}, and g_1, Figure 4.2(a), a single-band filter that exhibits a passband at 2.4 GHz has been realized. The parameter L_{16} is used for fine tuning of the position of the passband since it is the end segment of the resonator. The parameters L_{12} and g_1 are used predominantly to tune the coupling between the resonators, i.e. to achieve a desired bandwidth and insertion loss.

The resonators consist of six segments whose geometrical parameters are given in Table 4.1 in the section Initial structure. Due to the folding of the structure and disturbance of the electric field in the quarter-wavelength resonator, its overall length is equal to 26.9 mm, which is somewhat greater than the initially estimated 22.1 mm.

In the second step, the single-band filter operating at 3.5 GHz is designed. Analogous to the filter operating at 2.4 GHz, the initial length of $\lambda/4$ resonators is calculated in *TXLine* tool and it is equal to 15.6 mm. In order to provide good passband characteristics and a sharp roll-off skirt, the parameter L_{21} should be as long as a quarter of the resonator. However, its length is also dictated by the parameter L_{11}, i.e. L_{21} should not be significantly shorter than L_{11} since that would lead to large g_2 and ultimately to a weak coupling and high insertion loss. Therefore, the length L_{21} is somewhat longer than a quarter of the resonator, and the transmission zero is not positioned as close to the passband as in the lower single-band filter and thus its selectivity is slightly worse.

Following the same steps as in the previous single-band filter, a filter operating at 3.5 GHz has been realized. The parameters L_{22}, L_{25}, and g_2 were used for the fine tuning – L_{25} to tune the position of the passband, and L_{22} and g_2 to tune the coupling between the resonators. The final dimensions of the filter are given in Table 4.1 in the section Initial

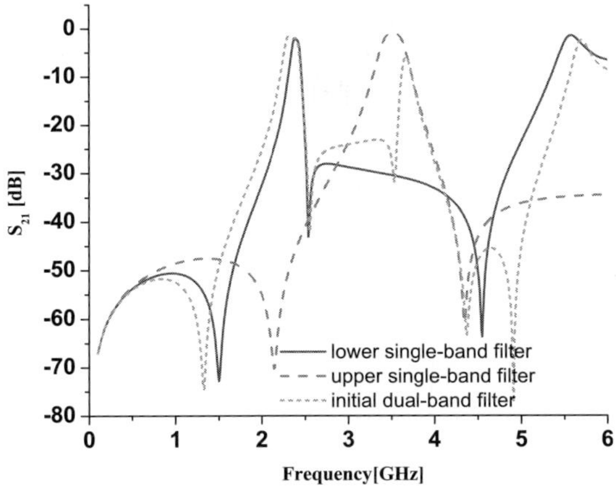

Figure 4.4 Responses of the single-band filters and the initial dual-band filter.

structure. Similar to the first single-band filter, the final length of the resonator is longer than the initial one, which is due to the folding of the resonator.

Figure 4.4 shows the responses of the single-band filters. Apart from the passbands that are positioned at predefined frequencies, it can be noted that at the higher part of each passband a transmission zero occurs, which is due to the antiparallel structures in the filters. In addition, there is a transmission zero at the lower part of the passbands, which will be discussed below.

Once the two filters are designed, they are put together to form a dual-band filter. The response of the initial dual-band filter is also shown in Figure 4.4 and it can be noted that it deviates from the desired response, especially in terms of the properties of the second passband. This is due to the fact that the two filters are closely positioned and share common vias and thus affect each other's properties even though they are separate structures.

Therefore, a fine tuning of the geometrical parameters is needed in order to obtain a response that meets the predefined specifications. Table 4.1 shows the geometrical parameters of the initial and final structures and it can be noted that the parameters L_{12}, L_{16}, g_1, L_{22}, L_{25}, and g_2 were finely tuned. Nevertheless, their values have not been significantly changed. Since six parameters have been changed, the optimization process may seem to be demanding. However, we should note that the parameters L_{12} and g_1 and L_{22} and g_2, and thus their tuning, are interrelated.

The scattering parameters of the final filter are shown in Figure 4.5 and its characteristics are summarized in Table 4.2. The proposed filter is characterized by a small insertion loss, 1.1 and 1.6 dB, as well as by excellent selectivity of both passbands owing to the four transmission zeros. At the same time, its dimensions of $0.13\lambda_g$ x $0.13\lambda_g$ makes it the most compact planar dual-band filter that operates at 2.4 and 3.5 GHz, which will be discussed in more detail in the Section 4.4.4.

In addition, the bandwidths of both passbands are equal to 100 MHz, which is in accordance with WLAN standards.

Table 4.2 Characteristics of the final dual-band filter

	f [GHz]	IL [dB]	RL [dB]	3 dB BW [MHz]	3 dB FBW [%]	overall dimensions [$\lambda_g x \lambda_g$]
first passband	2.4	1.1	15.3	100	4.6	0.13x0.13
second passband	3.5	1.6	15	100	2.85	

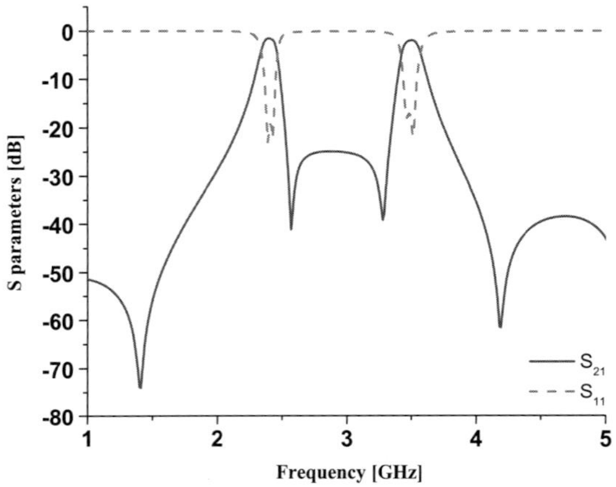

Figure 4.5 Response of the final dual-band filter.

We should also note that besides the two transmission zeros, which are due to the antiparallel coupling between the resonators, another two transmission zeros occur in the response of the filter and they are positioned at the lower side of each passband.

The first transmission zero is due to the presence of the vias in the filter structure. As was stated in the Section 4.2, vias have finite inductance, i.e. they represent inductors that are mutually coupled due to their proximity in the filter configuration. In that manner, another signal path is formed, and when it is out-of-phase with the main signal path, it gives a transmission zero. The position of this transmission zero is predominantly determined by the via inductance, i.e. by its dimension d, as well as by the strength of the coupling between the vias which depends on the distance between them.

Figure 4.6 shows the responses of dual-band filters that differ in the via dimension d and it confirms that an increase in the via dimensions, which corresponds to the decrease in the inductance, leads to a shift of the transmission zero towards higher frequencies. A change in the via inductance also causes a change in the resonant frequency of the resonators, which can be compensated through variation of other geometrical parameters.

The via dimension equal to 400 μm has been chosen since it is a good trade-off between the complexity of the fabrication procedure and selectivity. Namely, a larger

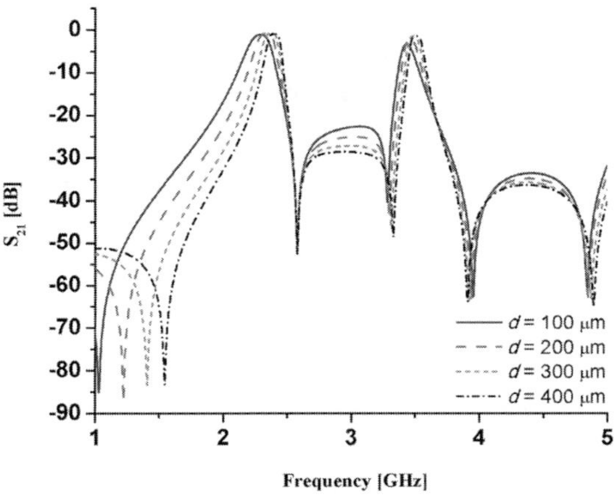

Figure 4.6 Influence of the via dimension on the filter response.

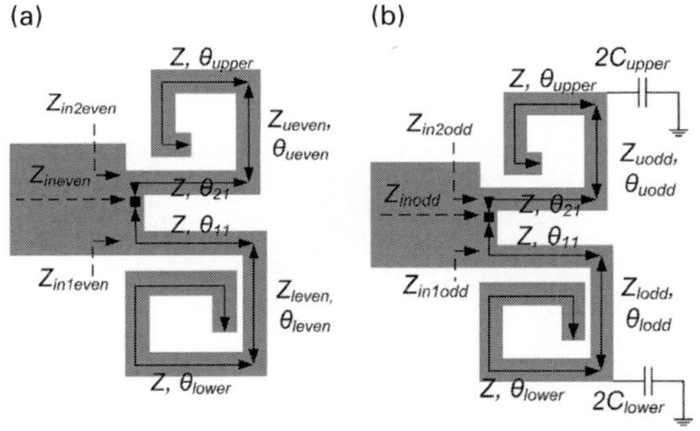

Figure 4.7 Equivalent circuit for (a) Even mode. (b) Odd mode.

via dimension positions the transmission zero closer to the passband. On the other hand, a larger dimension of a via eases the fabrication procedure, but the via should not have an exceedingly large dimension so as not to lose its inductive properties.

As for the transmission zero that is positioned at the lower side of the second passband, it occurs due to the specific configuration of the filter, i.e. due to a specific position of the feeding lines with respect to the resonators. Since the dual-band filter has a symmetrical structure, even-/odd-mode analysis can be applied for analysis of the behavior of the filter.

The even- and odd-mode equivalent circuits are shown in Figure 4.7. For the sake of a more precise analysis, even and odd impedances Z_{leven}, Z_{ueven}, Z_{lodd}, and Z_{uodd} have been assigned to the coupled segments and they represent the impedances of the coupled lines in even and odd modes [19].

The total input impedance for the even and odd mode represents a parallel combination of the input impedances of the lower and upper part of the filter, i.e. a parallel combination of the input impedances of the single-band filters. Thus, the even and odd input impedances can be expressed as

$$\frac{1}{Z_{ineven}} = \frac{1}{Z_{in1even}} + \frac{1}{Z_{in2even}}, \tag{4.18}$$

and

$$\frac{1}{Z_{inodd}} = \frac{1}{Z_{in1odd}} + \frac{1}{Z_{in2odd}}, \tag{4.19}$$

where $Z_{in1even}$, $Z_{in2even}$, Z_{in1odd}, and Z_{in2odd} represent the input impedances of the lower and upper part of the filter for the even and odd mode. Taking into consideration the notification in Figure 4.7, the input even and odd impedances of the upper and lower parts $Z_{in1even}$, $Z_{in2even}$, Z_{in1odd}, and Z_{in2odd} can be expressed in the following manner

$$Z_{in1even} = jZ\frac{Z_{leven}(Z_{leven}\tan\theta_{leven}\tan\theta_{lower} - Z) + Z\tan\theta_{11}(Z\tan\theta_{leven} + Z_{leven}\tan\theta_{lower})}{Z(Z\tan\theta_{leven} + Z_{leven}\tan\theta_{lower}) - Z_{leven}(Z_{leven}\tan\theta_{leven}\tan\theta_{lower} - Z) + \tan\theta_{11}}, \tag{4.20}$$

$$Z_{in1odd} = jZ\frac{Z_{lodd}\left(Z_{lodd}(\tan\theta_{lodd} + Z\omega 2C_{lower})\tan\theta_{lower} - Z\right) + Z\tan\theta_{11}\left(Z(\tan\theta_{lodd} + Z\omega 2C_{lower}) + Z_{lodd}\tan\theta_{lower}\right)}{Z\left(Z(\tan\theta_{lodd} + Z\omega 2C_{lower}) + Z_{lodd}\tan\theta_{lower}\right) - Z_{lodd}\left(Z_{lodd}(\tan\theta_{lodd} + Z\omega 2C_{lower})\tan\theta_{lower} - Z\right) + \tan\theta_{11}}, \tag{4.21}$$

$$Z_{in2even} = jZ\frac{Z_{ueven}(Z_{ueven}\tan\theta_{ueven}\tan\theta_{upper} - Z) + Z\tan\theta_{21}(Z\tan\theta_{ueven} + Z_{ueven}\tan\theta_{upper})}{Z(Z\tan\theta_{ueven} + Z_{ueven}\tan\theta_{upper}) - Z_{ueven}(Z_{ueven}\tan\theta_{ueven}\tan\theta_{upper} - Z) + \tan\theta_{21}}, \tag{4.22}$$

$$Z_{in2odd} = jZ\frac{Z_{uodd}\left(Z_{uodd}(\tan\theta_{uodd} + Z\omega 2C_{upper})\tan\theta_{upper} - Z\right) + Z\tan\theta_{21}\left(Z(\tan\theta_{uodd} + Z\omega 2C_{upper}) + Z_{uodd}\tan\theta_{upper}\right)}{Z\left(Z(\tan\theta_{uodd} + Z\omega 2C_{upper}) + Z_{uodd}\tan\theta_{upper}\right) - Z_{godd}\left(Z_{uodd}(\tan\theta_{uodd} + Z\omega 2C_{upper})\tan\theta_{upper} - Z\right) + \tan\theta_{21}}. \tag{4.23}$$

Due to the folds of the resonators, there are fringing fields at the corners of the resonators that influence their resonant frequency. In order to account for this in the even/odd analysis, fringing capacitances are introduced. For the sake of simplicity, the fringing capacitances at the corners between the segments are modeled by only two parameters, C_{lower} and C_{upper}. Since the symmetry plane in the even mode is a magnetic wall, there is no fringing capacitances in the even equivalent circuit.

From the mathematical point of view, S_{21} is equal to zero, i.e. a transmission zero occurs at a frequency at which the condition $Z_{ineven} = Z_{inodd}$ is met. Being positioned at the lower part of the second passband, it provides excellent selectivity as well as good performance in the stopband region between the two passbands.

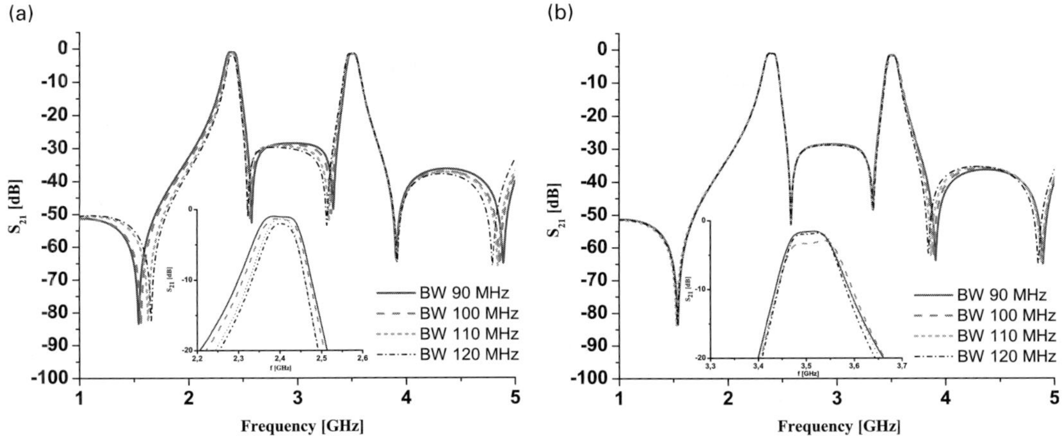

Figure 4.8 Independent control of: (a) First passband. (b) Second passband.

Besides compactness and good performance, the advantage of the proposed dual-band structure is the possibility of independent control of the passbands, which implies not only control of the positions, but also control of the bandwidths. Varying the parameters g_1, g_2, L_{12}, and L_{22}, the couplings between the resonators can be changed and the bandwidths can be tuned from 90 to 120 MHz, whilst the other filter characteristics are preserved. The independent varying of the two bandwidths is illustrated in Figure 4.8.

From the previous analysis, it can be concluded that the proposed structure and design approach allow greater design freedom as well as the realization of dual-band filters with excellent performance and small dimensions.

4.4.2 Illustration of the applicability of the proposed design approach

In practice, it is often needed to realize a filter of order higher than two, i.e. to realize a filter that comprises more than two resonators so as to obtain better passband performance. In order to demonstrate the applicability of the proposed approach and its potential for higher-order filter design, a dual-band filter of the fourth order has been realized.

To realize the fourth-order filter, the procedure presented in [20] is used. It relies on classical filter design theory, and external quality factors and coupling coefficients need to be determined. Since the passbands are realized with single-band filters, the external quality factors and coupling coefficients have been determined for each single-band filter separately, Table 4.3. Using the calculated values, the geometrical parameters for each single-band filter have been determined.

Similar to the dual-band filter of the second order, merging of the two single-band filters does not give a dual-band filter with the desired performance, but a fine tuning of the geometrical parameters L_{12}, L_{18}, L_{22}, g_1, g_2, g_3, and g_4 is needed. In this case, the tuning is more demanding since there are four resonators in each single-band filter.

Table 4.3 Quality factor and coupling coefficients for the fourth-order filter

	k_{12}	k_{23}	k_{34}	Q_e
f_1	0.034652	0.021804	0.034652	21.95
f_2	0.018017	0.0134	0.018017	46.8

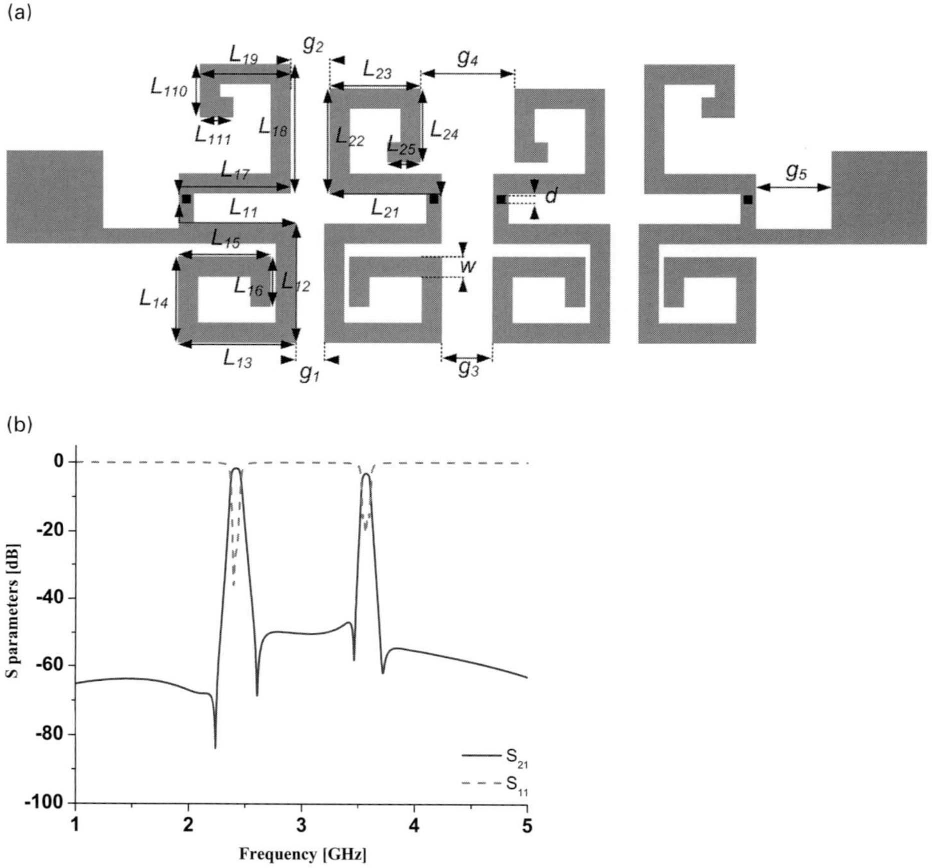

Figure 4.9 Dual-band filter of the fourth order: (a) Configuration. (b) Response.

Figure 4.9 shows the structure and the response of the final dual-band filter of the fourth order that operates at 2.4 and 3.5 GHz. Table 4.4 and Table 4.5 show the dimensions of the final filter and its characteristics, respectively.

The two outer resonators are inductively coupled to the feeding lines and their external quality factor is lower than that of the second-order dual-band filter since the tapped input lines are narrower. The resonators are capacitively coupled through the gaps g_1, g_2, g_3, and g_4. The upper part of the two inner resonators are folded in a somewhat different manner than the outer ones in order to obtain the desired coupling coefficient and good performance in the second passband.

Table 4.4 Geometrical parameters of the fourth-order dual-band filter

parameter [mm]	L_{11}	L_{12}	L_{13}	L_{14}	L_{15}	L_{16}	L_{17}	L_{18}
	7.4	6.2	6.2	4.5	4.9	2.6	7	6.5
	L_{19}	L_{110}	L_{111}	L_{21}	L_{22}	L_{23}	L_{24}	L_{25}
	4.4	3.6	1.7	7	5.5	4.7	3.8	1.7
	g_1	g_2	g_3	g_4	g_5	d	w	
	0.7	0.9	1.2	1.6	1.4	0.4	1	

Table 4.5 Characteristics of the fourth-order dual-band filter

	f [GHz]	IL [dB]	RL [dB]	3 dB BW [MHz]	3 dB FBW [%]
first passband	2.4	1.75	22	90	3.75
second passband	3.5	3	17	70	2

Table 4.6 Geometrical parameters of the dual-band filter operating at 3.5/5.2 GHz

parameter [mm]	L_{11}	L_{12}	L_{13}	L_{14}	L_{15}	g_1
	6.3	4.9	4.8	3.1	1.5	1.5
	L_{21}	L_{22}	L_{23}	g_2	d	w
	4.8	4.3	3.6	0.7	0.4	1

As for the response of the fourth-order filter, it can be noted that the higher-order filter also exhibits excellent performance in terms of insertion loss and selectivity. In comparison to the second-order filter, its selectivity is better since the transmission zeros are positioned closer to the passbands.

The excellent in-band and out-of-band characteristics of the fourth-order filter confirms the applicability of the proposed approach.

Another way to illustrate the applicability of the proposed approach is to realize a dual-band filter that operates at the frequencies other than 2.4 and 3.5 GHz. To that end, a dual-band filter of the second order operating at 3.5 and 5.2 GHz has been designed.

The filter has been designed in the same manner as the filter operating at 2.4 and 3.5 GHz. The first step was to design a single-band filter with the passband at 3.5 GHz and afterwards a single-band filter operating at 5.2 GHz. After the two filters had been put together, an additional optimization was done in order to obtain the desired response. The dimensions and the characteristics of the final filter are shown in Table 4.6 and Table 4.7, respectively, whilst Figure 4.10 shows its structure and response.

The upper part of the filter, which provides the passband at 5.2 GHz, consists of only three conductive segments since its resonators are significantly shorter than the resonators in the lower single-band filter, and the parameter L_{21} needs to be approximately as long as the parameter L_{11}. Therefore, the parameter L_{21} is longer than a quarter of the resonator and the highest transmission zero is positioned far from the second passband.

Table 4.7 Characteristics of the dual-band filter operating at 3.5/5.2 GHz

	f [GHz]	IL [dB]	RL [dB]	3 dB BW [MHz]	3 dB FBW [%]	overall dimensions [$\lambda_g x \lambda_g$]
first passband	3.5	1.5	17	160	4.57	0.16x0.15
second passband	5.2	1.3	17	300	5.77	

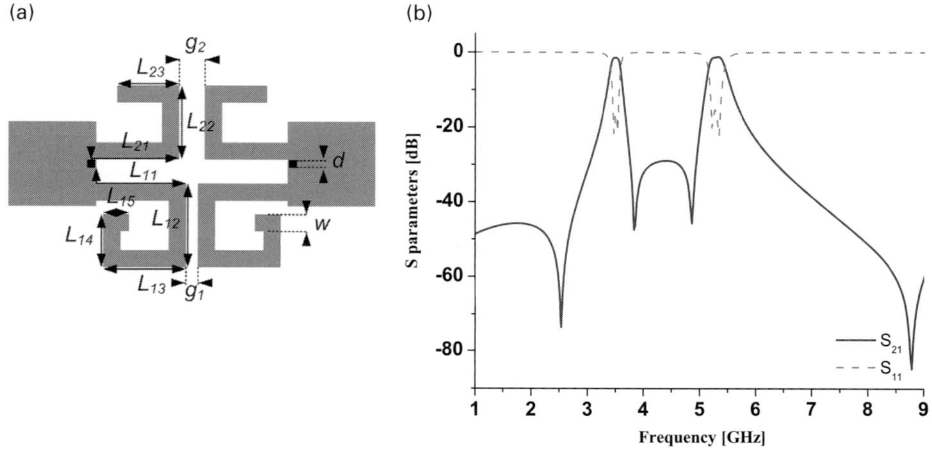

Figure 4.10 Dual-band filter operating at 3.5/5.2 GHz: (a) Configuration. (b) Response.

Nevertheless, the filter exhibits a very small insertion loss of both passbands, excellent selectivity, as well as small dimensions of $0.16\lambda_g$ x $0.15\lambda_g$, which is another confirmation of the applicability of the proposed approach to dual-band filter design.

4.4.3 Fabrication and measurement results

The proposed dual-band filter of the second order that operates at 2.4 and 3.5 GHz has been fabricated using standard printed circuit board technology. The circuit has been realized on the substrate Rogers RT/Duroid 5880 whose thickness is equal to 1.575 mm, dielectric constant $\varepsilon_r = 2.2$, and loss tangent $\tan\delta = 0.0009$.

A photograph of the fabricated dual-band filter is shown in Figure 4.11(a), whilst the simulated and measured results are compared in Figure 4.11(b).

The proposed filter comprises two vias and thus its fabrication might be sensitive to fabrication discrepancies, which is of great importance since the positions of both passbands and the lowest transmission zeros depends on the vias and their inductances. Nevertheless, Figure 4.11(b) shows that there is good agreement between the simulated and measured response.

Table 4.8 Characteristics of the simulated and fabricated dual-band filters

	f_1/f_2 [GHz]	IL [dB]	RL [dB]	3 dB BW [MHz]	3 dB FBW [%]
simulated circuit	2.4/3.5	1.1/1.6	15.3/15	100/100	4.6/2.85
fabricated circuit	2.41/3.55	1.6/2	10/12	170/120	7.1/3.4

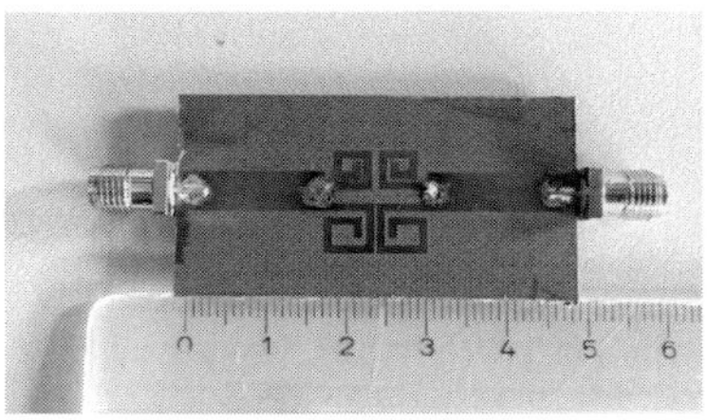

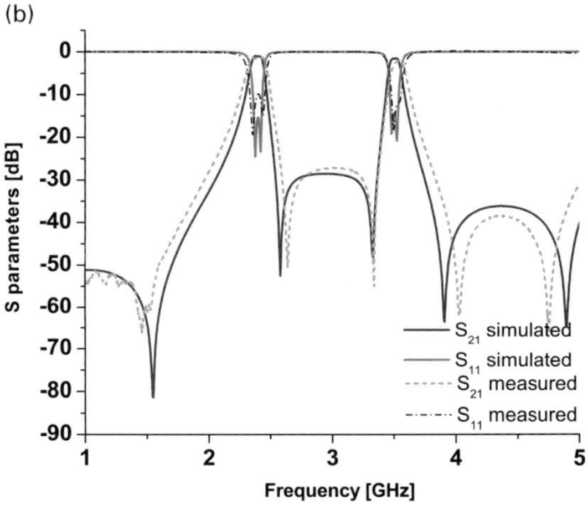

Figure 4.11 Dual-band filter operating at 2.4/3.5 GHz: (a) Photograph of the fabricated circuit. (b) Simulated and measured responses.

Table 4.8 gives a comparison between the characteristics of the simulated and fabricated circuits and it also confirms good agreement between the simulated and measured results.

4.4.4 Comparison to other published dual-band filters

The proposed dual-band filter has been designed to operate at 2.4 and 3.5 GHz. As was discussed in the previous chapter, a very small number of dual-band filters operating at 2.4 and 3.5 GHz have been published so far. Table 4.9 gives a comparison of the characteristics of the proposed filter and the characteristics of other dual-band filters operating at the same frequencies. Also, Table 4.9 summarizes the characteristics of filters that operate at different frequencies yet have good performance and represent the best published dual-band filters in terms of performance and size. The parameter λ_g is the guided wavelength at the central frequency of the first passband.

The proposed filter exhibits a small insertion loss in both passbands, 1.1 and 1.6 dB, good return loss, excellent selectivity, and very compact dimensions of only $0.13\lambda_g$ x $0.13\lambda_g$.

In comparison to the filters operating at the same frequencies, the proposed structure exhibits significantly better properties in terms of in-band and out-of-band

Table 4.9 Comparison of the characteristics of the proposed filter and other dual-band filters

	f_1/f_2 [GHz]	IL [dB]	RL [dB]	transmission zeros	dimensions [λ_gxλ_g]	3 dB FBW [%]
proposed filter	2.45/3.5	1/1.6	15.3/15	4	0.13x0.13	4.6/2.8
[21]	1/1.5	1.58/1.54	16/15	3	0.2x0.2	8/5.7
[22]	2.37/5.8	0.55/1.31	25/30	5	0.083x0.15	19/15
[23]	2.4/3.8	1.45/1.46	20/18	3	0.227x0.113	13/8.3
[24]	2.5/3.5	1.8/2.4	20/15	3	0.24x0.33	6.3/4.4
[25]	2.45/5.25	1.1/2.34	17/18	4	0.064x0.235	11/7.2
[26]	2.45/5.7	1/1.2	14/22	3	0.092x0.22	4.5/2.5
[27]	2.4/5.2	1.5/1.4	16/19	5	0.1x0.1	6.7/5.8
[28]	2.35/3.2	1.8/3	25/30	3	0.18x0.19	3.9/2.8
[29]	2.4/3.3	1.2/1.5	20/20	3	0.21x0.18	8.1/4.2

characteristics and overall dimensions. In addition, the performance of the proposed configuration is better than that of most other filters operating at different frequencies. The structures presented in [22] [25], and [27] occupy smaller areas in comparison to the proposed filter, but it should be noted that they have been designed to operate at 2.4 and 5.2 GHz, which eases the design procedure and reduces the overall dimensions. In addition, each of the three structures comprises resonators that are mutually coupled and thus their independent control of the passbands is limited. On the other hand, the proposed filter is comprised of two separate filters and thus allows independent control of the passbands.

It can be concluded that the proposed configuration is the most compact planar filter that operates at 2.4 and 3.5 GHz. Also, it is characterized by small insertion losses and excellent selectivity, as well as by the possibility of independent control of the passbands.

4.5 Tri-band filter based on $\lambda/4$ resonators

The presented dual-band structure can be used as a basis for the design of a tri-band filter with excellent performance and independent control of the passbands. In this section, a tri-band filter based on $\lambda/4$ resonators will be presented and analyzed, and its performance compared to those of other published tri-band filters.

4.5.1 Configuration and analysis of the behavior of tri-band filter

If a pair of conductive segments $L_{16} + L_{17} + L_{18}$ long is introduced in the structure of the proposed dual-band filter, a pair of ring-shaped $\lambda/2$ resonators is obtained, Figure 4.12. In that manner, another single-band filter, i.e. another passband, is formed and the structure exhibits tri-band behavior.

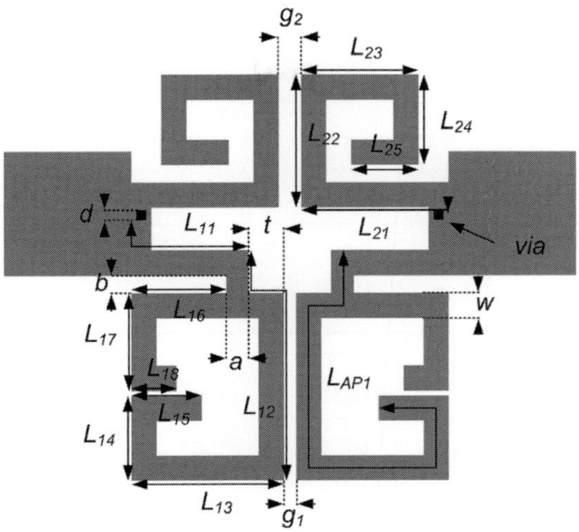

Figure 4.12 Configuration of the tri-band filter.

Design of the tri-band filter is based on the design procedure of a dual-band filter. The first step is to design a dual-band structure, i.e. the single-band filters that operate at the frequencies f_1 and f_2. Afterwards, the third single-band filter operating at the frequency f_3 is designed.

The single-band filter operating at f_2 is designed in the manner described in the previous section. Design of the filter that gives a passband at f_1 differs from the corresponding one in the dual-band filter due to the fact that the resonators are folded only to some extent to allow space for $\lambda/2$ resonators. In order to decrease losses in the first and third passbands, the filter configuration should be as simple as possible, i.e. excessive folding of the resonators has to be avoided.

Once the filters operating at f_1 and f_2 are designed, a new conductive segment $L_{16} + L_{17} + L_{18}$ long is introduced into the structure so that the dimensions are kept as small as possible. In that manner, a pair of $\lambda/2$ resonators, i.e. a single-band filter is formed. The filter operates at the frequency f_3 that is higher than the frequencies f_1 and f_2. At the same time, a strong coupling between the segments L_{11} and $L_{16} + L_{17} + L_{18}$ needs to be avoided to decrease the coupling between the filter that provides the passband at the frequency f_1 and the one that provides the passband at the frequency f_3.

Following the proposed procedure, a tri-band filter for WLAN frequencies 2.4, 3.5, and 5.2 GHz has been designed. As was stated previously, after the single-band filters operating at 2.4 and 3.5 GHz are designed and put together, a fine tuning is needed to obtain the desired response. In the case of the upper filter, the parameters L_{22}, L_{25}, and g_2 are used for fine tuning, whilst in the case of the lower filter the parameters L_{12}, L_{15}, and g_1 are used.

The length of the segment that is introduced in the structure is determined using previously determined parameters L_{12}, L_{13}, L_{14}, and L_{15}. Namely, the sum of those parameters, and $L_{16} + L_{17} + L_{18}$ should correspond to the electrical length of $\lambda/2$ at

Table 4.10 Geometrical parameters of the tri-band filter

parameter [mm]	L_{11}	L_{12}	L_{13}	L_{14}	L_{15}	L_{16}	L_{17}	L_{18}	g_1	w
	5.9	10.6	6.1	3.4	2.8	3.8	3.9	1.8	0.5	1
	L_{21}	L_{22}	L_{23}	L_{24}	L_{25}	g_2	a	b	t	d
	7	5.3	4.7	3.6	2.7	0.9	0.9	0.7	1.4	0.4

Table 4.11 Characteristics of the tri-band filter

	f [GHz]	IL [dB]	RL [dB]	3 dB BW [MHz]	3 dB FBW [%]	overall dimensions [$\lambda_g x \lambda_g$]
first passband	2.4	0.75	31.1	130	5.8	
second passband	3.5	1.55	18	90	2.6	0.17x0.133
third passband	5.2	1	19.5	220	4	

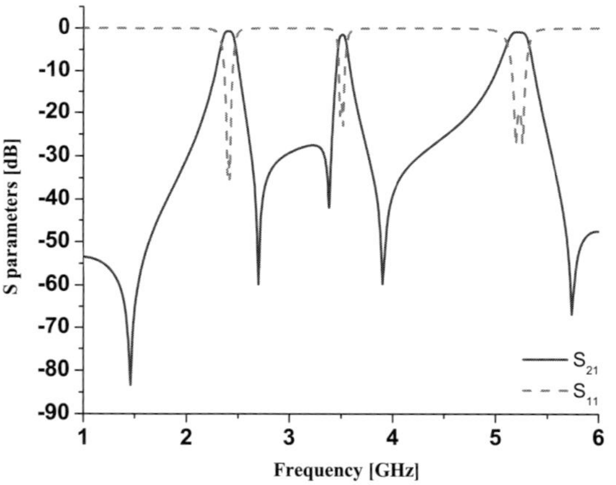

Figure 4.13 Response of the tri-band filter.

5.2 GHz on the given substrate. Since the position of the third passband is determined by the segment $L_{16} + L_{17} + L_{18}$ that does not influence the other two passbands, it is clear that the position of the third passband can be independently controlled.

As in the case of the dual-band filter, after the filters are put together the desired response is not obtained and thus fine tuning of the geometrical parameters is needed. The fine tuning has been done using the parameters L_{17}, L_{18}, a, and b. Table 4.10 gives the geometrical parameters of the final structure.

Figure 4.13 shows the response of the final filter, and Table 4.11 shows its characteristics. The filter exhibits small insertion loss, excellent return loss, and very compact dimensions.

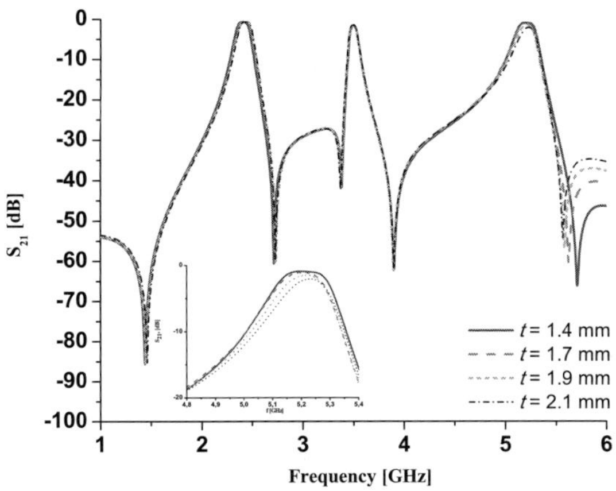

Figure 4.14 Influence of the parameter t on the filter response.

In the configuration of the filter, it can be noted that the parameter t offers an additional degree of freedom in tri-band filter design. Namely, the ring-shaped resonators may be treated as the structures that are inductively coupled to the feeding lines through the segment a. The point at which the resonator is coupled to the feeding line determines the external quality factor, i.e. the strength of the coupling between them. The more the coupling point is closer to the virtual grounding, the less is the coupling factor, i.e. the greater the external factor.

By a change of the parameter t, the characteristics of the third passband can be varied and the performance of the first passband preserved, providing that t does not have a very large value.

Figure 4.14 shows the responses of the filters that differ in the parameter t. In order to ensure that all passbands have the same central frequency, the parameter L_{18} is finely tuned. It can be noted that a change in t influences the bandwidth of the third passband and the range of the bandwidths are from 160 to 220 MHz.

As was stated, the proposed filter is characterized by excellent selectivity, which is due to five transmission zeros. Similarly as in the dual-band filter, the lowest transmission zero occurs due to the presence of the vias in the configuration, which act as coupled inductors, whilst the zeros positioned at the upper part of the first and second passbands are due to antiparallel coupling between the resonators.

As for the zero positioned at the lower part of the second passband and the zero positioned at the upper part of the third passband, they occur due to the specific position of the feeding lines in respect to the resonators. Since the filter has a symmetrical structure, even-/odd-mode analysis can be applied for the analysis of these transmission zeros. The equivalent even and odd circuits are shown in Figure 4.15. Similar to the dual-band filter, the total input impedance for the even and odd mode represents a parallel combination of the input impedances of the lower and upper part of the filter. Thus, the total even and odd input impedances can be expressed using the expressions (4.18) and (4.19).

The even and odd input impedance for the upper part of the structure can be expressed using the same expressions as in the case of the dual-band filter, (4.22) and

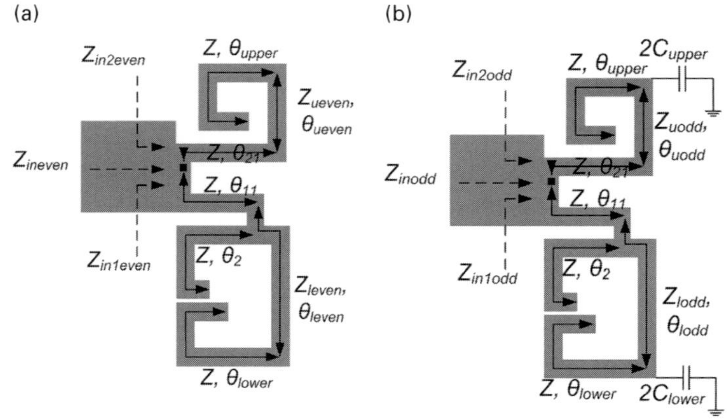

Figure 4.15 Equivalent circuit for (a) Even mode. (b) Odd mode.

(4.23). As for the lower part of the filter, it comprises an additional conductive segment in comparison to the dual-band filter, and thus the expressions for its even and odd input impedances are more complex than in the case of the dual-band filter

$$Z_{in1even} = Z\frac{\frac{Q_{even}P}{Q_{even}+P} + j\tan\theta_{11}}{Z + j\frac{Q_{even}P}{Q_{even}+P}\tan\theta_{11}}, \tag{4.24}$$

$$Z_{in1odd} = Z\frac{\frac{Q_{odd}P}{Q_{odd}+P} + j\tan\theta_{11}}{Z + j\frac{Q_{odd}P}{Q_{odd}+P}\tan\theta_{11}}, \tag{4.25}$$

where

$$\begin{aligned} P &= \frac{Z}{j\tan\theta_2}, \\ Q_{even} &= Z_{leven}\frac{Z - Z_{leven}\tan\theta_{leven}\tan\theta_{lower}}{jZ_{leven}\tan\theta_{lower} + jZ\tan\theta_{leven}}, \\ Q_{odd} &= Z_{leven}\frac{Z - Z_{lodd}\tan\theta_{lodd}(\tan\theta_{lower} + 2\omega ZC_{lower})}{jZ_{lodd}(\tan\theta_{lower} + 2\omega ZC_{lower}) + jZ\tan\theta_{lodd}}. \end{aligned} \tag{4.26}$$

At the frequencies at which the condition $Z_{ineven} = Z_{inodd}$ is met, S_{21} is equal to zero, i.e. transmission zeros occur.

The previous analysis shows that the approach for tri-band filter design also allows greater design freedom, but also that it results in compact filters with excellent performance which enable independent control of the passbands.

4.5.2 Fabrication and measurement results

The proposed tri-band filter has been fabricated using a standard printed circuit board. The circuit has been realized on the substrate Rogers RT/Duroid 5880 whose thickness is equal to 1.575 mm, dielectric constant $\varepsilon_r = 2.2$, and loss tangent $\tan\delta = 0.0009$.

Table 4.12 Characteristics of the simulated and fabricated tri-band filters

	$f_1/f_2/f_3$ [GHz]	IL [dB]	RL [dB]	3 dB BW [MHz]	3 dB FBW [%]
simulated circuit	2.4/3.5/5.2	0.75/1.55/1	31.1/18/19.5	130/90/220	5.8/2.6/4
fabricated circuit	2.42/3.55/5.19	1.7/2.1/1.7	12/10/22	210/150/170	8.75/4.22/3.27

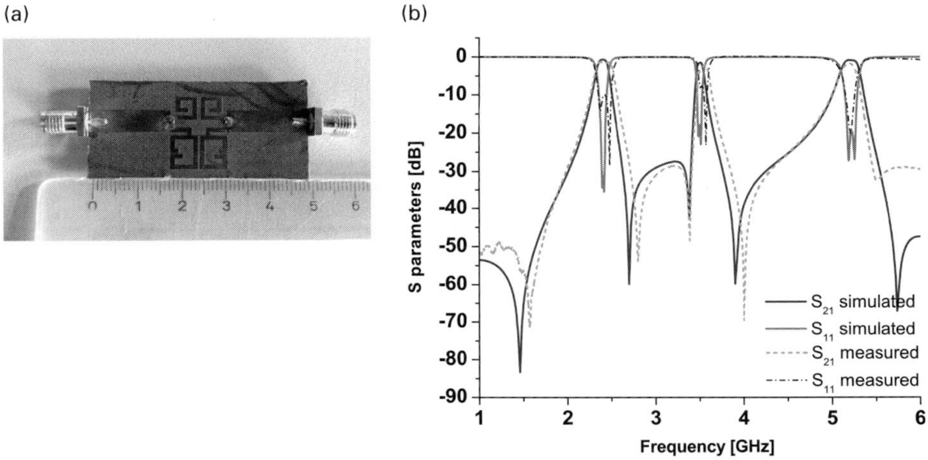

Figure 4.16 Tri-band filter: (a) Photograph of the fabricated circuit. (b) Simulated and measured responses.

A photograph of the fabricated tri-band filter is shown in Figure 4.16(a), whilst the simulation and measured results are compared in Figure 4.16(b).

Similar to the dual-band filter, the tri-band filter comprises two vias and thus its fabrication is also sensitive to fabrication discrepancies, which is important since the positions and properties of the two passbands and the two transmission zeros depend on the vias and their inductances. However, Figure 4.16(b) shows good agreement between the simulated and measured results.

Table 4.12 summarizes the characteristics of the simulated and measured results and it also confirms that there is good agreement between the two results.

4.5.3 Comparison to other published results

The proposed tri-band filter has been designed to operate at 2.4, 3.5, and 5.2 GHz. The most proposed tri-band filters have been designed to operate at the same frequencies. In Table 4.13, comparison between the characteristics of the proposed filter and those of other state-of-the-art tri-band filters is given.

Table 4.13 Comparison of the characteristics of the proposed filter and other tri-band filters

	$f_1/f_2/f_3$ [GHz]	IL [dB]	RL [dB]	transmission zeros	dimensions [λ_gxλ_g]	3 dB FBW [%]
proposed filter	2.4/3.5/5.2	0.75/1.55/1	31.1/18/19.5	5	0.17x0.133	5.8/2.6/4
[30]	1.57/2.4/3.5	0.77/1.51/1.8	20/30/25	6	0.35x0.25	12.5/8/6
[31]	1.57/2.4/3.5	1.6/1.5/2.3	9/18.9/13.5	5	0.38x0.34	5.2/3.8/4.6
[32]	2.4/3.5/5.8	0.9/1.3/1.8	30/15/20	6	0.22x0.35	7.6/5.8/3.71
[33]	2.4/3.5/5.2	1.2/1.1/1.5	16.5/18/14.5	8	0.183x0.264	5/3.7/4.2
[34]	2.4/3.5/5.25	1.9/1.42/1.51	14.3/15/16.8	4	0.11x0.2	6.2/12/11.8

The proposed filter shows small insertion losses in all three passbands, 0.75, 1.55, and 1 dB, excellent return loss, 31, 18, and 19 dB, excellent selectivity, and very compact dimensions of $0.17\lambda_g$ x $0.133\lambda_g$.

In comparison to other configurations, the proposed filter shows similar characteristics in terms of insertion loss, return loss, and bandwidths. When it comes to the overall size of the circuit, the proposed filter is also superior to all other published tri-band filters with the exception of the filter proposed in [34]. However, the selectivity of the structure in [34] is poor, especially in the third passband.

Taking into consideration the performance of all filters, it can be concluded that the proposed filter represents the most compact planar tri-band structure, which, at the same time, is characterized by excellent selectivity and small insertion loss. In addition, the three passbands can be independently controlled.

In Chapter 6, another tri-band bandpass filter with high-performance, excellent selectivity, and compact size will be presented. The filter proposed in this chapter together with the filter to be proposed in Chapter 6 represent the most compact tri-band structures proposed so far.

4.6 Conclusion

In this chapter, two multi-band bandpass filters have been proposed. The filters are based on $\lambda/4$ resonators, which are twice as short as conventional $\lambda/2$ resonators, so they represent good candidates for the realization of compact filtering circuits. At the same time, $\lambda/4$ resonators have better potential for minimizing interferences at higher resonances.

The proposed dual-band filter consists of two single-band filters based on $\lambda/4$ resonators, which enables independent control of the passbands. It was shown that the folding of the resonators results in not only size reduction, but also in a transmission zero that occurs at the upper part of the passband.

The proposed design approach of the dual-band filter consists of a separate design of single-band filters, and optimization of the geometrical parameters once the dual-band structure is formed.

Using the design procedure, a dual-band filter of the second order that operates at 2.4 and 3.5 GHz has been proposed. Apart from small insertion loss and excellent return loss, the realized filter is characterized by excellent selectivity which is due to four transmission zeros. Two zeros occur due to the antiparallel coupling between the resonators, whilst the other two occur due to the presence of the vias in the structure and the specific position of the feeding lines in respect to the resonators.

The proposed filter has been fabricated in printed circuit board technology and there is good agreement between the measured and simulated results. It was shown that in comparison to other dual-band filters, the proposed dual-band filter is the most compact planar dual-band filter which operates at 2.4 and 3.5 GHz, and is also characterized by excellent performance and the possibility of independent control of the passbands.

In order to illustrate the applicability of the proposed approach, a dual-band filter of the fourth order for 2.4 and 3.5 GHz has been realized, as well as a filter of the second order that operates at 3.5 and 5.2 GHz. Both filters exhibit very good performance.

By introducing a conductive segment in the structure of the proposed dual-band filter, a pair of $\lambda/2$ resonators is formed and it provides another passband. In that manner, a tri-band filter is obtained. It was shown how a tri-band filter is designed, and the influence of the geometrical parameters on the response was discussed.

Applying the proposed approach to a tri-band filter that operates at 2.4, 3.5, and 5.2 GHz has been realized. The filter has been fabricated using standard printed circuit board technology and the comparison between the simulated and measured results showed good agreement.

It was concluded that in comparison to other tri-band filters the proposed filter represents the most compact tri-band filter published so far. At the same time, it is characterized by excellent performance and the possibility of independent control of the passbands.

References

[1] N. Janković, V. Radonić, R. Geschke, and V. Crnojević-Bengin, "A compact dual-band bandpass filter using folded quarter-wavelength resonators," *Proceedings of 42nd European Microwave Conference*, 2012, pp. 360–363.

[2] N. Janković, V. Radonić, and V. Crnojević-Bengin, "Compact tri-band bandpass filter based on quarter-wavelength resonators," *Journal of Electromagnetic Waves and Applications*, vol. 27, no. 6, pp. 750–757, April 2013.

[3] *AWR Microwave Office* 2008.

[4] *Sonnet ver.* 13.

[5] D. M. Pozar, *Microwave Engineering*, John Wiley & Sons, New Jersey, 1998.

[6] G. Matthaei, L. Young, and E. M. T Jones, *Microwave Filters, Impedance Matching Networks, and Coupling Structures*, Artech House, Norwood, 1985.

[7] Q. Chu, L. Ye, and F. Chen, "Design of compact dual-band bandpass filter using $\lambda/4$ resonators," *Proceedings of Asia-Pacific Microwave Conference*, 2008, pp. 1–4.

[8] M. Doan and W. Che, "Dual-band bandpass filter using quarter-wavelength resonator with controllable bandwidths and good selectivity," *Proceedings of International Conference on Advanced Technologies for Communications*, 2011, pp. 199–202.
[9] H. Wang and Q. Chu, "A compact dual-band filter with adjustable transmission zeros," *Proceedings of European Microwave Conference*, 2009, pp. 117–120.
[10] M. Farhat and B. Virdee, "Dual-band bandpass filter using embedded short-circuited resonators in the open-loop c-shape resonator," *Proceedings of European Microwave Conference*, 2011, pp. 759–762.
[11] G. Dai, Y. Guo, and M. Xia, "Dual-band bandpass filter using parallel short-ended feed scheme," *IEEE Microwave and Wireless Components Letters*, vol. 20, no. 6, pp. 325–327, June 2010.
[12] R. Geschke, B. Jokanovic, and P. Meyer, "Filter parameter extraction for triple-band composite split-ring resonators and filters," *IEEE Transactions on Microwave Theory and Techniques*, vol. 59, no. 6, pp. 1500–1508, June 2011.
[13] F. Chen and Q. Chu, "Design of compact tri-band bandpass filters using assembled resonators," *IEEE Transactions on Microwave Theory and Techniques*, vol. 57, no. 1, pp. 165–171, Jan. 2009.
[14] X. Zhang, Q. Xue, and B. Hu, "Planar tri-band bandpass filter with compact size," *IEEE Microwave and Wireless Components Letters*, vol. 20, no. 5, pp. 262–264, May 2010.
[15] M. Doan, W. Che, K. Deng, and W. Feng, "Compact tri-band bandpass filter using stub-loaded resonator and quarter-wavelength resonator," *Proceedings of China–Japan Joint Microwave Conference Proceedings*, 2011, pp. 1–4.
[16] Q. Chu, X. Wu, and F. Chen, "Novel compact tri-band bandpass filter with controllable bandwidths," *IEEE Microwave and Wireless Components Letters*, vol. 21, no. 12, pp. 655–657, Dec. 2011.
[17] X. Lin, "Design of compact tri-band bandpass filter using $\lambda/4$ and stub loaded resonators," *Journal of Electromagnetic Waves and Applications*, vol. 24, no. 14, pp. 2029–2035, Oct. 2010.
[18] M. Makimoto and S. Yamashita, "Bandpass filters using parallel coupled stripline stepped impedance resonators," *IEEE Transactions on Microwave Theory and Techniques*, vol. 28, no. 12, pp. 1413–1417, Dec. 1980.
[19] C. Tsai, S. Lee, and H. Lee, "Transmission-line filters with capacitively loaded coupled lines," *IEEE Transactions on Microwave Theory and Techniques*, vol. 51, no. 5, pp. 1517–1524, May 2003.
[20] J. Hong and M. J. Lancaster, *Microstrip Filters for RF/Microwave Applications*, John Wiley & Sons, New Jersey, 2001.
[21] C. Tang and P. Wu, "Design of a planar dual-band bandpass filter," *IEEE Microwave and Wireless Components Letters*, vol. 21, no. 7, pp. 362–364, July 2011.
[22] Z. Zhang, Y. Jiao, X. Wang, and S. Cao, "Design of a compact dual-band bandpass filter using opposite hook-shaped resonator," *IEEE Microwave and Wireless Components Letters*, vol. 21, no. 7, pp. 359–361, July 2011.
[23] H. Wang and Q. Chu, "A compact dual-band filter with adjustable transmission zeros," *Proceedings of European Microwave Conference*, 2009, pp. 117–120.
[24] C. Karpuz and A. Gorur, "Dual-mode dual-band microstrip filters," *Proceedings of European Microwave Conference*, 2009, pp. 105–108.

[25] P. Singh, S. Basu, and Y. Wang, "Miniature dual-band filter using quarter wavelength stepped impedance resonators," *IEEE Microwave and Wireless Components Letters*, vol. 18, no. 2, pp. 88–90, Feb. 2008.

[26] C. Hsu, C. Chen, and H. Chuang, "A miniaturized dual-band bandpass filter using embedded resonators," *IEEE Microwave and Wireless Components Letters*, vol. 21, no. 12, pp. 658–660, Dec. 2011.

[27] X. Luo, H. Qian, J. Ma, K. Ma, and K. Yeo, "Compact dual-band bandpass filters using novel embedded spiral resonator (ESR)," *IEEE Microwave and Wireless Components Letters*, vol. 20, no. 8, pp. 435–437, Aug. 2010.

[28] B. Wu, C. Liang, Q. Li, and P. Qin, "Novel dual-band filter incorporating defected SIR and microstrip SIR," *IEEE Microwave and Wireless Components Letters*, vol. 18, no. 6, pp. 392–394, June 2008.

[29] L. Ren and H. Huang, "Dual-band bandpass filter based on dual-plane microstrip/interdigital DGS slot structure," *IET Electronics Letters*, vol. 45, no. 21, pp. 1077–1079, Oct. 2009.

[30] C. Hsu, C. Lee, and Y. Hsieh, "Tri-band bandpass filter with sharp passband skirts designed using tri-section SIRs," *IEEE Microwave and Wireless Components Letters*, vol. 18, no. 1, pp. 19–21, Jan. 2008.

[31] M. Doan, W. Che, and W. Feng, "Tri-band bandpass filter using square ring short stub loaded resonators," *IET Electronics Letters*, vol. 48, no. 2, pp. 106–107, Jan. 2012.

[32] Q. Yin, L. Wu, L. Zhou, and W. Yin, "A tri-band filter using tri-mode stub-loaded resonators (SLRs)," *Proceedings of IEEE Electrical Design of Advanced Packaging and Systems Symposium*, 2010, pp. 1–4.

[33] W. Chen, M. Weng, and S. Chang, "A new tri-band bandpass filter based on stub-loaded step-impedance resonator," *IEEE Microwave and Wireless Components Letters*, vol. 22, no. 4, pp. 179–181, April 2012.

[34] Q. Chu, X. Wu, and F. Chen, "Novel compact tri-band bandpass filter with controllable bandwidths," *IEEE Microwave and Wireless Components Letters*, vol. 21, no. 12, pp. 655–657, Dec. 2011.

5 Dual-band filters based on grounded patch resonators

Vasa Radonić, Riana Helena Geschke, and Vesna Crnojević-Bengin

5.1 Introduction

In Chapter 2, it was shown that a number of multi-band microstrip-based filters have recently been published based on the use of various innovative geometric designs. Nevertheless, grounded patch resonators (GPR) or mushroom structures are pervasively exploited in antenna design, have not been widely adopted for multi-band filter applications, despite a number of advantages over other commonly used planar resonators.

This chapter's main focus is the design of compact dual-band filters using the grounded patch resonator (GPR). Section 5.2 is devoted to the design of the square patch resonator, analysis of its topology, and possible modifications. The first modification to be analyzed is application of the Hilbert fractal curve for purposes of miniaturization. Since the square GPR is inherently a non-degenerate dual-mode resonator, this property is also exploited and it will be shown how the two fundamental modes can be independently controlled by changing the width and length of the resonator. Introduction of a slot perturbation in the rectangular GPR gives the perturbed grounded patch resonator (PGPR) and the behavior of such a structure is analyzed in detail. It is demonstrated that the inter-resonator couplings of the PGPR at two centre frequencies can be adjusted separately, which allows the design of dual-band filters with independent passband centre frequencies and respective bandwidths. Unlike many other realizations and approaches, this method allows passbands to be spaced further apart than is typically possible with narrow-band coupled-resonator methods.

Section 5.3 addresses filters based on the resonators introduced in Section 5.2. To highlight the properties of the grounded patch resonator, bandstop filters were first designed using different geometrical positions of the patch resonators. The grounded Hilbert patch resonator embedded in the microstrip line is used to design a super-compact highly selective bandstop filter and compared to other state-of-the-art bandstop filters available in the literature. Afterwards, single-band bandpass filters based on GPRs in which only the first resonant mode of the GPR is taken into consideration are proposed.

Finally, dual-band filters based on the proposed resonators are presented and analyzed in detail. Unlike other published dual-band filters based on dual-mode resonators in which the limiting factor is the fact that two centre frequencies cannot be adjusted separately, in this chapter we describe dual-mode resonators where the two modes are decoupled and quite different in resonant frequency. This allows the design of dual-band filters with wide spacing between passbands and independent passband characteristics, which will be demonstrated.

5.2 Grounded patch resonator

5.2.1 Resonator design

As was stated previously, the GPR have been popular in the context of antenna design [1]. In the antenna field, high impedance surfaces were created using grids of grounded patches. Surprisingly, this structure has not been widely exploited in filter design, despite a number of advantages over other commonly used planar resonators. The fabrication of the basic grounded patch is less sensitive to dimension tolerances, in contrast to many resonators previously reported in the literature. Generally, it does not require narrow conductive lines or small gaps. It is very compact in size and exhibits low insertion loss and suppressed spurious response, which makes the GPR suitable for implementation in the design of compact filters with good performance. The implementation of different modifications of the grounded patch resonator in the realization of single-band and dual-band microstrip filters were described in [2–10].

The GPR is a metal square etched on the top side of the dielectric substrate connected by a metal via to the ground plane on the lower substrate side, Figure 5.1. The resonator is realized on 1.27 mm thick Taconic CεR-10 substrate with a dielectric constant of ε_r = 9.8, and dielectric losses equal to 0.0035. The conductor losses are modeled using bulk conductivity for copper ($\sigma = 5.88 \cdot 10^7$ S/m). All resonators and filters presented in this chapter have been designed on this substrate, unless otherwise is specified.

The influence of the various geometrical parameters of the GPR on its performance can be analyzed in the case where the resonator is excited by the microstrip line, i.e. in bandstop filter configuration. The typical configuration of narrowband bandstop filters consists of a transmission line coupled electrically or magnetically to the resonators, and in the microstrip architecture, three different feeding topologies of the patch resonator can be used to obtain the stopband characteristic. The patch can be positioned either next to or under the microstrip line, or the patch can be embedded in the microstrip,

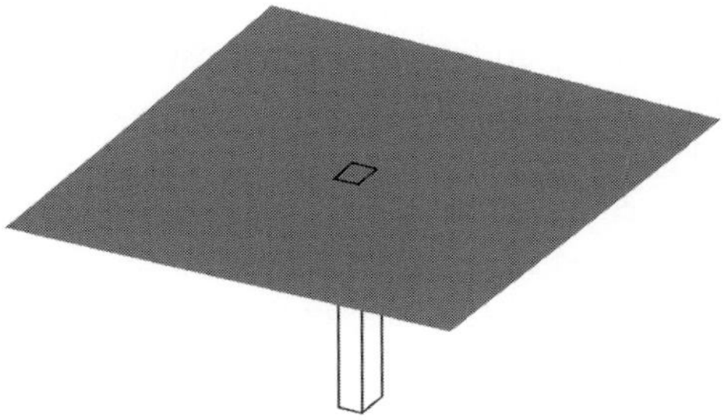

Figure 5.1 Grounded patch resonator, GPR [2].

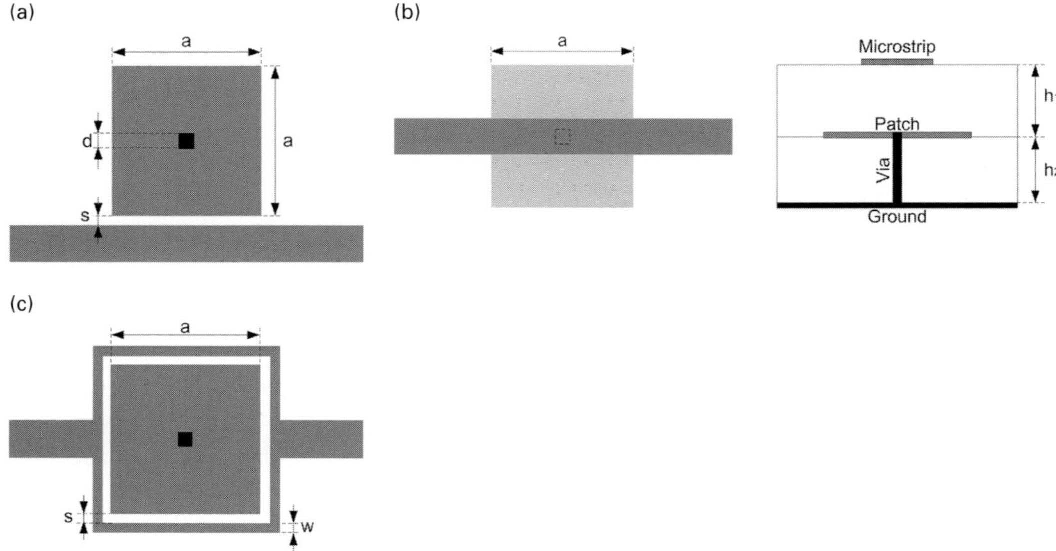

Figure 5.2 Excitation of the GPR: (a) Patch positioned next to microstrip line [6]. (b) Patch placed under microstrip line. (c) Patch embedded in microstrip line [6].

Figure 5.2. Since the coupling configuration has a strong influence on the response of the resonator, we first analyze this phenomenon.

The configuration in which the GPR is positioned next to the microstrip line is shown in Figure 5.2(a). To enhance the coupling, the distance between the resonator and the microstrip line, s, is chosen to be the minimum achievable in standard PCB technology, which is equal to 0.1 mm. In the second case, where the patch is positioned under a microstrip line, the patch is located halfway between the ground plane and the top layer, Figure 5.2(b), at a distance equal to $h_2 = 0.5$ mm above the ground plane. In the third case, a microstrip line encloses a patch, Figure 5.2(c), i.e. the microstrip line is extended and the patch resonator is implemented inside. The ring width around the patch, w, and distance between patch and ring, s, are set to 0.1 mm for the same reason as mentioned above.

In all simulations, via is modeled using a square cross section. In order to increase the inductance of the resonator, the via has a minimum achievable width, with dimension d equal to 0.1 mm. The initial length of the side of the square patch was set to a = 5 mm. The S_{21} responses for three different excitations of the GPR illustrated in Figure 5.2 are compared in Figure 5.3. All simulations were performed using Sonnet Software v13.56 [11].

The patch resonator placed under the microstrip line has the highest resonant frequency while the resonator positioned next to the microstrip line has the smallest insertion loss. It can be seen that the patch embedded in the microstrip has the lowest resonant frequency, the highest insertion losses, and the widest stopband for the same dimensions of the resonator. Thus, the GPR resonator embedded in the microstrip line is suitable for the design of bandstop filters with good characteristics.

Table 5.1 Simulation results forthe GPR positioned next to microstrip line

a (mm)	5	3.1	2.5	1.5	0.7
f_{s1} (GHz)	3.06	4.68	5.58	8.2	13.35
BW (MHz)	40.2	82.5	120.8	263.8	464
FBW (%)	1.3	1.76	2.16	3.21	3.47
S_{21} (dB)	–18.83	–21.94	–24.4	–24.05	–24.85

Table 5.2 Simulation results forthe GPR positioned under microstrip line

a (mm)	5	3.1	2.5	1.5	0.7
f_{s1} (GHz)	3.12	4.82	5.84	9.23	17.25
BW (MHz)	122.4	264.1	374.9	614.1	1190
FBW (%)	3.92	5.48	6.42	6.65	6.91
S_{21} (dB)	–26.1	–29.73	–30.1	–32.4	–34.96

Table 5.3 Simulation results for the GPR embedded in the microstrip line

a (mm)	5	3.1	2.5	1.5	0.7
f_{s1} (GHz)	2.94	4.44	5.3	7.8	13.1
BW (MHz)	183	397	562	1030	1564
FBW (%)	6.22	8.93	10.6	13.2	11.9
S_{21} (dB)	–35.53	–36.12	–37.68	–36.73	–40.48

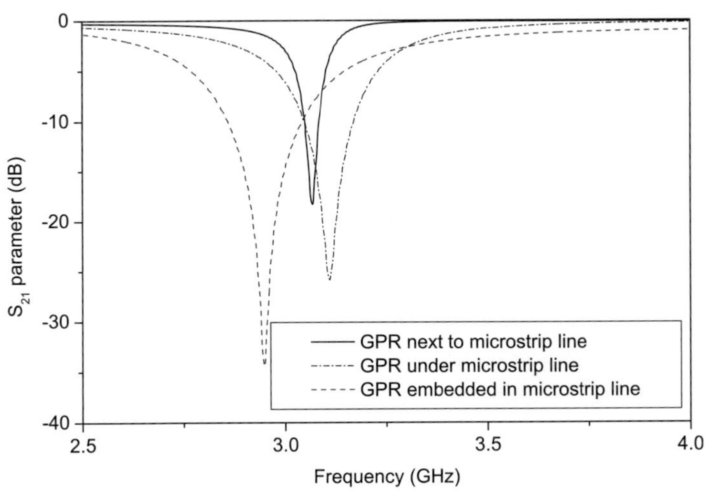

Figure 5.3 Responses of the structures illustrated in Figure 5.2.

Apart from the coupling configuration, the resonant frequency of the resonator predominantly depends on the dimensions of the resonator. To investigate the influence of the patch size on its performance, the outer dimensions of the patch, a, were varied. Tables 5.1, 5.2, and 5.3 show simulation results for all three structures, presented in

Table 5.4 Simulation results for a GPR placed under microstrip for different heights h_2

h_2 (mm)	1.17	1	0.5	0.27	0.1
f_{s1} (GHz)	1.8	2.38	3.12	3.44	3.76
BW (MHz)	231	136	122.4	67	na
FBW (%)	12.83	5.7	3.92	1.94	na
S_{21} (dB)	–43.91	–35.27	–24	–17.01	–7.02

Figure 5.2, where f_{s1} denotes the first resonant frequency, S_{21} is the insertion loss at f_{s1}, *BW* is the 10 dB bandwidth, and *FBW* denotes the fractional bandwidth.

By decreasing the size of the resonator, the resonant frequency and the bandwidth increase. As is expected, the patch embedded in the microstrip line has the lowest resonant frequency and the highest insertion losses. In addition, it exhibits the widest stopband for all dimensions of the grounded patch due to the stronger coupling between the microstrip line and the resonator.

The influence of the coupling between the microstrip line and the patch can be analyzed by changing the distance between them in the case when the patch is positioned under the microstrip line. Table 5.4 shows simulation results for different patch heights h_2 (distance between patch and ground plane), where f_{s1} denotes the first resonant frequency, S_{21} is the insertion loss at f_{s1}, and *BW* presents the 10 dB bandwidths. The outer dimension of the patch were kept constant at 5 mm x 5 mm, while the via dimension was set to $d = 0.1$ mm. The substrate thickness was constant at 1.27 mm.

Due to the increase in the coupling strength and extension of the via, i.e. increase of the via inductance, the resonant frequency is decreased, and the bandwidth and insertion loss increased.

Despite the fact that the patch placed under the microstrip line can provide a lower resonant frequency and larger bandwidth in regard to the patch embedded in the microstrip line, one of the main limitations of this geometry is the impossibility to implement it using the standard PCB technology. Considering that substrates are only available in standard thicknesses, it is very difficult to control the thicknesses of the substrates in the multi-layer technologies and to obtain a perfect alignment between layers without air gaps. This directly influences filter performance.

Another important aspect of the resonator configuration is the position of the via and its influence was investigated in the case when the patch is positioned next to the microstrip line. Four different edge via positions were analyzed, Figure 5.4. The distance between the patch and the microstrip line is again chosen to be the minimal achievable in standard PCB technology. Simulation results are compared in Table 5.5 and in Figure 5.5, where f denotes resonant frequency, S_{21} is the insertion loss at f, and *BW* is the 10 dB bandwidth. As expected, changes in the positions of the via result in a significant change of the resonant frequency. For instance, a maximal change in the via position provides a reduction of more than 12% of the resonant frequency. The lowest resonant frequency exhibits the structure in which the via is positioned in the upper position, while the highest relates to the case where the via is placed in the centre

Table 5.5 Influence of via position on the performance

Via Position	Centre	Left/Right	Down	Up
f **(GHz)**	3.06	2.702	2.717	2.692
f **(%)**	na	11.7	11.21	12
S_{21} **(dB)**	–18.83	–17.26	–16.43	–17.25
BW **(MHz)**	40.2	27.6	24.8	27.9
BW **(%)**	na	31.3	38.31	30.6

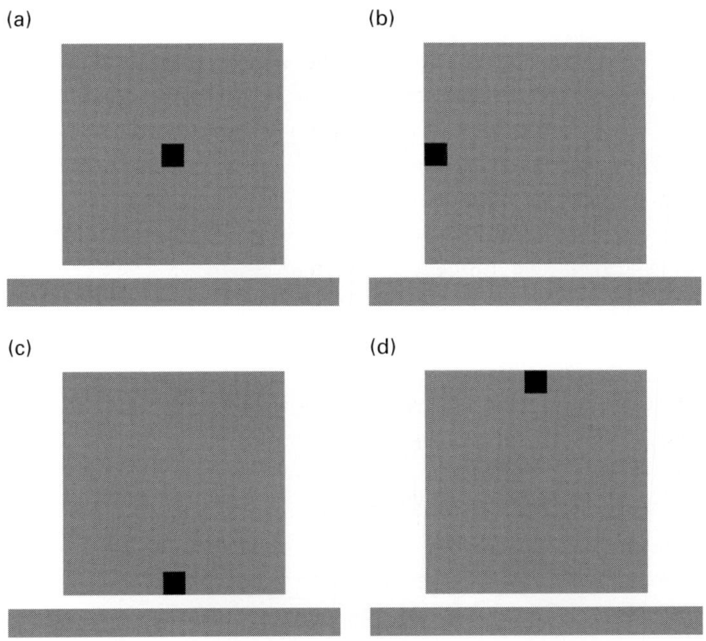

Figure 5.4 Via position related to microstrip line: (a) Centered. (b) Left. (c) Next to the microstrip (bottom). (d) As far as possible from the microstrip (top).

position. On the other hand, by moving the via to the side of the patch, more than a 30% reduction of the bandwidth is observable.

The via diameter is a parameter that could be very difficult to control in a fabrication process and thus the influence of via diameter is also investigated. Three different dimensions of the square via were analyzed in the case where the patch is embedded in the microstrip line: 0.1 mm, 0.3 mm, and 0.5 mm. Simulation results are depicted in Figure 5.6. In Table 5.6, the simulated resonant frequency of the patch, f_r, and corresponding calculated results for the via inductance, L, are shown. The inductance of the via is calculated according to standard analytical expressions for via inductance (5.2). It can be seen that small changes in via dimensions result in relatively high changes of the inductance, which ultimately cause large changes of the resonant frequency. For example, variation of the via size from 0.1 mm to 0.3 mm results in a

Table 5.6 Influence of the via dimensions on the performance

Via dimensions(mm)	0.1 x 0.1	0.3 x 0.3	0.5 x 0.5
L (nH)	0.786	0.533	0.431
L (%)	na	–32	–45
f_r (GHz)	2.94	3.66	4.22
f_r (%)	na	24	43.5

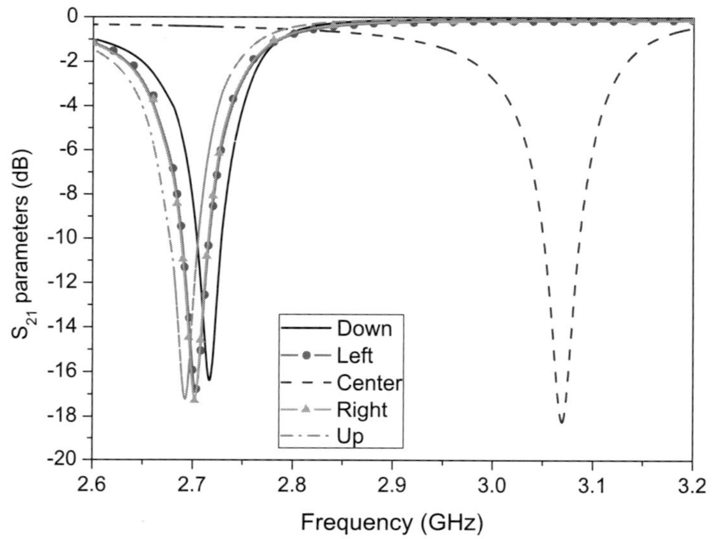

Figure 5.5 Influence of via positions on the response.

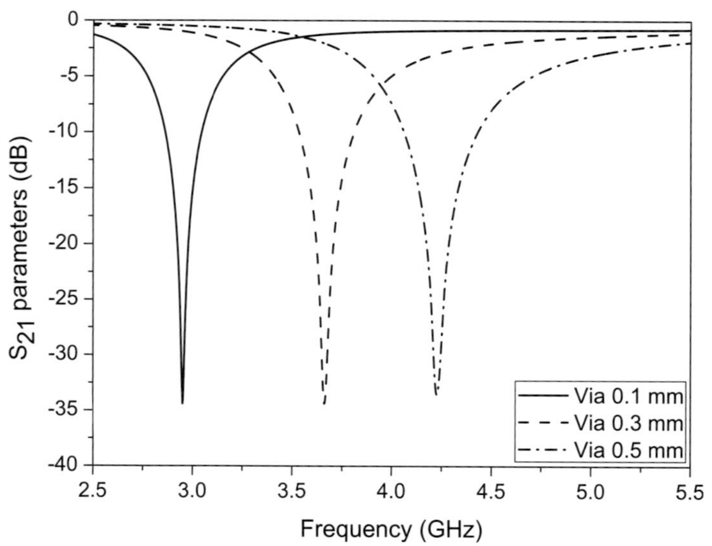

Figure 5.6 Influence of via dimensions on the performance in the case where the GPR is embedded in microstrip line.

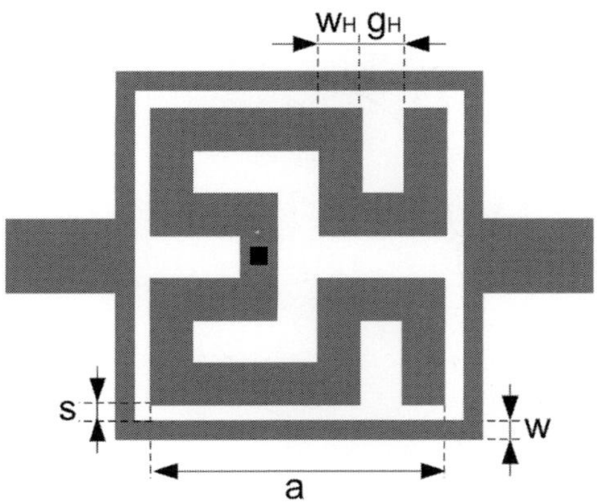

Figure 5.7 Layout of the Hilbert grounded resonator [7].

change of the via inductance by 32%, resulting in a change in the resonant frequency of about 24%.

The previously analyzed patch resonator has a conventional square shape. In order to obtain more design freedom as well as miniaturization, the square patch is replaced by a Hilbert fractal patch resonator. A novel resonator based on the grounded Hilbert patch embedded in the microstrip line is proposed in Figure 5.7. Due to their space-filling property, fractal curves allow the design of electrically long lines on a finite substrate area, and they have been applied for miniaturization of different passive components and circuits [12–14].

The proposed resonator consists of a grounded Hilbert patch embedded in the microstrip line, where s denotes the gap between the microstrip line and the resonator, a is length of the side of the patch, w is line width of the ring, and d denotes side dimensions of the square via positioned in the middle of the Hilbert patch. Dimensions w_H and g_H denote the line width and the spacing of the Hilbert fractal patch, respectively. Two different configurations of the Hilbert resonator with constant outer dimensions a = 2.9 mm and different line-to-spacing ratio w_H/g_H were analyzed using Sonnet software: one in which w_H = 0.5 mm and g_H = 0.3 mm and other in which w_H = 0.3 mm and g_H = 0.7 mm. In order to enhance the coupling between the patch and the microstrip, the width of the microstrip line around the patch as well as the spacing between the ring line and the patch are set to minimal values achievable in standard PCB technology, i.e. to 0.1 mm. The via is modeled with a square cross section, with dimensions equal to 0.1 mm by 0.1 mm. Simulated responses of the proposed Hilbert resonators are shown in Figure 5.8. Additionally, they are compared to those of a conventional square resonator whose dimensions were set to a = 2.9 mm and d = 0.1 mm.

The Hilbert grounded patch, similar to the GPR, gives a notch response when it is properly excited by the electromagnetic field of the microstrip line. It can be seen that the conventional GPR exhibits stronger rejection at resonance, but its resonant frequency is significantly higher than that of the grounded Hilbert patch. The proposed Hilbert fractal modification results in a reduction of the resonant frequency by about

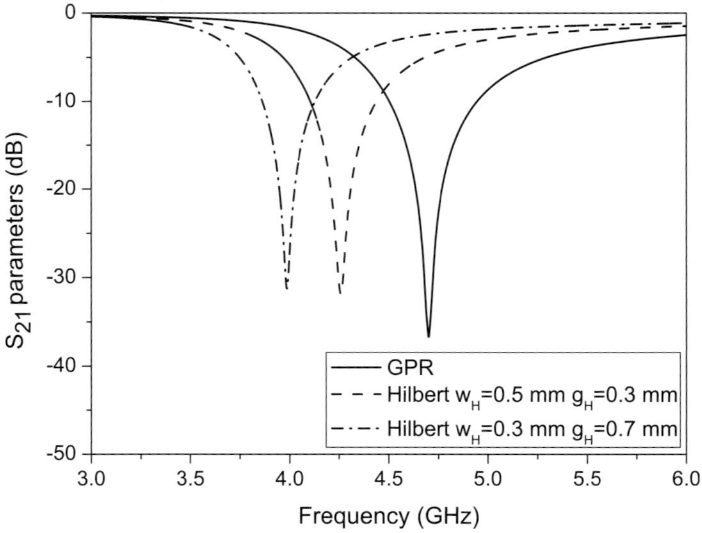

Figure 5.8 Comparison of the simulation results for Hilbert grounded resonators with different line-to-spacing ratios and the GPR.

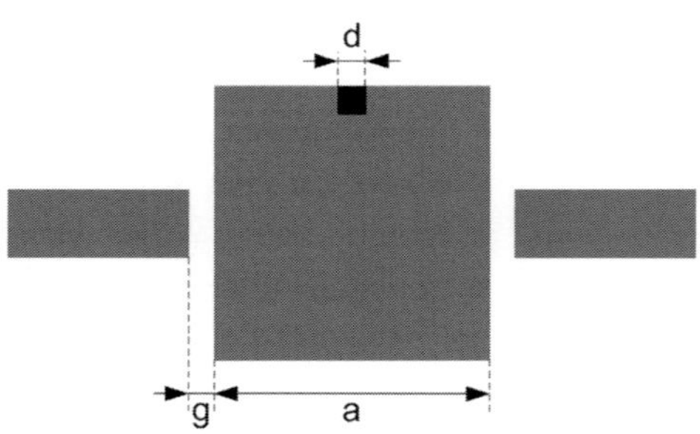

Figure 5.9 GPR capacitively coupled to the microstrip feeds [8].

15.2% in comparison with a conventional square patch. Additional miniaturization can be obtained by reducing the line-to-spacing ratio of the Hilbert fractal curve or increasing the fractal order of the Hilbert curve. According to these results, it is clear that the grounded Hilbert patch embedded in the microstrip is more suitable for the design of compact bandstop filters.

The behavior of the square patch resonator is particularly interesting in terms of application in dual-band filters, i.e. it is interesting to analyze its fundamental and first higher resonance. To that end, a square patch resonator in passband configuration, capacitively coupled to the feed lines through the gap g is investigated, Figure 5.9. The size of the square patch was chosen to be $a = b = 6$ mm, while the via dimension d is set to 0.4 mm and the gap is $g = 0.1$ mm. In the proposed case, the via is placed on the end side of the patch. The response of the proposed resonator is shown in Figure 5.10. The resonator exhibits the first resonance at $f_r = 2.85$ GHz, while the second one occurs at

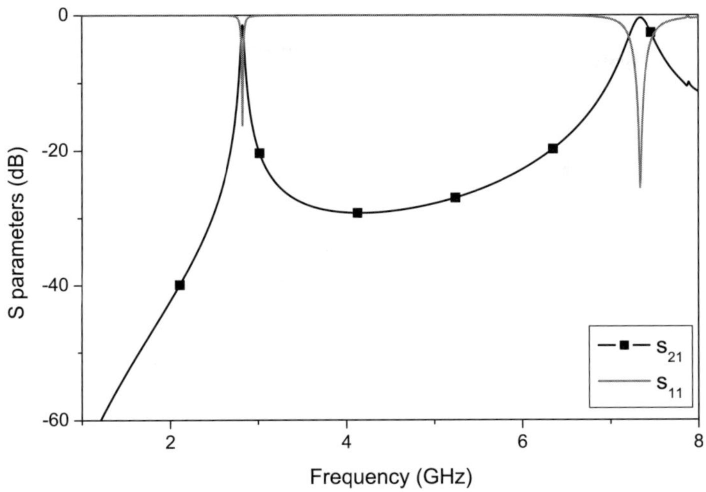

Figure 5.10 The full-wave simulated response of the GPR capacitively coupled to the microstrip feeds [8].

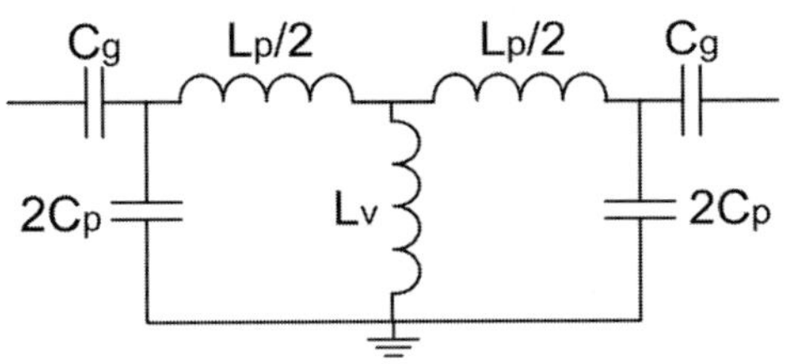

Figure 5.11 Equivalent circuit of the proposed resonator [8].

approximately $2.6f_r$. The square GPR is inherently a non-degenerate dual-mode resonator whose two resonant frequencies may be independently controlled by varying its dimensions.

To fully understand its behavior, the resonator has been modeled using an LC equivalent circuit shown in Figure 5.11, where L_v represents the via inductance, C_p is the patch capacitance towards the ground, L_p is the series inductance of the patch, and C_g models the capacitance between the feeds and the resonator. The values of the patch capacitance toward the ground and via inductance that determine the first resonance can be obtained using equations (5.1) and (5.2), respectively, derived in [15]

$$C_P = \frac{0.264 \cdot 10^{-12}(\varepsilon_r + 1.41)}{\text{In}\left(\frac{5.98h}{0.8a+t}\right)}. \tag{5.1}$$

$$L_{via} = 2 \cdot 10^{-7} h \left[\text{In}\left(\frac{4h}{d}\right) + 0.5\left(\frac{d}{h}\right) - 0.75\right]. \tag{5.2}$$

Table 5.7 Parameters of the circuit model

	L_v (nH)	L_p (nH)	C_p (pF)	C_g (pF)
Calculated	0.495	0.585	6.479	0.1362
Tuned	0.416	0.59	6.105	0.205

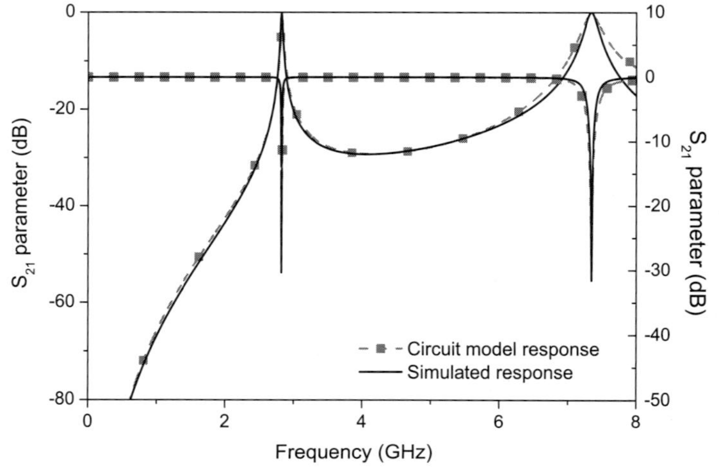

Figure 5.12 Comparison of the full-wave simulated response of the GPR and the response of the circuit model [8].

The second resonance is determined by the patch inductance $L_{p,}$ and the edge capacitance C_g. The patch inductance L_p and the edge capacitance C_g can be calculated using equation (5.3) and (5.4), derived in [16] and [17], respectively

$$L_p = 2a\cdot 10^{-7}\left[\ln\left(\frac{a}{b+t}\right) + 1.193 + 0.2235\frac{b+t}{a}\right]\left(0.57 - 0.145\ln\frac{b}{h}\right) \quad (5.3)$$

$$C_g = 500\,h\cdot Q_1\cdot e^{-1.86\frac{g}{h}}\left[1 + 4.19\left(1 - e^{-0.785\sqrt{\frac{h}{w}\frac{a}{w}}}\right)\right], \quad (5.4)$$

where w is the width of the feeding line, t is the thickness of the metallization, b is the width of the patch (in the case of the GPR $b = a$), and Q_1 can be calculated using

$$Q_1 = 0.04598\cdot\left(0.03 + \frac{w}{h}\right)^{\frac{1.23}{1+0.12(a/w-1)^{0.9}}}\cdot(0.272 + 0.07\varepsilon_r). \quad (5.5)$$

All variables are in meters. For the square patch realized on a 1.27 mm thick Taconic CεR-10 substrate, with dimensions of a = 6 mm and a via size of d = 0.4 mm, the initial calculated values for the circuit model are shown in Table 5.7. The response of the proposed electrical model had been additionally tuned. The response of the final electric model and the electromagnetic response obtained using Sonnet software are compared in Figure 5.12, while the tuned values are presented in Table 5.7. The response of the model, indicated by a dashed line, exhibits good agreement with the

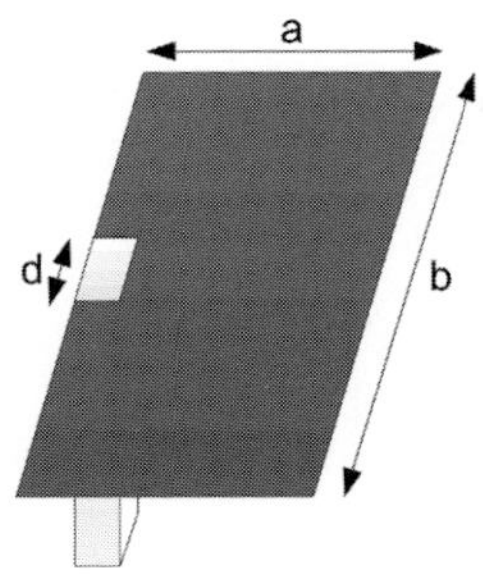

Figure 5.13 Rectangular grounded patch resonator, the RGPR [10].

electromagnetic one. Considering that the proposed formulas are valid for conventional microstrip line segments, the presence of the via causes small variations between tuned and calculated values especially in the case of patch edge capacitance and capacitance toward the ground. The patch capacitance is slightly smaller, by 6%. while the edge capacitance is 30% larger.

The square patch resonator can be modified to a rectangular patch resonator to allow for more design freedom and considerable spacing between resonances. The rectangular grounded patch resonator (RGPR) is presented in Figure 5.13. Similar to its square counterpart, this resonator supports two non-degenerate resonances. The first resonance is created by the capacitance to ground and the via inductance, whilst the second resonance is modeled by the patch edge capacitance and the series inductance. Clearly, the two resonances are positioned at different frequencies by selecting the appropriate geometrical parameters, namely a, b, and d in Figure 5.13.

The patch capacitance to ground is determined by the size of the patch, i.e. by the parameters a and b and the substrate characteristics. These parameters also determine the patch inductance. The via inductance is related to the via diameter, whilst the edge capacitance is related to the parameter a. This is illustrated in Figure 5.14, where the dimensions a and b of the RGPR, capacitively coupled to 50 Ω feed lines, have been varied. Here, the via diameter $d = 0.4$ mm was kept constant, and the dimensions of the patch were varied while maintaining a constant patch area. By doing this, the first resonance was fixed at 2.4 GHz and the second resonance was shifted by almost 100%. This demonstrates the first of two criteria for full control of the dual-band magnitude response, i.e. control over the individual passband positions.

Alternatively, by increasing the ratio between resonances, single bandpass filters with extended stopband regions can be realized. However, when two RGPRs are used to form a filter, the inter-resonator coupling at two resonant frequencies cannot be adjusted separately, and two passbands of the filter cannot be separately optimized, as will be shown in Section 5.3.3. To change the coupling characteristics, the dual-band resonator shape, or dimensions, usually need to be changed to obtain the proper couplings at both resonances. In general, this has also been the case with all other dual-mode resonators presented so far, and it significantly limits their applicability in the realization of dual-band filters.

In Figure 5.15, the current distributions of the first two modes supported by a RGPR are shown for the resonator dimensions a = 9.3 mm, b = 5.5 mm, and d = 0.4 mm.

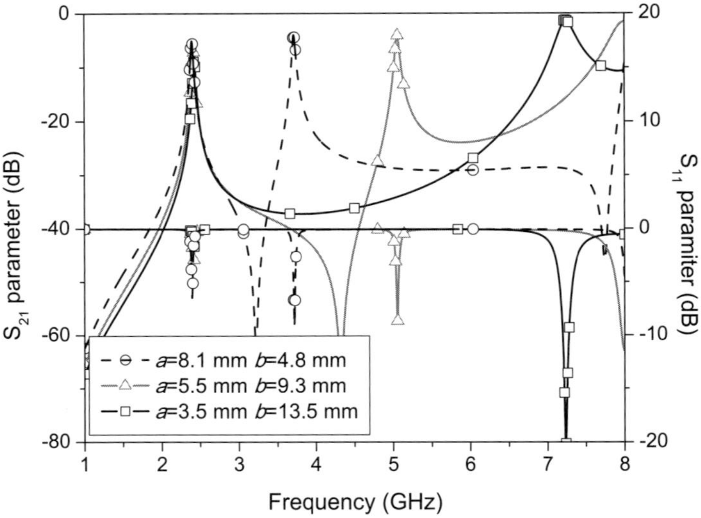

Figure 5.14 Influence of RGPR dimensions on its transmission and reflection characteristics [10].

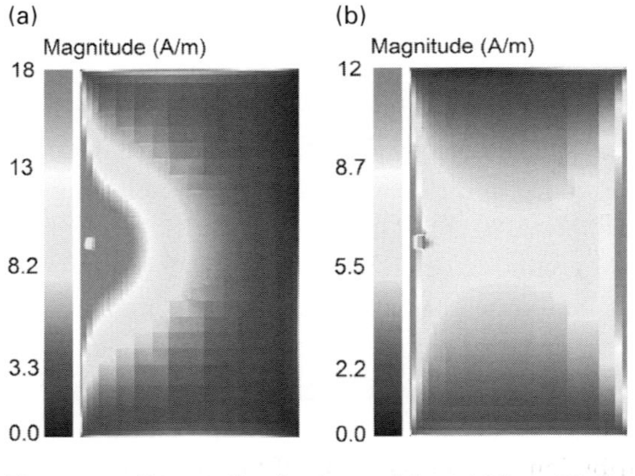

Figure 5.15 Current distributions of the (a) First. (b) Second resonant mode of the RGPR with dimensions a = 9.3 mm, b = 5.5 mm, and d = 0.4 mm [10].

These graphical presentations of current magnitude show that the via region of the patch supports the strongest current at the first resonant frequency. At the second resonance, the currents are strongest on the two patch edges. This led to the idea to introduce a slot perturbation at the edge opposite to the via, i.e. to obtain a mechanism to influence only the second mode. This method is described in detail in the following section.

5.2.2 Perturbed grounded-patch resonator

As was stated in the previous section, introduction of a small perturbation to the edge of the resonator opposite the via gives a resonator with a mechanism to control only the second mode. The configuration of such a resonator, a perturbed grounded patch resonator (PGPR), is shown in Figure 5.16. Parameters s_l and s_w denote the length

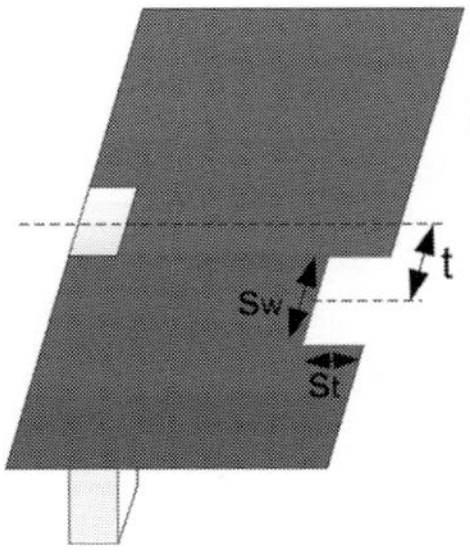

Figure 5.16 Perturbed grounded patch resonator, PGPR [10].

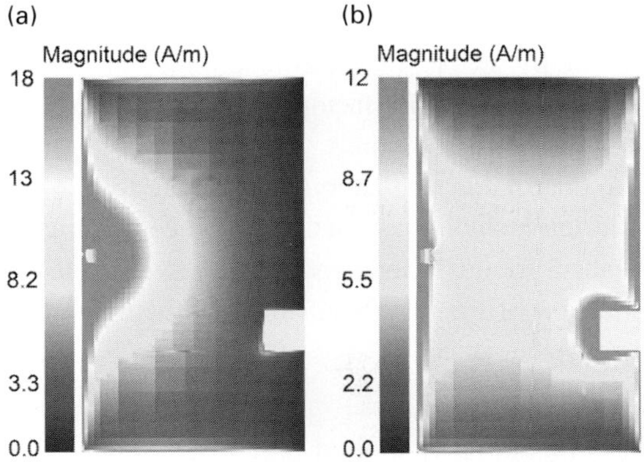

Figure 5.17 Current distributions of the (a) first and (b) second resonant mode of PGPR with dimensions $a = 9.3$ mm, $b = 5.5$ mm, $d = 0.4$ mm, $s_l = 1.1$ mm, $s_w = 1$ mm, and $t = 1.3$ mm [10].

and the width of the slot, respectively, while t denotes the slot offset with respect to the symmetry axis of the PGPR (dashed line).

Due to the current distribution, the introduction of the edge slot causes a negligible shift in the resonant frequencies of the resonator, and these remain positioned at 2.4 GHz and 5.2 GHz, as was the case for the RGPR of the same dimensions. The explanation for this surprising result is that the slot does not interrupt the flow of the current for the first mode, while it negligibly interrupts the flow of the current for the second mode, due to the fact that the slot size is much smaller than the length of the patch a. This is also evident from Figure 5.17, where the current distributions of two resonant modes of the PGPR are shown.

Nevertheless, when two or more PGPR resonators are coupled to realize a filter, edge slot perturbations play a significant role, as will be shown in Section 5.3.3. In that case, the width, length, and position of the slots can be optimized to provide the desired coupling coefficient and external quality factor at the second operating frequency, independently from the coupling coefficient at the first operating frequency leading to a design dual-band filter with independently controlled passband characteristics.

5.3 Application of grounded patch resonators in filter design

In this section, single-band bandstop and bandpass filters as well as dual-band filters based on the GPR and its modification are proposed. The synthesis procedure based on coupled-resonator theory is used to design different types of filters based on coupled GPR resonators.

In Section 5.3.1, a compact bandstop microstrip filter is proposed, based on the GPR and Hilbert grounded patch embedded in the microstrip line. The superiority of the proposed filters is confirmed by comparison with state-of-the-art bandstop filters of similar type in terms of size, performance, and fabrication complexity.

In the following section, the highly selective compact single-band bandpass filter of third order and the cross-coupled filter based on the GPR are presented. Their design is based on conventional filter theory, i.e. on the determination of the design curves for coupling coefficients and the external quality factors of the first resonant mode of the GPR from the electromagnetic (EM) simulations.

Section 5.3.3 is devoted to dual-mode resonators based on the RGPR and the PGPR. It presents a detailed filter design including estimation of the coupling coefficients and quality factors at both resonances for dual-mode resonators based on the RGPR and the PGPR, modeling, and fabrication of the dual-band filters with straightforward design and independent passbands characteristics.

5.3.1 Single-band bandstop filters based on grounded-patch resonators

In this section, a novel compact bandstop microstrip filter is proposed, based on the grounded patch resonator and Hilbert grounded patch embedded in the microstrip line. The characteristics of the proposed filter are compared with state-of-the-art stopband filters of similar type in terms of size, performance, and fabrication complexity.

To illustrate the potential of the proposed GPR in an arbitrary filtering characteristic, bandstop filters of the fourth order were designed with different geometrical arrangements of GPRs. The layouts of the filters with the patch resonators positioned next to and under the microstrip line, and the layout of the filter with the patch resonators embedded in a microstrip line are shown in Figure 5.18. The outer dimensions of patch resonators were set to 5 mm x 5 mm. The distance between patches is 0.3 mm, the distance between microstrip line and the patch is 0.1 mm, and via dimensions are equal to 0.1 mm x 0.1 mm. The frequency responses of the structures are shown in Figure 5.19, while Table 5.8 compares the responses of the proposed filters, where the f_s denotes centre bandstop frequency, $BW_{3\text{dB}}$ and $BW_{10\text{dB}}$ are 3 dB and 10 dB bandwidths, respectively, and S_{21_0} is the insertion loss in the passband. In Table 5.8, filter dimensions are given in millimeters and guided wavelength λ_g.

All proposed filters are characterized by very low insertion loss in the passband and relatively wide bandwidth. In the filter where the patches are positioned next to the microstrip line, the 3 dB fractional bandwidth is equal to 15.2% centered at 3.1 GHz. Nevertheless, the insertion loss in the passband is not sufficient and it is equal only to 8 dB. The filter with patches positioned under a microstrip line demonstrates sharper

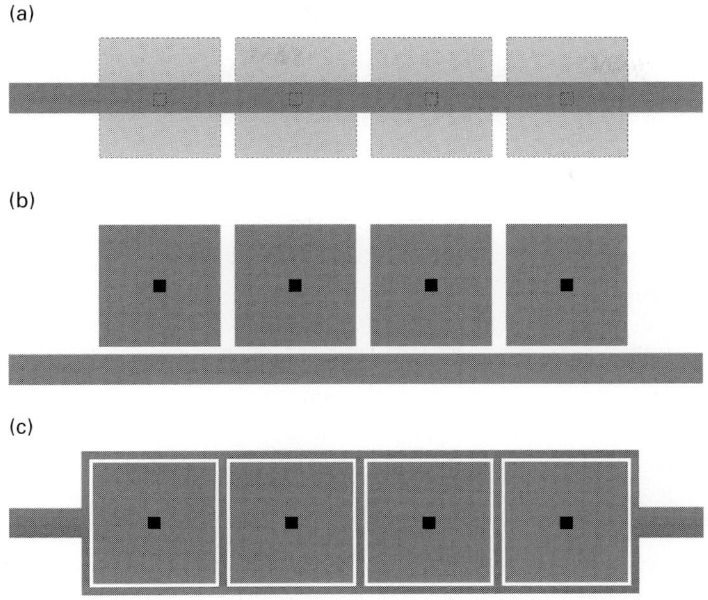

Figure 5.18 Layout of the proposed bandstop filter of the fourth order with: (a) Patch positioned under microstrip line. (b) Patch placed next to microstrip line. (c) Patch embedded in microstrip line [6].

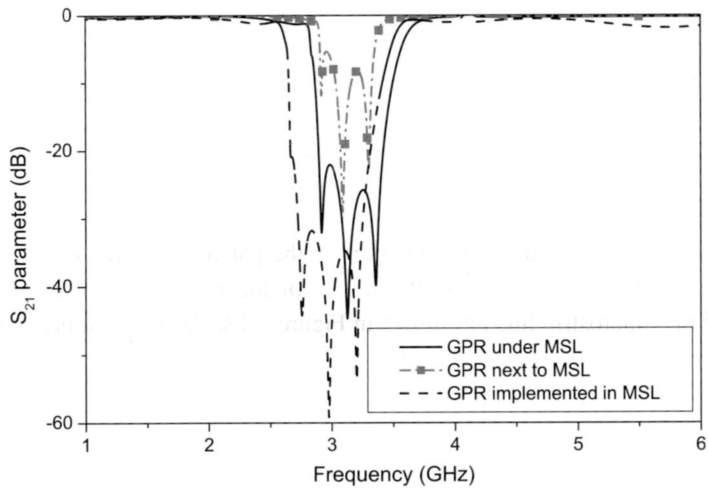

Figure 5.19 Simulation results for bandstop filters illustrated in Figure 5.18.

transitions for both sides of the stopband with reflection loss of more than 20 dB. The filter with patch resonators implemented in the microstrip line shows a slightly lower resonant frequency than the other two filters. It has 32% 10 dB fractional bandwidth centered at 3 GHz with an insertion loss of more than 30 dB. Stronger coupling between

Table 5.8 Comparison of the characteristics of the bandstop filters based on the GPR

Filter	Patch under MSL	Patch next to MSL	Patch implemented in MSL
f_s (GHz)	3.1	3.1	3
BW_{3dB} (MHz)	772.7	472	970
FBW_{3dB} (%)	25.75	15.2	32.3
BW_{10dB} (MHz)	644	Na	900
FBW_{10dB} (%)	21.45	Na	30
S_{21_0} (dB)	–22	–8	–32
Dimensions (mm)	21.5 x 5	21.5 x 6.5	21.5 x 5.4
Dimensions (λ_g)	0.52 x 0.12	0.52 x 0.16	0.52 x 0.13

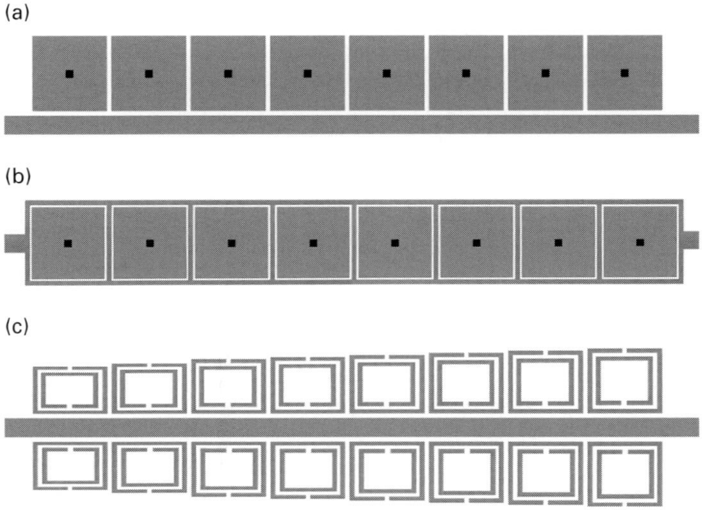

Figure 5.20 Comparison of the eighth-order filter based on the GPR and filter based on SRR: (a) Filter with GPR resonator placed next to microstrip line. (b) Filter with patch resonators implemented in microstrip line. (c) Filter based on SRRs [18].

patches and the microstrip line in the case of the embedded patch produces deeper stopbands and causes small ripples just outside the stopband.

In order to compare the characteristics of the proposed filter configurations based on the GPR, the eighth-order stopband filters are compared with a stopband filter based on the Split Ring Resonator (SRR), published in [18]. The layouts of the filters are shown in Figure 5.20. The external dimensions of the patch are optimized to 3.3 mm x 3.3 mm, in order to obtain the same central frequency as in the case of the filter based on the SRR. The distance between the patch resonators (in Figure 5.20a) and the distance between the microstrip line and patch resonators (in Figure 5.20b) are set to 0.1 mm, while in both cases square vias have dimensions equal to 0.1 mm x 0.1 mm.

The total lengths of the filters based on patch resonators are 28.9 mm, which correspond to $0.54\lambda_g$ on the given Taconic substrate. In the case of the filter proposed in [18], a microstrip line is loaded with SRRs on both sides. The dimensions of the

Table 5.9 Comparison of the characteristics of the eighth-order stopband filters based onthe GPR and filter based on SRRs

Filters	Microstrip loaded with GPR	GPR implemented in Microstrip	Microstrip loaded with SRR
f_s **(GHz)**	4.55	4.35	4.55
BW_{10dB} **(MHz)**	820	1410	714
FBW_{10dB} **(%)**	18	32.4	15.7
S_{21_0} **(dB)**	–21.8	–60	–18
Dimensions (mm)	28.5 x 4.8	28.9 x 3.7	39.3 x 9.97
Dimensions (λ_g)	1.02 x 0.17	1.04 x 0.13	1.6 x 0.4

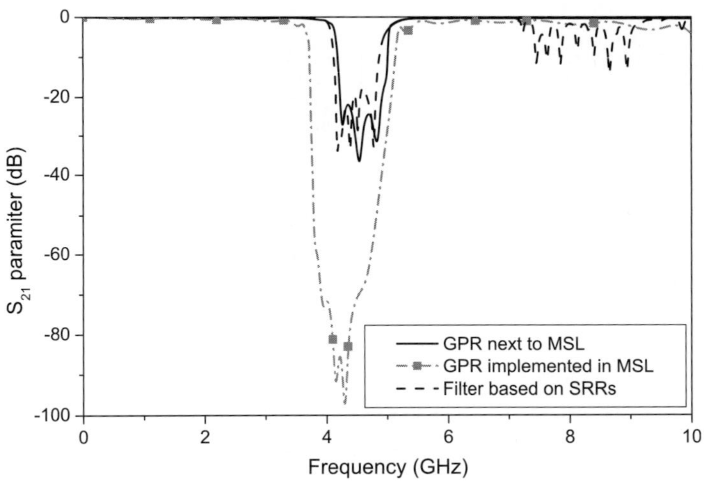

Figure 5.21 Comparison of the filters based on the GPR and on SRR illustrated in Figure 5.20.

SRRs are precisely optimized to achieve a broadband response. The SRR dimensions vary between 4.1 mm x 2.7 mm for the largest SRR resonator and 3.55 mm x 2.7 mm for the smallest one, resulting in a total length of the filter of 39.3 mm, which correspond to $1.6\lambda_g$. The line width of the SRR, the distance between rings, the distance between the SRR, and the distance between the rings and microstrip line were set to 0.3 mm. The filter in [18] is realized on a Rogers RO3010 substrate with a thickness of 1.27 mm, a relative permittivity of 10.2, and a loss tangent of 0.0022. Figure 5.21 compares the simulation responses of the proposed filters, while the Table 5.9 summarizes the characteristic filter parameters, where f_s is the central bandstop frequency, BW_{10dB} is the 10 dB bandwidth, and s_{21_0} is the insertion loss in the passband.

It can be noted that the filter with the GPR positioned next to the microstrip line shows characteristics similar to the filter based on the SRR in terms of bandwidth and return losses. The filter with patch resonators implemented in the microstrip lines shows significantly higher suppression in the stopband, more than 60 dB and 17% wider fractional bandwidth. In comparison to the filter based on the SRR, the filter with the embedded GPR resonator has a bandwidth extension of more than twice as much higher

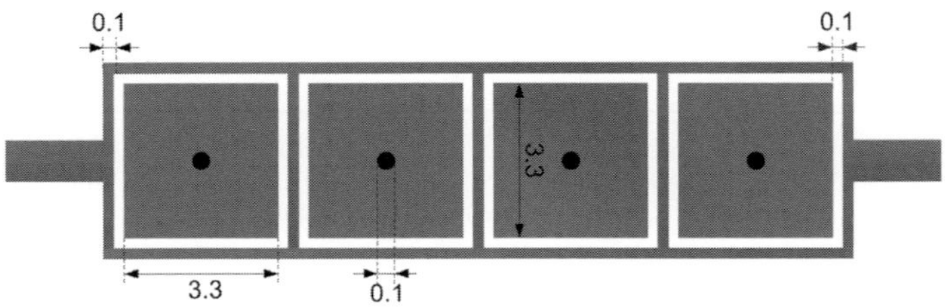

Figure 5.22 Layout of the fourth-order bandstop filter with the GPR embedded in a microstrip line, dimensions in mm [6].

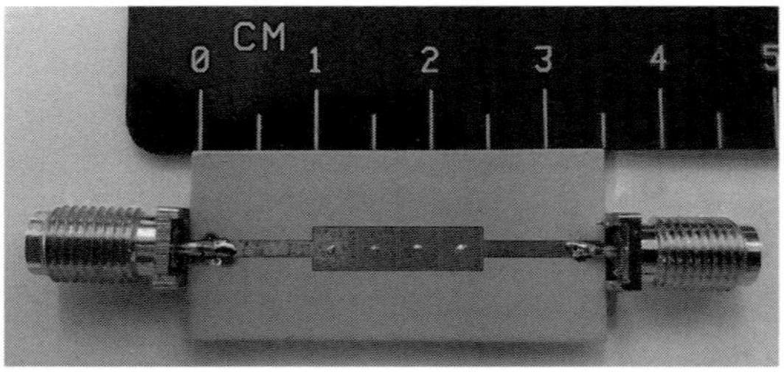

Figure 5.23 Photograph of the fabricated fourth-order bandstop filter based on GPRs embedded in a microstrip line [6].

suppression in the stopband. Another advantage of the filters based on the patch implemented in the microstrip line is the position of the second harmonic. In the case of the filter based on the SRRs, the second harmonic occurs at frequency of $2f_s$, while for filters based on patch resonators, the second harmonic occurs at approximately $3f_s$. In contrast to the filter based on SRRs, for filters based on GPRs there is no need for further optimization of the size of patches in the filter. Moreover, the total dimensions of the filters are smaller in the case of the filters based ona GPR.

Based on the proposed grounded patch embedded in the microstrip, a super compact high selectivity stopband filter of fourth order was designed, fabricated, and measured [6]. The layout of the filter is shown in Figure 5.22 for the outer dimension of the patches of 3.3 mm x 3.3 mm. All resonators are identical, and no time-consuming optimization is needed in the process of filter design. All the spacings between patches and the microstrip are equal to 0.1 mm, and size of all vias is 0.1 mm. The overall filter dimensions are only 14.5 mm x 3.7 mm, i.e. approximately $0.52\lambda_g$ x $0.13\lambda_g$, where λ_g is the guided wavelength.

The photograph of the fabricated prototype is shown in the Figure 5.23, while simulated and measured responses of the filter are compared in Figure 5.24. Good agreement can be observed. The fabricated filter exhibits a 35% 10 dB fractional bandwidth centered at 4.36 GHz with rejection of more than 30 dB and a maximum reflection loss of about 1 dB in the stopband. The fabricated fourth-order filter has an overall dimension of $0.52\lambda_g$ x $0.13\lambda_g$. The second stopband appears at approximately 10 GHz.

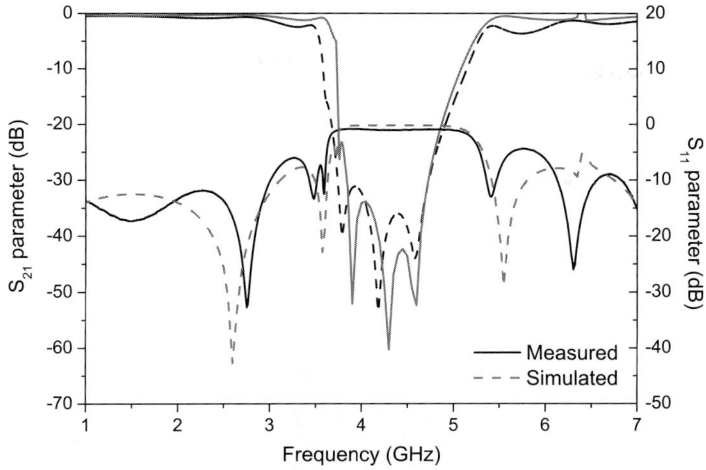

Figure 5.24 Simulated and measured responses of the fourth-order bandstop filter based on GPRs embedded in microstrip line [6].

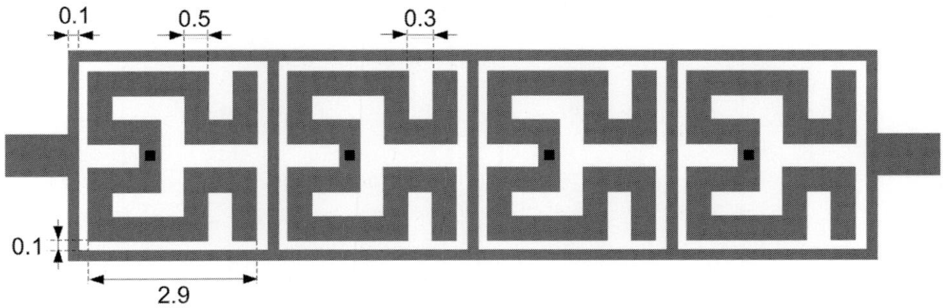

Figure 5.25 Layout of the bandstop filter of the fourth order with Hilbert grounded patch embedded in microstrip line, all dimensions in mm [7].

Based on the previously proposed Hilbert grounded patch resonator embedded in the microstrip, another fourth-order super compact highly selective bandstop filter was designed [7]. The layout of the proposed filter for the outer dimensions of the patches equal to 2.9 mm x 2.9 mm is shown in Figure 5.25. All resonators are identical, and no time-consuming optimization is needed in the process of filter design. The spacing between patches and the microstrip are equal to 0.1 mm and the size of all the vias is 0.1 mm. The overall filter dimensions are 12.9 x 3.3 mm, i.e. approximately $0.47\lambda_g$ x $0.12\lambda_g$, where λ_g is the guided wavelength.

Simulated response of the filter is shown in Figure 5.26. The filter is characterized by a 10 dB fractional bandwidth of 30% at 4.43 GHz, with rejection of more than 30 dB and the maximal insertion losses around 1 dB in the passbands.

To benchmark the new proposed filter topology based on the Hilbert grounded patch resonator, a comparison with five other recently published stopband filters is made: one based on conventional GPRs [6], one based on an L-shaped resonator [16], and another

Table 5.10 Comparison of the characteristics of the filter based on the Hilbert grounded patch embedded in the microstrip and other bandstop filters [7]

Filters	Hilbert GPR filter	Filter1 [19]	Filter2 [18]	Filter3 [19]	Filter 4 [20]	Filter 5 [16]	GPR filter [6]
f_c (GHz)	4.43	9.25	4.5	9.25	2.44	3.4	4.36
***BW* (MHz)**	1330	≈333	760	3000	60	142	1360
***FBW* (%)**	30	3.6	17	33	2.45	4.2	35
Rejection (dB)	32	35	25	45	20	50	30
Dimensions (mm)	12.9x3.3	19.5x7.75	39.3x9.97	19.5x3	23.7x23.7	44.5x17.4	3.7x14.5
Dimensions (λ_g)	0.47x0.12	1x0.4	1.6x0.4	1x0.15	0.5x0.5	1.36x0.53	0.52x0.13
Filter size, ($100*\lambda_g^2$)	5.64	40	64	15	25	72	6.76
Filter order	4	6	8	6	4	5	4

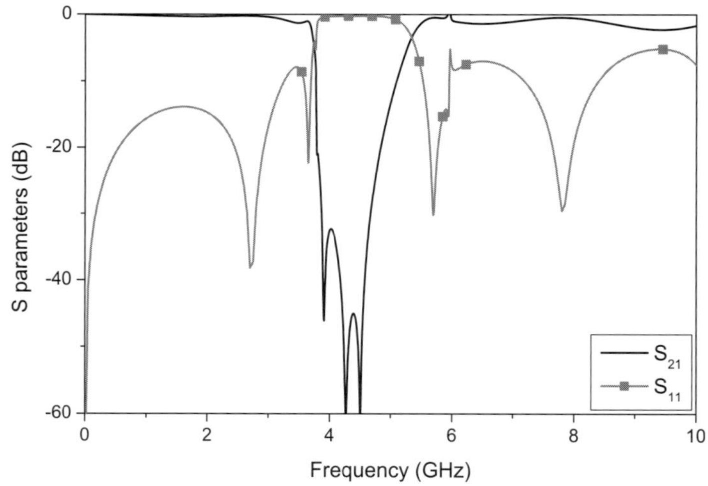

Figure 5.26 Simulated response of the proposed bandstop filter of the fourth order with the Hilbert grounded patch embedded in the microstrip line [7].

based on different configurations of SRRs [18–20]. The comparison is detailed in Table 5.10, where f_c is the centre frequency, *FBW* is the 10 dB fractional bandwidth, and the *Rejection* is the rejection in the stopband. Table 5.10 also gives the order of different filters, and their overall dimensions and chip area in terms of guided wavelengths. In [18–20], the values are obtained from the measured responses presented in the papers, as exact numbers for the parameters in the table were not presented in [18–20].

Filter 1 [19] of Table 5.10 consists of a microstrip line coupled to Split Ring Resonators (SRRs) on each side of the microstrip line. The filter is of the sixth order and its footprint is seven times that of the Hilbert GPR filter. Furthermore, it exhibits a narrow stopband of 3.6%.

Filter 2 [18] is similar to Filter 1 but the dimensions are optimized and the SRRs are non-synchronously tuned to provide a larger stopband, which is 17% in this case.

Eight unit cells were required and the footprint is eleven times larger than the Hilbert GPR filter presented in this chapter.

Filter 3 [19] uses Complementary Split Ring Resonators (CSRRs). These are etched into the ground plane below a microstrip feed line. The ground plane etching complicated the manufacturing process due to the requirement for very careful alignment of the etching on two conductive layers. Filter 3 has similar performance to the Hilbert GPR, although the footprint is three times larger and the filter exhibits increased insertion loss at frequencies above the stopband.

Filter 4 [20] has the smallest bandwidth and the worst stopband rejection of all the examples listed in Table 5.10, and furthermore has a large footprint.

Filter 5 [16] is a fifth-order microstrip bandstop filter based on L-shape resonators. The filter exhibits a narrow stopband equal to 4.2% but with higher suppression in the stopband. On the other hand, its footprint is almost 13 times larger than the filter based on the Hilbert resonator.

Instead of using the square GPR, the Hilbert grounded patch is used in [6], where the length of filter is reduced by 9.6% and its width by 8.3%, without degrading the performance. The performance of this filter is detailed in the last column of Table 5.10. The response compares well to the Hilbert GPR filter (first example of Table 5.10), although its footprint is 16.5% larger. The miniaturization of the Hilbert GPR filter is enhanced by either changing the order of the Hilbert curve, or more practically, by changing the line-to-spacing ratio of the fractal.

5.3.2 Single-band bandpass filters based on grounded-patch resonators

In this section, a compact bandpass microstrip filter based on grounded patch resonators is proposed. Only the first mode of the GPR is taken into account in the design of single-band filters. The coupling model of the filter is established and the coupling coefficients of the coupled pair of resonators and the external quality factors of the basic feeding configurations are investigated using full-wave electromagnetic simulations. As an example, a third-order direct-coupled filter and cross-coupled filter based onthe GPR are designed.

5.3.2.1 Direct coupled filters based on grounded-patch resonators

A square grounded patch is used to design compact filters following classical filter design theory and the coupling coefficient method [21–22, 16]. As a demonstration, a third-order Chebyshev bandpass filter for Wireless LAN applications that operates at 2.4 GHz is designed. The proposed configuration of the third-order direct-coupled filter is shown in Figure 5.27. The filter has been designed to exhibit a Chebyshev response centered at 2.4 GHz, with a fractional bandwidth equal to $FBW = 3.96\%$ and a passband transmission ripple equal 0.05 dB.

First, the resonator is designed to operate at 2.4 GHz. The dimensions of the square GPR resonator were found to be 7 mm x 7 mm with the via size equal to 0.4 mm.

The circuit model of the third-order bandpass filter is shown in Figure 5.28. Each resonator has been modeled by a parallel RLC circuit, where the parallel R_p models the effect of a finite resonator unloaded Q-factor. The coupling between resonators is

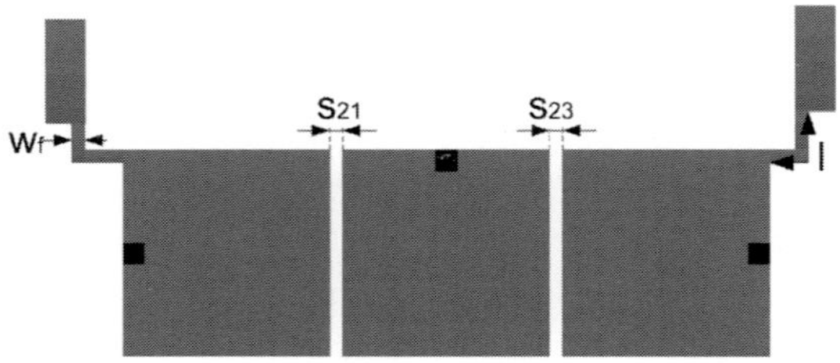

Figure 5.27 Configuration of the proposed third-order direct-coupled filter [9].

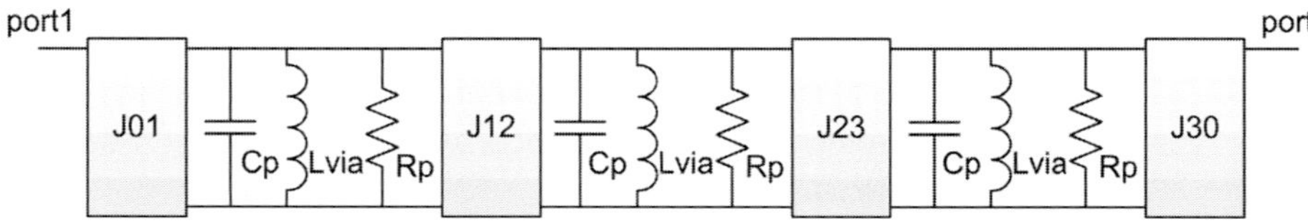

Figure 5.28 The third-order bandpass filter prototype [9].

modeled by frequency independent J-inverters, where the J_{ij} $(i,j = 0,1,2,3)$ in Figure 5.28 represents frequency-independent admittance inverters that can be calculated from the filter specifications. It should be noted that these models are only accurate in the vicinity of the resonant frequency. The filter parameter values for the prototype, found in [21], are: $g_0 = g_4 = 1$, $g_1 = g_3 = 1.031$, $g_2 = 1.147$.

The related coupling coefficients and external quality factors of the direct-coupled filter can be determined by the following equations [16]

$$k_{ij} = \frac{FBW}{\sqrt{g_i g_j}}. \tag{5.6}$$

$$Q_{e1} = \frac{g_0 g_1}{FBW}. \tag{5.7}$$

The coefficients k_{ij} specify the coupling between adjacent resonators i and j of the filter. Because of the symmetry of the structure, two coupling coefficients must be equal, $k_{12} = k_{23}$. The external quality factors Q_{e1} and Q_{e3} that specify the input and output couplings, respectively, are also equal. For the defined filter specifications, the following values have been calculated: $k_{12} = k_{23} = 0.0363846$, $Q_{e1} = Q_{e3} = 26.046$.

The characteristic admittances for the frequency independent inverters are then

$$J_{01} = \sqrt{\frac{FBW \cdot b}{Z_0 g_0 g_1}}, \tag{5.8}$$

$$J_{30} = \sqrt{\frac{FBW \cdot b}{Z_0 g_3 g_4}}, \tag{5.9}$$

$$J_{ij} = \frac{FBW \cdot b}{\sqrt{g_i g_j}}, \tag{5.10}$$

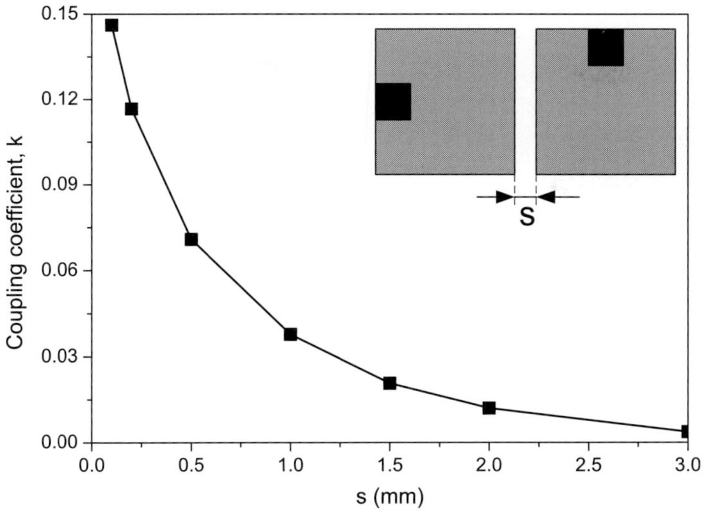

Figure 5.29 Coupling coefficients versus resonator-to-resonator distance s for orientation as shown [9].

where Z_0 is the characteristic line impedance of the resonator and b is the susceptance slope parameter. The resistance R_p can be determined from the unloaded quality factor of the resonator using equation

$$R_p = \frac{Q_u}{w_0 C_P}, \tag{5.11}$$

where Q_u is the external quality factor of the resonator given by

$$Q_u = \frac{f_0}{BW_{3\ dB}}. \tag{5.12}$$

In Eq. (5.12), BW_{3dB} denotes the 3 dB bandwidth of the single grounded patch resonator. It should be mentioned that the proposed resonator has an unloaded quality factor equal to Q_u =100 at the first resonance.

The coupling coefficient k can be calculated from the two split resonant frequencies f_1 and f_2, obtained from full-wave EM simulations of two identical synchronously tuned coupled resonators [16], using

$$k = \frac{f_2^2 - f_1^2}{f_2^2 + f_1^2}. \tag{5.13}$$

Due to the specific resonator design, where the via is shifted to one side of the metal patch, several basic coupling scenarios can be realized for the first resonant mode [8]. They stem from different orientations of otherwise identical grounded patch resonators, which are separated by spacing s and may be shifted for a lateral offset. The obtained coupling coefficient as a function of resonator-to-resonator distance s and for the orientation chosen in the proposed filter is shown as an inset in Figure 5.29. It can be seen that the coupling coefficient decreases monotonously with distance s.

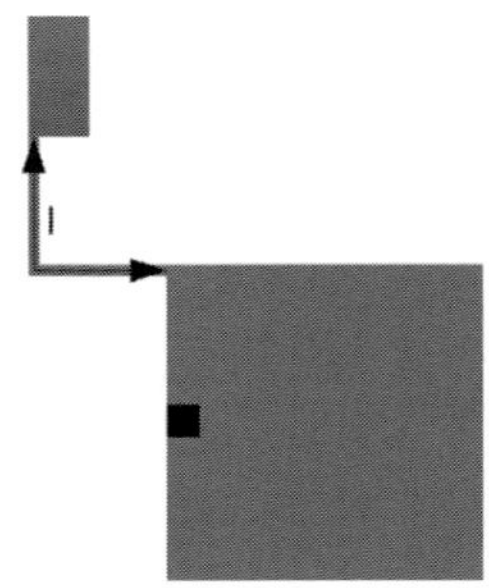

Figure 5.30 Tapped line structure for determination of the Q-factor of the GPR [9].

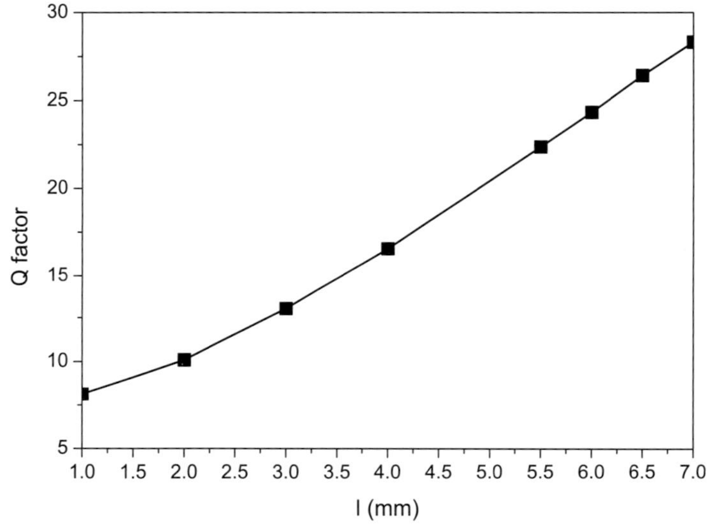

Figure 5.31 External quality factor as a function of the coupling line length l [9].

In the proposed case, the 50 Ω feeding line is directly tapped to the input/output resonator and the external quality factor is controlled by the line length, l, as detailed in Figure 5.30. The external quality factor is extracted from the group delay at resonance using equation (5.14), reprinted from [16], where w_0 denotes resonant frequency and $\tau_{s11}(w_0)$ presents group delay at w_0. The external quality factor was determined as a function of the feed line length l, for a fixed line width of $w_f = 0.1$ mm, since the narrow line width results in the smaller external quality factor of the resonator. The external quality factor in the function of the line length l is plotted in Figure 5.31.

$$Q_e = \frac{\omega_0\, \tau_{s11}(\omega_0)}{4}. \tag{5.14}$$

The initial dimensions of the filter, such as resonator-to-resonator spacing and the feed length, were determined using the obtained design curves calculated for the coupling coefficient and Q-factor. Spacing between the resonators was set to $s_{12} = s_{23} = 1.1$ mm

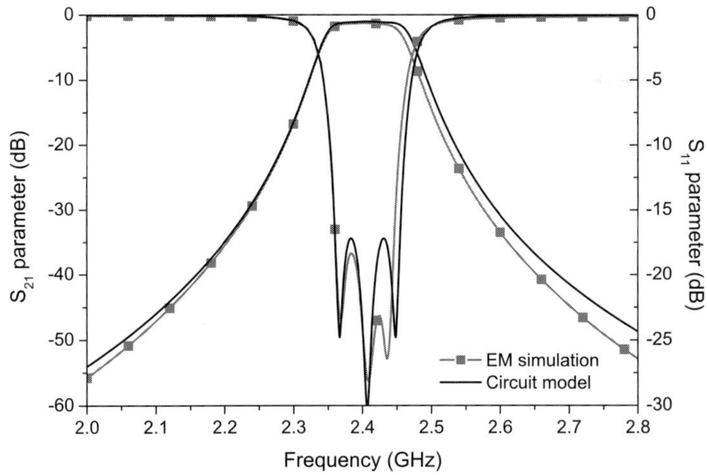

Figure 5.32 Comparison of electromagnetic response and response of coupling model [9].

and the lengths of the feeds to $l = 6.3$ mm. This filter was simulated in Sonnet software and it exhibited larger bandwidth and passband ripple than expected. In order to obtain the required response, a numerical tuning procedure based on the filter model was performed, in which the filter and the corresponding model were divided into individual elements and analyzed separately.

We have concluded that the type of feed used slightly influences the resonances of the patches. To compensate for the frequency shift effect because of the feeding lines, the first and the third patch have to be larger, in order to resonate at the same frequency as the middle one. In addition, the Q-factor has to be higher; therefore, the length of the feeds has to be slightly longer.

The final dimensions of the filter are as follows: dimensions of the first and third patch are 7.3 mm x 7 mm, while the middle one is 7 mm x 7 mm; the distances between patches are $s_{12} = s_{23} = 1.05$ mm and the length of the feeds is $l = 6.4$ mm. The overall size of the filter together with feeds is only $0.31\lambda_g$ x $0.64\lambda_g$, where λ_g denotes the guided wavelength at an operating frequency of 2.4 GHz.

In Figure 5.32, a comparison of the electromagnetic and the circuit model response is shown, where very good agreement with the theoretical model can be observed. The electromagnetic response of the filter in the wide frequency range is shown in Figure 5.33. The filter is characterized by small insertion loss in the passband equal to 1.25 dB, and passband ripple smaller than 0.07 dB. The fractional bandwidth is equal to 3.91%, which conforms well to the designed specifications. The proposed filter exhibits a wide stopband region up to second harmonic that exists at 6.25 GHz, with a more than 60 dB rejection in the stopband.

5.3.2.2 Cross-coupled filter based on grounded-patch resonators

In this section, the design, fabrication, and measurements of a passband cross-coupled filter of fourth order based on the GPR are presented. The filter has been designed

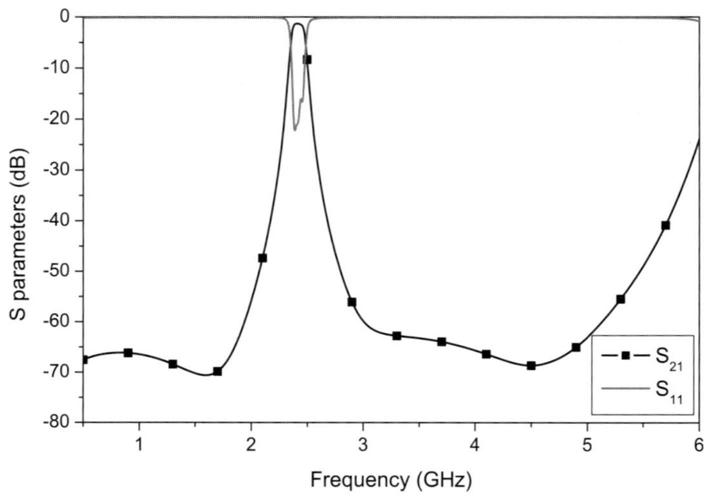

Figure 5.33 Electromagnetic response of the third-order filter based on the GPR over a wide frequency range [9].

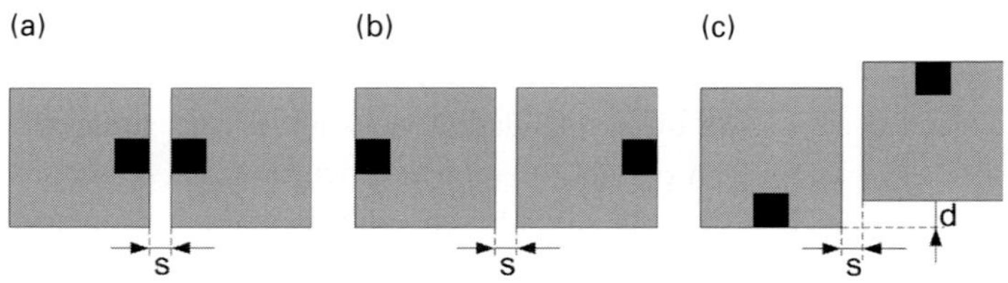

Figure 5.34 Basic coupling structures of the coupled GPR [8]: (a) Magnetic coupling. (b) Electric coupling. (c) Mixed coupling.

according to the following specifications: central frequency $f_c = 2.85$ GHz and fractional bandwidth of $FBW = 5\%$, with two transmission zeros positioned at 2.85 ± 0.15 GHz.

In the first step, the dimension of the resonator that operates at 2.85 GHz was found. The dimensions of the square GPR resonator were found to be 6 mm x 6 mm with a via size of 0.4 mm.

Typically quasi-elliptic response filters are realized by exploiting coupling between non-adjacent resonators. Due to the specific resonator design, where the via is shifted to one side of the metal patch, three basic coupling structures can be realized usingthe GPR, as is shown in Figure 5.34. They result from different orientations of identical patch resonators, which are separated by spacing s and may be shifted by an offset d.

The nature and the intensity of the fringing fields in those three geometrical arrangements determine the nature and the strength of the coupling. The maximum magnetic coupling is obtained when the vias are placed close to each other, Figure 5.34(a), while the electric coupling exists when "free" sides of the patches are in proximity, i.e. when the vias are placed on the opposite sides of the patches, Figure 5.34(b). Mixed coupling refers to the structure proposed in Figure 5.34(c), since both electric and magnetic coupling occur in that case.

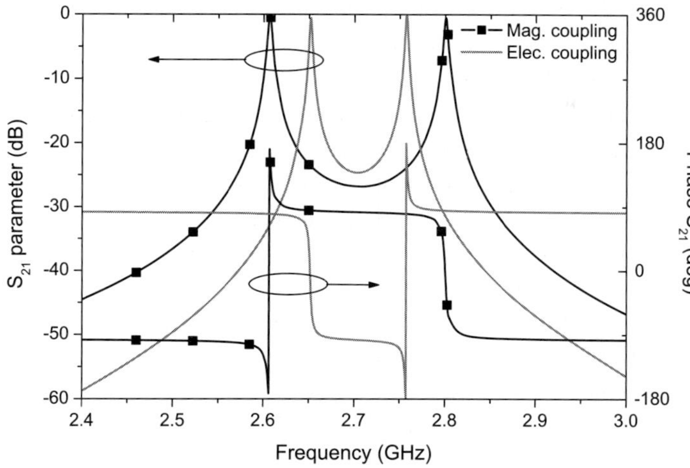

Figure 5.35 Comparison of electric and magnetic coupling responses [8].

Coupling curves of three basic coupling structures encountered in this class of filters are investigated using full-wave electromagnetic simulations. Figure 5.35 shows transmission characteristics for structures with electric and magnetic coupling, together with corresponding phase responses. In both cases, the distance s is equal to 1.2 mm. It can be seen that the magnetic coupling is stronger than the electric one, for the same distance s. In addition, it can be seen that the phase characteristics are out of phase confirming that two coupling coefficients will have the opposite signs [16]. To design cross-coupled filters based on the proposed resonator, the values of the coupling coefficients are needed for various resonator-to-resonator distances, s. The design curves for the coupling coefficients and the external quality factor are calculated using equations (5.13) and (5.14), respectively.

The obtained coupling coefficients as a function of resonator-to-resonator distance s and offset d are shown in Figure 5.36. Mixed coupling in the function of resonator-to-resonator distance s and resonator offset d is shown in Figure 5.36(c). As expected, the mixed coupling results in the smallest values of the coupling coefficient. Also, it can be noted that in the case of mixed coupling, small variation of offset d has very small influence on the coupling coefficient.

Similarly, the external quality factor was determined as a function of the coupling line length l, Figure 5.37. For $l<a$, the coupling line lies only above the resonator, while for $a<l<2a$, it extends to the resonator's side as well, as shown in Figure 5.37(a). Figure 5.38 shows the external quality factor as a function of the coupling line length l. Fine-tuning of the external quality factor can be obtained by changing the feed line position t, Figure 5.37(b). Fine-tuning of the external quality factor by changing the position of the feed line, t, for a fixed length of the coupling line set at l = 11.2 mm is shown in Figure 5.39.

Using the designed curve, a bandpass filter of fourth order with the coupling structure shown in Figure 5.40 has been designed. The generalized equivalent network of the prototype is shown in Figure 5.41 where J and g denote admittance inverters and

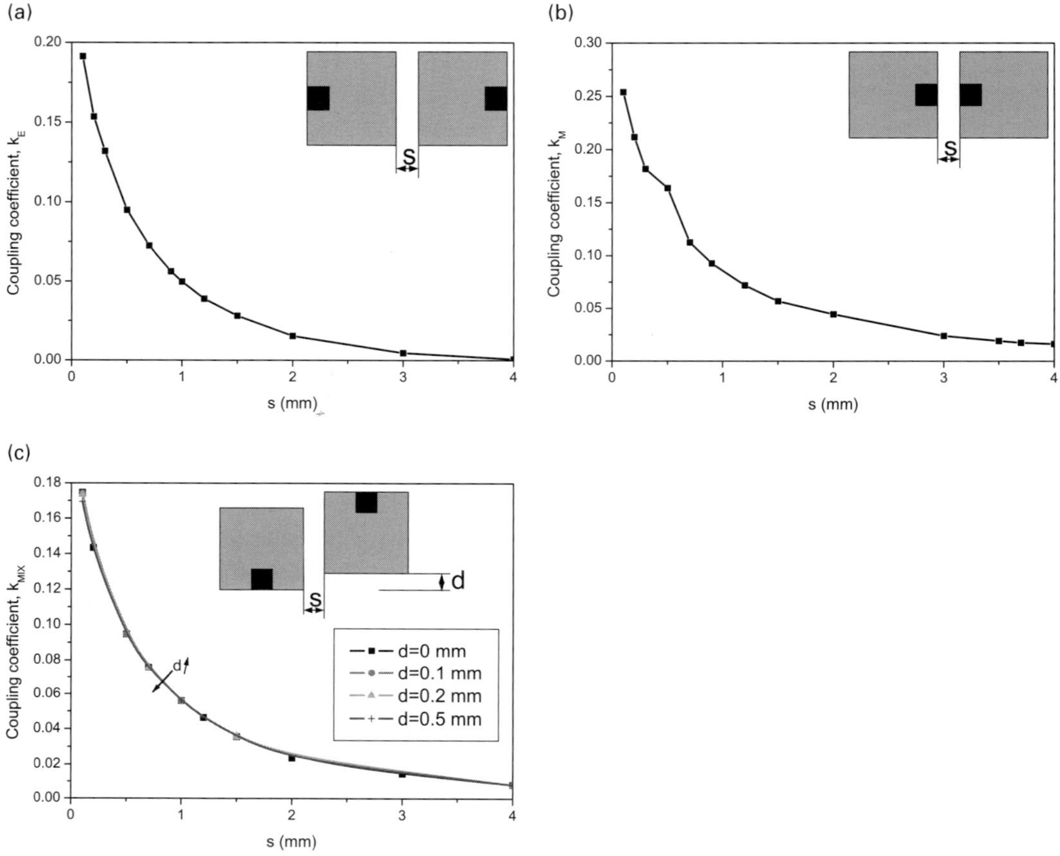

Figure 5.36 Coupling coefficients versus resonator-to-resonator distance *s* for: (a) Electric coupling. (b) Magnetic coupling. (c) Mixed coupling (dependence from an offset *d* is also shown) [8].

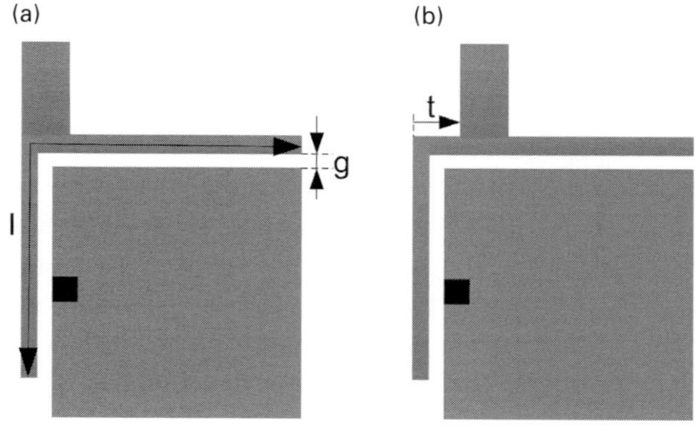

Figure 5.37 Determination of the Q-factor of the proposed resonator [8]: (a) External quality factor as a function of the coupling line length l. (b) Fine tuning of the external quality factor by changing the position of the feed line, t, for fixed l = 11.2 mm.

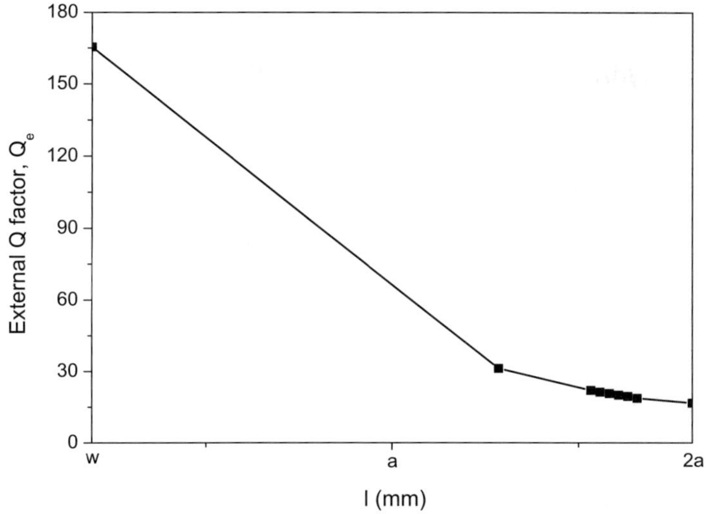

Figure 5.38 External quality factor as a function of the coupling line length *l* [8].

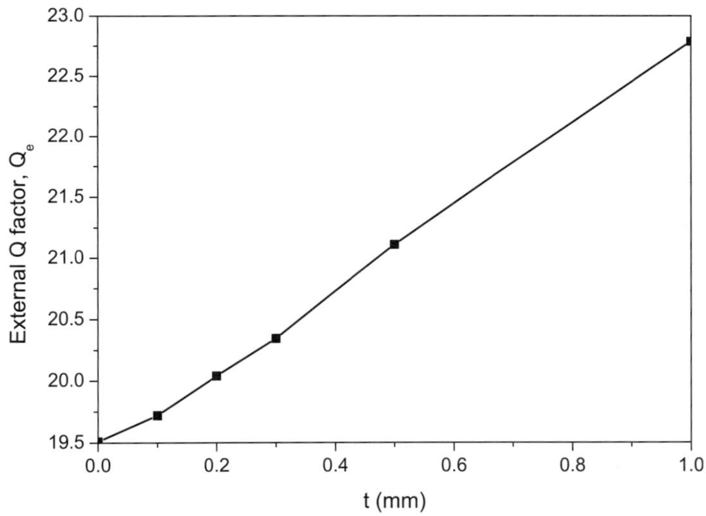

Figure 5.39 Fine-tuning of the external quality factor by changing the position of the feed line, *t*, for fixed length l = 11.2 mm [8].

capacitances, respectively. The values of g_i and J_i have been calculated using synthesis equations defined in [16]. For the fourth-order filter with return loss $L_R = -20$ dB and frequency allocation of the transmission zeros at $\Omega = 1.82$, the calculated values are: $g_1 = 0.95914$, $g_2 = 1.4169$, $J_1 = -0.20509$, and $J_2 = 1.1105$.

Those values can be used to calculate the external quality factor and coupling coefficients using the following expressions [16]

$$Qei = Qeo = \frac{g_1}{FBW}, \tag{5.15}$$

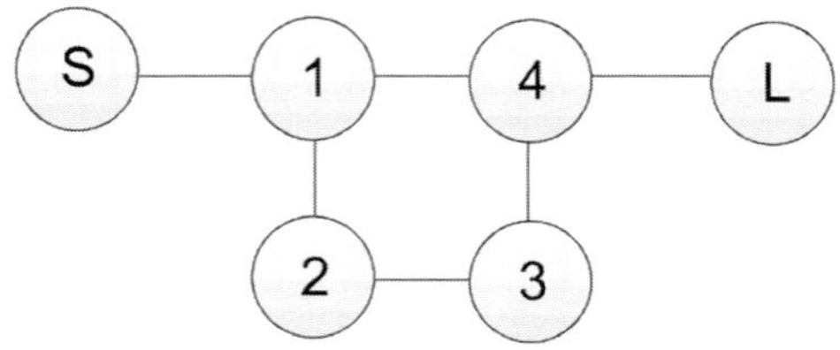

Figure 5.40 Coupling structure of the cross-coupled filter [8].

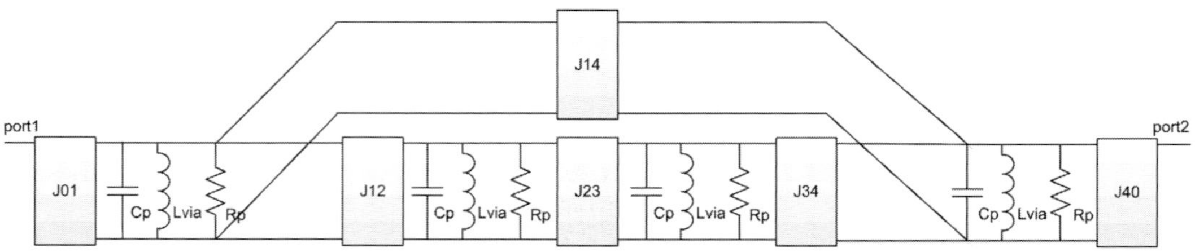

Figure 5.41 Generalized equivalent circuit network of the cross-coupled filter [8].

$$M_{1,2} = M_{3,4} = \frac{FBW}{\sqrt{g_1 g_2}}, \tag{5.16}$$

$$M_{2,3} = \frac{FBW{\cdot}J_2}{g_2}, \tag{5.17}$$

$$M_{1,4} = \frac{FBW{\cdot}J_1}{g_1}. \tag{5.18}$$

The calculated coupling values are $M_{12} = M_{34} = 0.0428903$, $M_{23} = 0.0391877$, and $M_{14} = -0.0106913$. To achieve the desired values of M_{14}, $M_{23,}$ and M_{12}, three coupled pairs with electric, magnetic, and mixed coupling are employed, respectively. Positive coupling coefficients $M_{12} = M_{34}$ and M_{23} are realized by the mixed and magnetic coupling, respectively, while the negative coupling coefficient M_{14} is realized using electric coupling.

Using the design curves given above, all dimensions of the filter, such as spacings and feed length and location were determined. The obtained spacings between the resonators are found to be $s_{14} = 2.51$ mm, $s_{23} = 2.27$ mm, and $s_{12} = 1.35$ mm for offset $d = 0.1$ mm. Since the actual line widths and spacing are determined by the resolution of the PCB technology, after the tuning procedure we set them to $s_{14} = 2.5$ mm, $s_{23} = 2.3$ mm, $s_{12} = 1.2$ mm, and $d = 0.1$ mm. To compensate for a frequency shift effect because of the feeding lines, the dimensions of the first and fourth resonators are set to 6.1 mm x 6.1 mm. The layout of the filter with all the relevant dimensions is shown in Figure 5.42, while the comparison between circuit model and electromagnetic simulation of the filter are compared in Figure 5.43. Good agreement can be observed.

The cross-coupled filter is fabricated using standard PCB technology. The photograph of the fabricated circuit is shown in Figure 5.44. The overall size of the

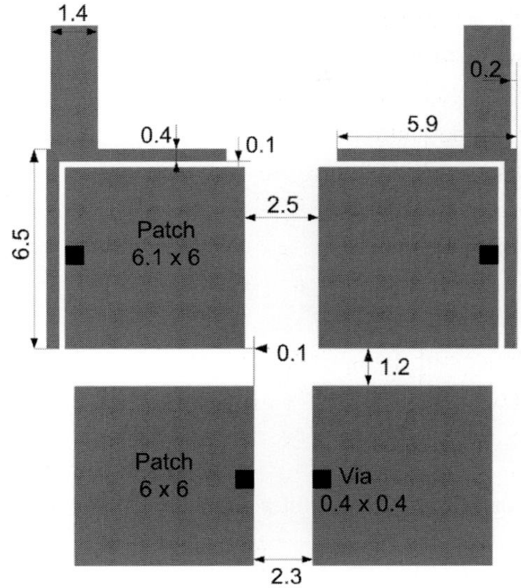

Figure 5.42 Layout of the filter with all dimensions in mm [8].

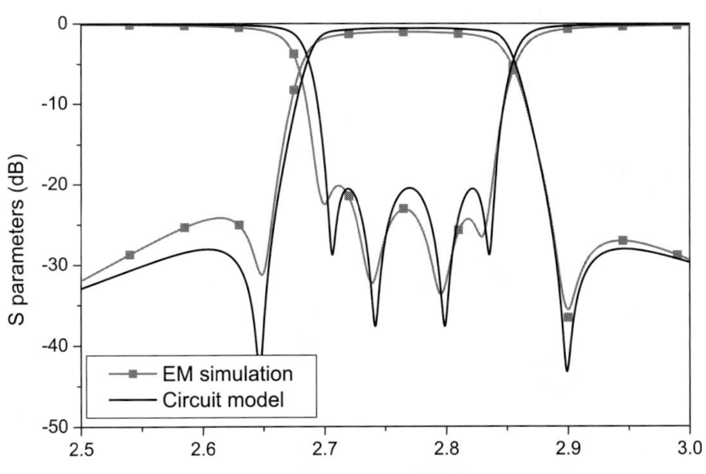

Figure 5.43 Comparison of electric circuit model and electromagnetic simulation.

realized filter is only $0.385\lambda_g$ x $0.34\lambda_g$, where λ_g denotes the guided wavelength at 2.85 GHz.

The simulated and measured responses of the filter are compared in Figure 5.45, where the specified transmission zeros near the edges of the passband can be clearly identified. Good agreement between the measured and simulated responses can be observed, except for a shift in frequency approximately equal to 1.4% and higher insertion losses in the passband. Since manufacturer specifications for substrate material

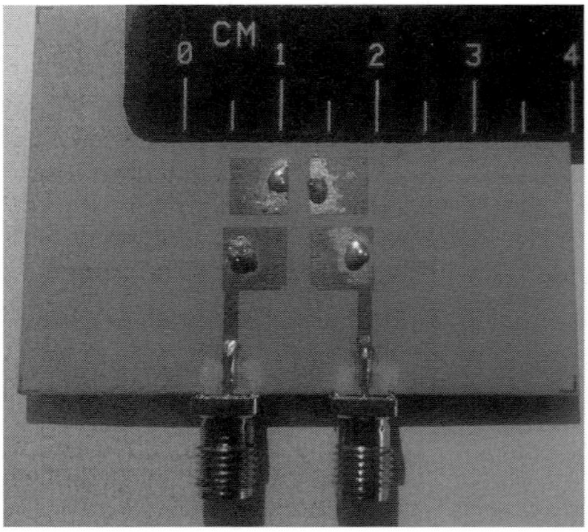

Figure 5.44 Photograph of the fabricated cross-coupled filter [8].

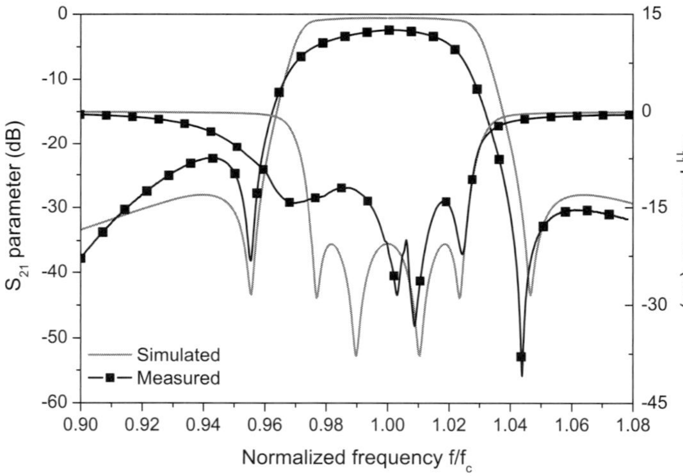

Figure 5.45 Comparison of simulated and measured responses [8].

allow variations in the range ±0.5, as well as variations in substrate thickness, this can be explained by the discrepancy between the actual and simulated values of the dielectric constant and the substrate thickness.

The measured central frequency is 2.88 GHz and the filter has stopband rejection up to 8 GHz. For a better comparison of the responses, a normalized frequency axis is used in Figure 5.45. Furthermore, the filter also shows small radiation loss from the left side of the passband.

5.3.3 Dual-band filters with independent passbands based on grounded-patch resonators

In this section, three dual-band filters based on the RGPR and the PGPR with independent control of passbands that operate according to the IEEE 802.11b/g/n standard are proposed. We first propose using a rectangular grounded patch resonator (RGPR) instead of the square one, as the rectangular patch is inherently a non-degenerate dual-mode resonator whose two fundamental resonant frequencies may be independently controlled, as was detailed in Section 5.2.1.

However, when two RGPRs are coupled, the inter-resonator coupling at two resonant frequencies cannot be adjusted separately. The filters based on the PGPR with an edge slot perturbation allow separate control of the couplings at both operating frequencies resulting in independent control of both the position and the filter bandwidths. The coupling models of the filters are established and the coupling coefficients of the coupled pairs of resonators and the external quality factors at both operating modes are derived using full-wave electromagnetic simulations. As an example, two dual-band filters of second order and one of fourth order have been designed, fabricated, and measured.

5.3.3.1 Dual-band filter based on the RGPR

To demonstrate the potential of the RGPR for dual-band application, a second-order dual-band bandpass filter is designed based on a previously presented resonator. The filter based on the RGPR has been designed according to the following specifications: central frequency of the first band is f_{c1} = 2.45 GHz with a 3 dB bandwidth of 160 MHz and ripple equal to 0.1 dB. The frequency of the second band is set to f_{c2} = 5.2 GHz. Unfortunately, the bandwidth and ripple in the second band cannot be specified since the bandwidths of the two bands are mutually dependent.

A common situation that occurs when dual-band filters are designed is that the inter-resonator couplings at two resonant frequencies cannot be independently controlled, since both coupling coefficients are controlled by the same parameter – inter-resonator spacing. Filters of this type are typically designed using a procedure which requires multiple tuning of dimensions and finding a set of design curves that meet specifications which in turn increases the time needed to design a filter.

We first determine the dimensions of the resonator in order to obtain resonating modes at the 2.45 and 5.2 GHz. For the selected Taconic substrate, the length and width of the resonator have been found as 5.5 mm and 9.3 mm, respectively, while the via dimension is set to 0.2 mm.

The coupling coefficients and loaded Q-factors were determined for both bands together because they are mutually dependent. The couplings between resonators and Q-factors depend of the orientations of the resonators. Two basic coupling structures can be realized, as is shown in Figure 5.46 and the coupling coefficient can be calculated from the EM simulation using equation (5.13) at both resonant modes separately. Figure 5.47 shows the obtained coupling coefficient for both orientations as the function of various resonator-to-resonator distances s for the patch dimensions equal to 5.5 mm x 9.3 mm.

The inter-resonator couplings depend on the orientation of the coupled resonators. It can be seen that the resonant frequency of the second mode is less dependent on the distance between resonators. In the case of Orientation A, the first mode has stronger coupling values for small distances, while for the greater distances the coupling of

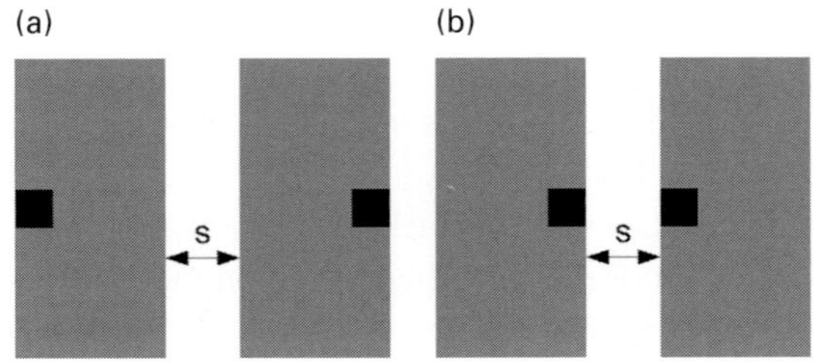

Figure 5.46 Basic coupling structures of coupled the RGPR: (a) Orientation A. (b) Orientation B.

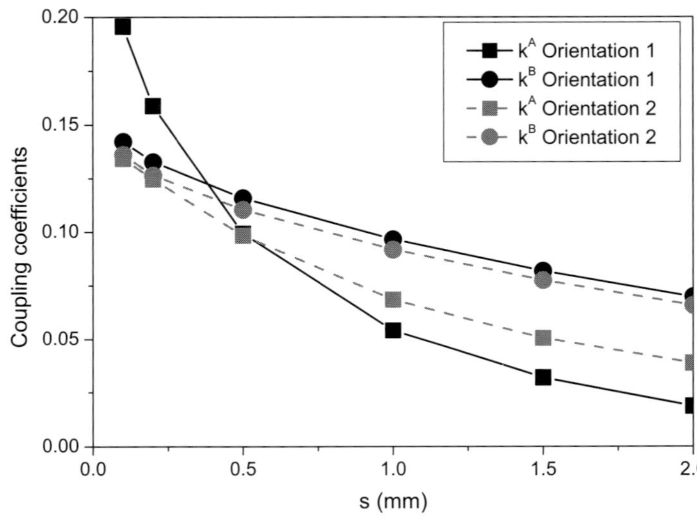

Figure 5.47 Coupling coefficients versus resonator-to-resonator distance *s* for RGPR resonator.

second mode is stronger. In the case of Orientation B, the coupling coefficient at the second band is always stronger than that of the first one. It should be noted that the coupling coefficients k_i^A and k_i^B have the opposite signs, since at f_1 the coupling is predominantly magnetic in Structure A and predominantly electric in Structure B (and the other way round at f_2).

Similarly, the external quality factors can be determined as a function of the coupling line length l for both modes simultaneously. Figure 5.48 shows the topologies to excite the input/output dual-band resonator. The dimension of the feed line width is set to w = 1.4 mm, the spacing between the resonator and feed is set to g = 0.1 mm, while the width of the line is w_L = 0.3 mm. Figure 5.49 shows the obtained Q-factor curves for both orientations as the function of various feed line lengths l. It should be mentioned that the proposed resonator has a high unloaded quality factor of the second mode equal to $Q_u{}^A = 190$. Unfortunately, the unloaded Q factor for the first mode is only $Q_u{}^B = 100$.

For a longer line, the input/output coupling is stronger, resulting in lower values of the external quality factor. In the case of both orientations, the first mode has a higher external Q-factor. It can be noted that there is a point where the two modes have the same couplings and a point where the values of the Q-factors are equal for both modes.

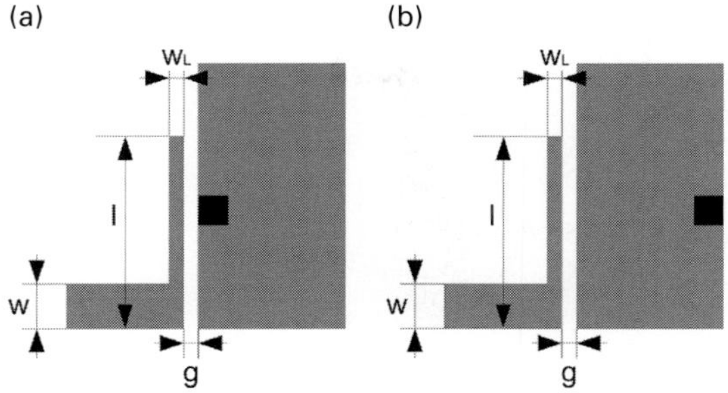

Figure 5.48 Basic feed topologies: (a) Orientation A. (b) Orientation B.

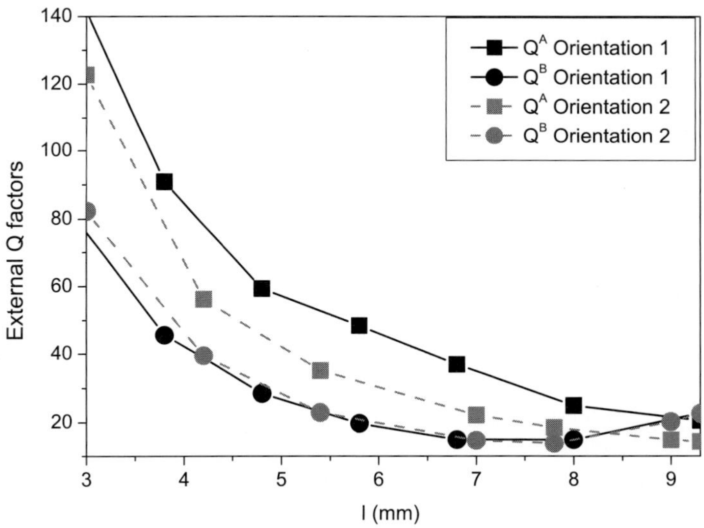

Figure 5.49 External quality factors as a function of the coupling feed line length *l*.

This suggests that the dual-band filter with identical characteristics in both bands can easily be designed using the intersection point of the curves for the coupling coefficients and Q-factors. Unfortunately, it leaves no freedom in the design of filters with arbitrary characteristics of passbands.

The standard filter theory and procedure described in [16] can be used to design a dual-band filter. Unfortunately, this procedure requires finding of a set of design curves, which increases the time needed to design the filter. In the case of the proposed resonator, shape or layout cannot be easily changed without affects on both resonant frequencies.

Since the coupling coefficients and loaded Q-factors of the different modes of the GPR were determined, a simplified coupling model of the second order may be used in the dual-band filter design. The dual-band filter is modeled as the superposition of two channels that operate at two different frequencies, Figure 5.50, while each dual-band resonator of the filter is modeled using two sub-resonators. Channel 1 is formed by two

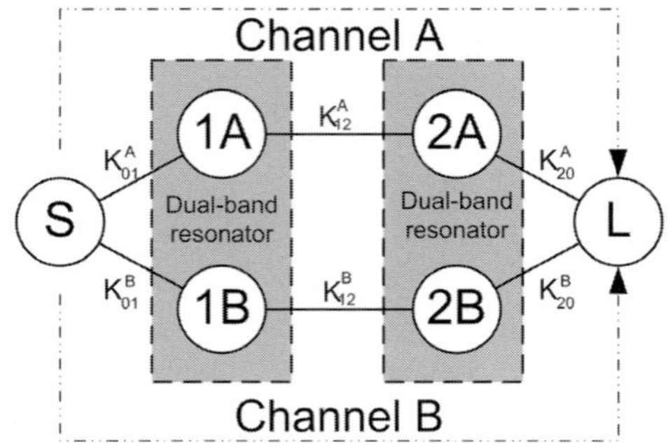

Figure 5.50 Coupling model for a second-order GPR dual-band filter.

sub-resonators, denoted 1.1 and 1.2, which operate at first resonance, f_1, while channel 2 is formed by another two sub-resonators, 2.1 and 2.2, which operate at the second resonance f_2. The source and load are denoted by S and L, respectively.

The proposed model includes only the main couplings that exist in the filter. Parasitic couplings, such as the couplings between sub-resonators operating at different frequencies and the couplings between source/load with non-adjacent resonators are not taken into account. This does not diminish the validity of the model as long as the resonant frequencies of the filter are significantly different, which is the case here. In such a case, parasitic couplings cannot be reliably extracted from the simulations as they are not defined for resonators that operate at significantly different frequencies. However, it should be noted that the model is valid only in the passbands, due to the narrow-band validity of coupled-resonator models around the central frequencies of the passbands. The proposed model can also be extended for filters of higher orders, as will be shown below.

For the predefined specification, calculated values for the Q-factor and coupling coefficient in the first band are $Q^A = 24.52$ and $k_{12}{}^A = 0.04756$, while the values for the second band obtained using designed curves are $Q^B = 14.6$ and $k_{12}{}^B = -0.1006$. The designed parameters can be found using the design curves proposed in Figure 5.47 and Figure 5.49. Since the standard PCB technology allows a resolution of 0.1 mm, all dimensions are rounded to the integer product of the minimal resolution. The filter's layout, with all the relevant determined dimensions satisfying the design specification, is shown in Figure 5.51.

Figure 5.52 shows a comparison of the electromagnetic response and circuit models responses – good agreement in both passbands was obtained. The model contains transmission zeros which existed in the transmission characteristics between the two passbands. They arise as a result of different signs of the coupling between resonators in different modes.

The parasitic coupling, such as the coupling between the source/load with the non-adjacent resonator and the coupling between the non-adjacent resonators at both resonances, is not taken into account. These values are very small and can slightly affect the zeros positions.

The proposed filter based on the RGPR is fabricated in standard PCB technology. A photograph of the fabricated circuit is shown in Figure 5.53, while the simulated and

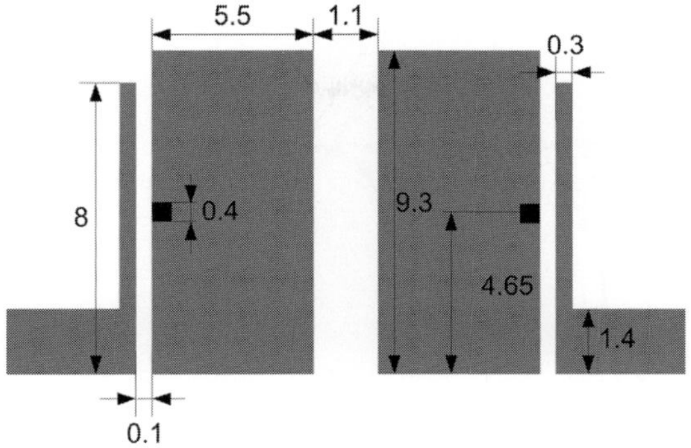

Figure 5.51 Layout of the RGPR dual-band filter with dimensions in millimeters.

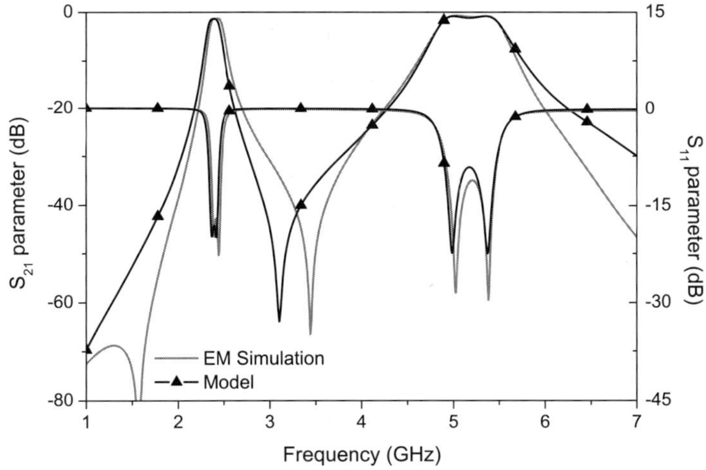

Figure 5.52 Comparison of the electromagnetic response and circuit model responses of the RGPR dual-band filter.

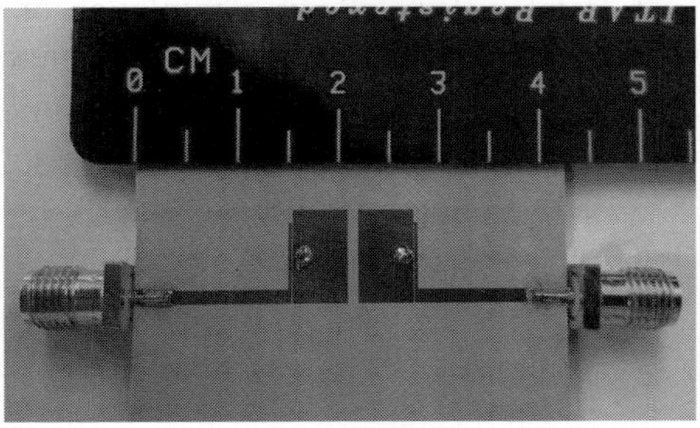

Figure 5.53 Photograph of the fabricated RGPR dual-band filter.

measured responses of the filter are compared in Figure 5.54. Good agreement between the measured and simulated responses can be observed, except for a small shift of the second band and the transmission zero. The passband centre frequencies obtained in the measurement are 2.406 GHz in the first band and 5.27 GHz in the second one. Measured

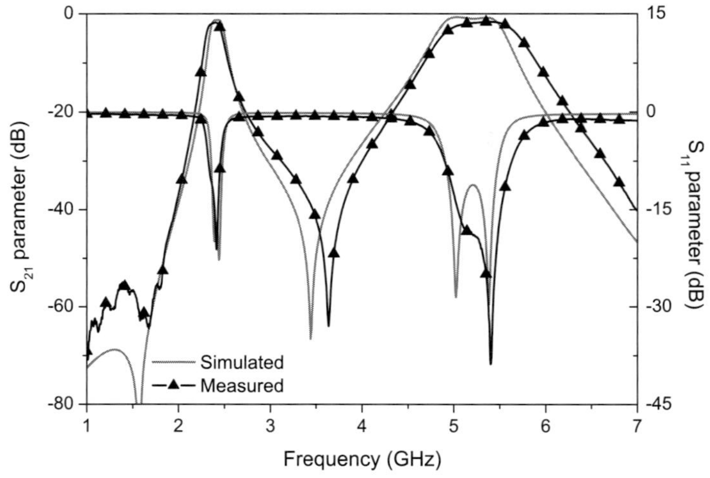

Figure 5.54 Comparisons of the simulated and the measured responses of the RGPR dual-band filter.

bandwidths are equal to 180 MHz and 840 MHz, while the insertion losses in the passbands are equal to –1.804 dB and –1.638 dB, respectively. The filter is very compact in size, with the overall dimension of only 12.9 mm x 9.3 mm, i.e. $0.268\lambda_g$ x $0.191\lambda_g$, where λ_g denotes the guided wavelength at the first resonant frequency.

5.3.2.2 Dual-band filter based on PGRP

The perturbed grounded patch resonator (PGPR) contains a slot perturbation that allows two fundamental resonant modes of the resonator to be used for dual-band filter design, with complete control over the passband magnitude response in each passband. The design procedure is presented in detail in this section.

In the previous section, it was shown that the PGPR allows independent control of the positions of the two fundamental resonant modes. In addition, the PGPR allows independent control of passband bandwidths, as will be demonstrated in the following.

Figure 5.55 shows two cases of coupled PGPRs, Structure A and Structure B, with design parameters s and t , whilst Figure 5.56 shows how the coupling coefficients change for different values of s and t at two fundamental resonances for both structures.

It can be seen that the parameter t influences the coupling coefficient of structure A at the first resonance, $k_1{}^A$, whilst it has no effect on the coupling coefficient at the second resonance, $k_2{}^A$. Nevertheless, this is not the case for the structure B, i.e. the parameter t has no effect on any of the coupling coefficients $k_1{}^B$ and $k_2{}^B$, Figure 5.56(b). It is interesting that the coupling coefficients k_i^A and k_i^B are opposite in sign. At the first resonance, magnetic coupling dominates in structure A and electric coupling dominates in structure B. This is reversed at the second resonance.

Figure 5.57 reveals another interesting fact – the slot dimensions s_l and s_w have a similar effect on the coupling coefficients of structure A as the parameter t, i.e. slot dimensions can be also used to control the bandwidths separately.

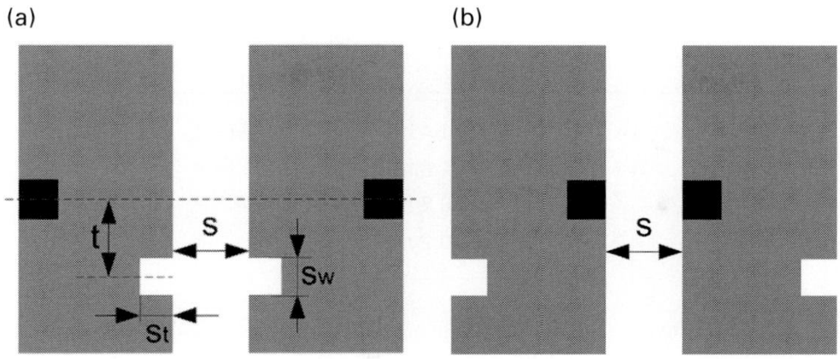

Figure 5.55 Basic coupling structures of coupled PGRP [10]: (a) Orientation A. (b) Orientation B.

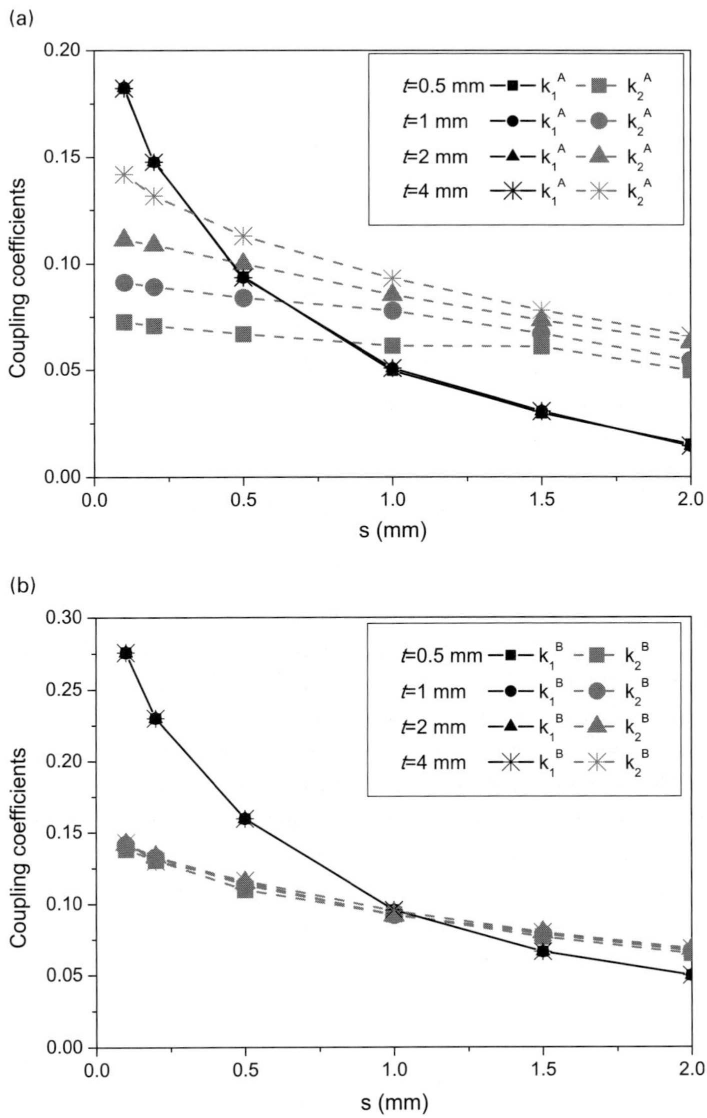

Figure 5.56 Coupling coefficients for (a) Structure A. (b) Structure B versus inter-resonator spacing s, for various offsets t. PGPR dimensions are $a = 9.3$ mm, $b = 5.5$ mm, $d = 0.4$ mm, $s_t = 1$ mm and $s_w = 1$ mm [10].

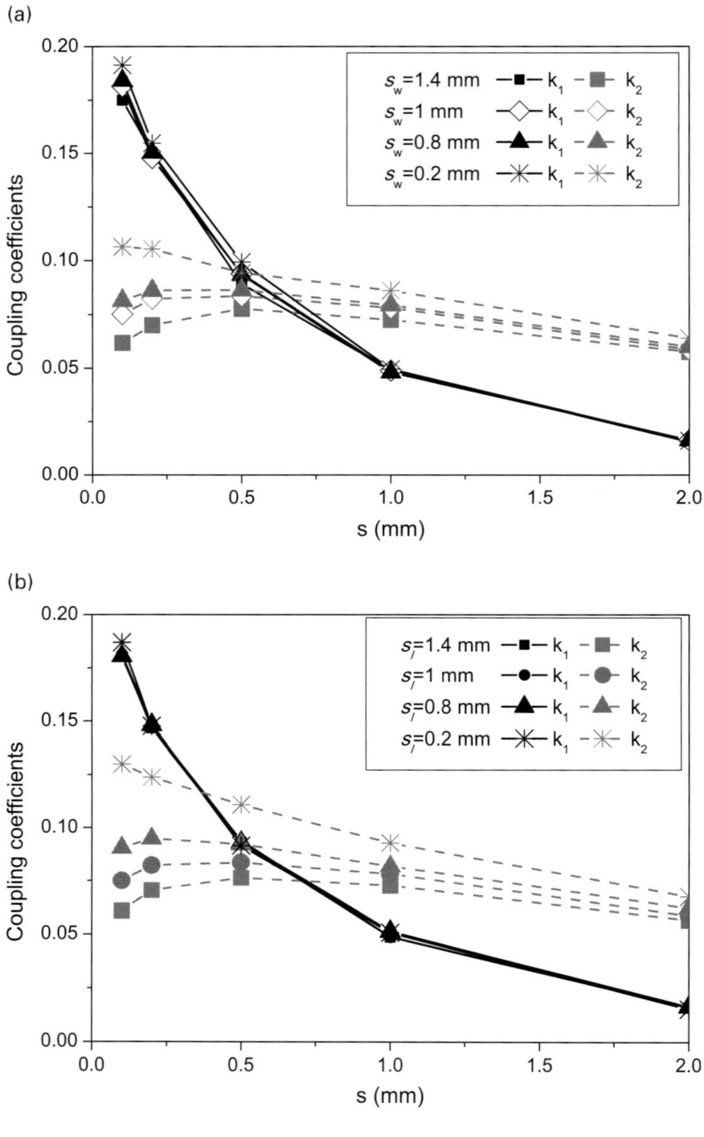

Figure 5.57 Coupling coefficients for Structure A versus inter-resonator spacing s, for various slot dimensions: (a) Varied s_w, for s_l =1 mm. (b) Varied s_l for s_w = 1 mm. Other PGPR dimensions are a = 9.3 mm, b = 5.5 mm, d = 0.4 mm, t = 2 mm [10].

The final option to independently control the bandwidths of the two passbands relies on the fact that the loaded Q-factors at the two fundamental resonances are set independently by the slot offset t – Figure 5.58 reveals that for the case where the edge with the via is closest to the feed line. At the same time, the change of slot dimensions does not have this effect on the quality factors. The consequence is that the Q-factors can be differentiated by changing the slot offset, whilst the coupling coefficients can be independently controlled using the slot dimensions.

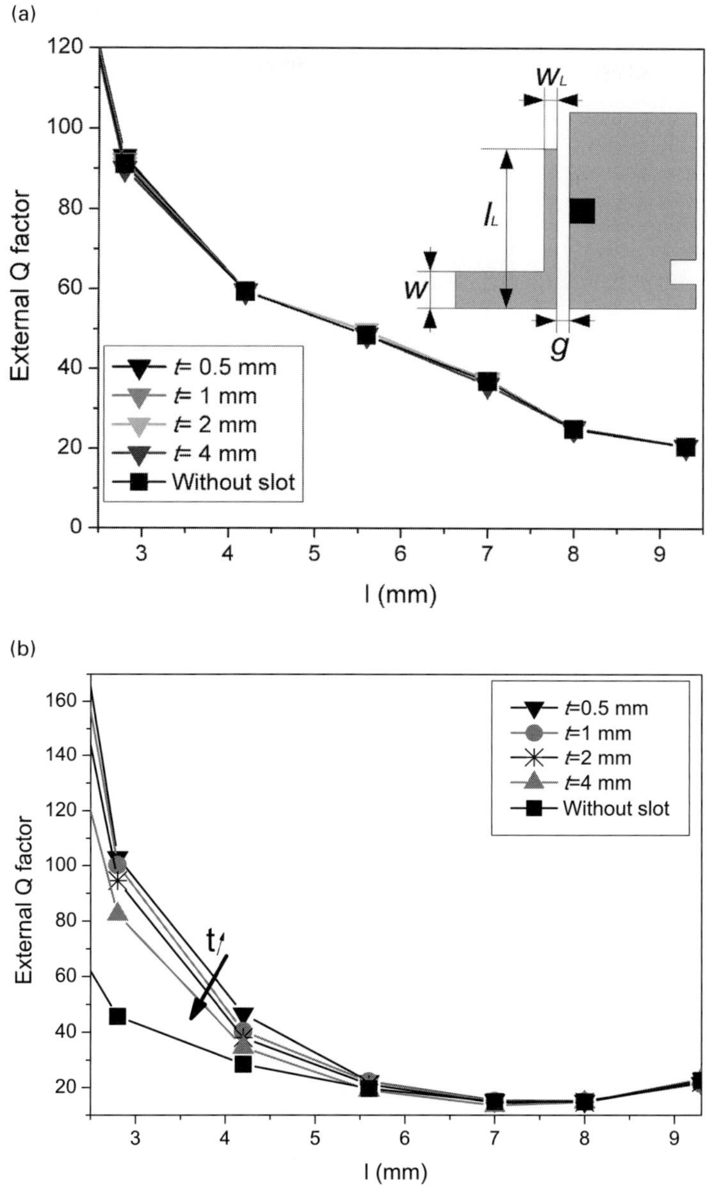

Figure 5.58 External Q-factors versus feed-line length l, for various slot offset t: (a) Q_1. (b) Q_2. The feed structure is shown in the inset and PGPR dimensions are $a = 9.3$ mm, $b = 5.5$ mm, $d = 0.4$ mm, $s_l = 1.1$ mm, $s_w = 1$ mm and $g = 0.1$ mm [10].

The analysis presented above illustrates that PGPR allows independent control of the coupling coefficients and Q-factors at two resonant frequencies, thus allowing the design of dual-band filters with completely independent passband central frequencies and bandwidths.

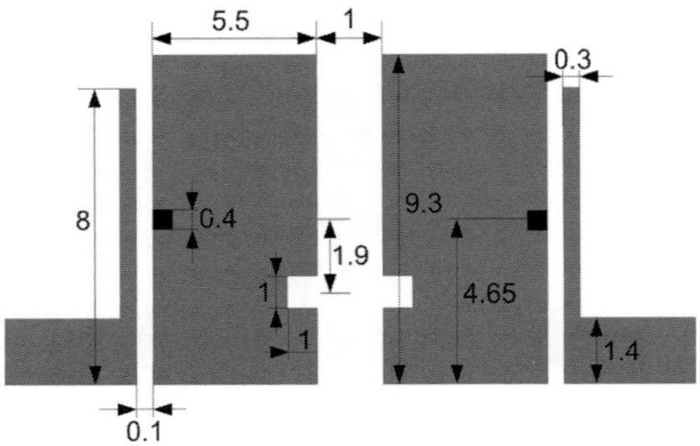

Figure 5.59 Layout of the proposed second-order dual-band bandpass filter based on PGPR, dimensions in mm [10].

To demonstrate the applicability of the proposed structure and mechanisms, two bandpass filters based on the PGPR have been designed, fabricated, and measured. The first filter is of second order and the second one is of fourth order. The fourth-order filter is strong evidence that the resonant modes are indeed independent, as one would not be able to optimize two passbands with exact specified passband responses if the resonances were not fully independent.

The first example based on the PGPR has required specifications of passbands at WLAN frequencies 2.45 and 5.1 GHz, with bandwidths of 150 and 590 MHz, respectively. Based on these requirements the values for the coupling coefficients and loaded Q-factors were found to be $k_1^{12} = 0.04956$, $k_2^{12} = -0.08367$, $Q_1 = 25.26$, and $Q_2 = 12.2$. Using the design curves presented in Figures 5.56 to 5.58, the required dimensions are $s = 1.1$ mm, $s_w = 1$ mm, $s_l = 1$ mm, $t = 1.9$ mm, and $l = 8$ mm. The filter layout is shown is Figure 5.59.

The EM simulation and the response of the circuit model are compared in Figure 5.60 and there is good agreement in the passbands. The coupling model is not accurate between the passbands, but the transmission zeros between passbands arise as a result of the different signs of the coupling at different resonant modes.

The simulated and measured results are compared in Figure 5.62, whilst a photograph of the fabricated circuit is shown in Figure 5.61. The filter footprint is 12.8 mm by 9.3 mm, which is $0264\lambda_g$ x $0.191\lambda_g$, where λ_g is the guided wavelength on the given substrate at the first resonant frequency of 2.45 GHz. The agreement between the results is good: the measured 3dB bandwidths are 193 MHz and 670 MHz, respectively. Discrepancies are attributed to the via manufacturing tolerances and these mainly impact the first passband as predicted by the theoretical analysis.

To illustrate the independence of the two patch modes, a fourth-order filter is designed, i.e. each of the passbands is based on a fourth-order response. This is a good demonstration that the two passbands are indeed independently designed, as a design with so many resonators would be extremely complex were the resonances not really independent. The design specifications were designed according to the IEEE 802.11b/g/n standard. The filter is designed to operate at 2.45/5.2 GHz, with 3dB bandwidths of $BW_1 = 150$ MHz and $BW_2 = 410$ MHz, respectively. Figure 5.63 shows the coupling scheme.

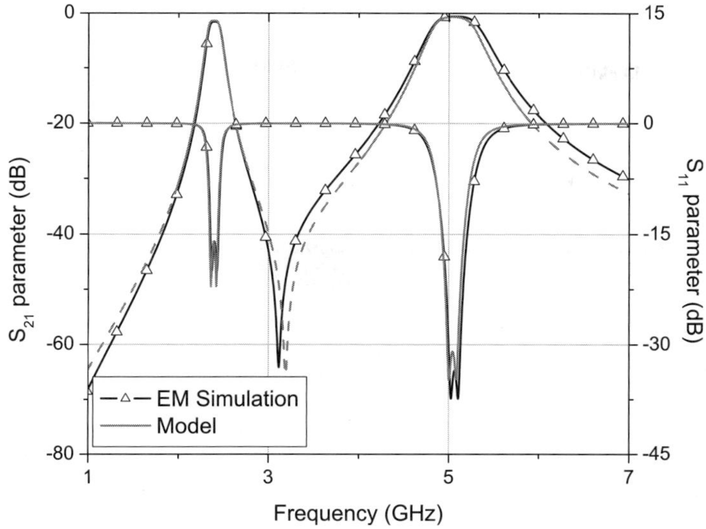

Figure 5.60 Comparison of the responses of the EM simulations and the circuit model for dual-band bandpass filter based on PGPR [10].

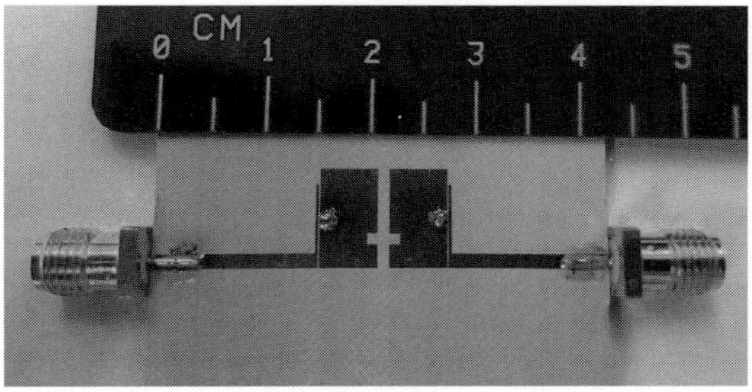

Figure 5.61 Photograph of the second-order dual-band PGPR filter [10].

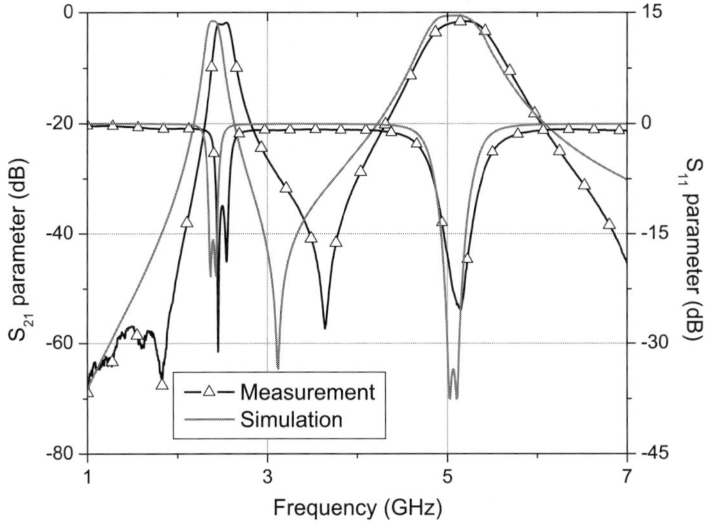

Figure 5.62 Comparison of the simulated and measured response of the second-order dual-band PGPR filter [10].

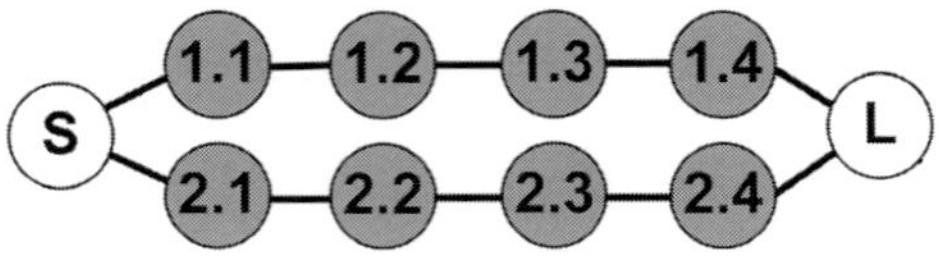

Figure 5.63 Coupling model scheme of the proposed fourth-order dual-band PGRP bandpass filter [10].

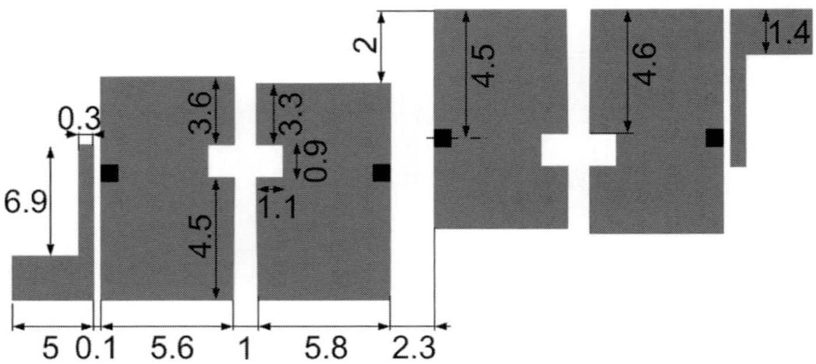

Figure 5.64 Layout of the proposed fourth-order dual-band filter based on PGPR (in mm) [10].

This filter was designed to be inside a metal box, to avoid radiation losses from the resonators and to obtain good selectivity. The design was done by including the box dimensions in the EM simulations. The model of the filter is shown in Figure 5.63, and it was used to fine-tune the dimensions of the filter after using the design curves presented in Figures 5.56 to 5.58. The filter dimensions were found using a standard numerical tuning method used with coupled-resonator filters. Figure 5.64 gives all the filter dimensions and Figure 5.65 details the dimensions of the metal box.

The N+2 coupling matrix for each passband are presented independently, since the passbands were designed entirely independently. The first passband has the centre frequency of 2.425 GHz and a bandwidth of 6%. The normalized coupling matrix is

$$m_1 = \begin{bmatrix} 0 & 0.847 & 0 & 0 & 0 & 0 \\ 0.847 & 0 & 0.7353 & 0 & 0 & 0 \\ 0 & 0.7353 & 0 & -0.5945 & 0 & 0 \\ 0 & 0 & -0.5945 & 0 & 0.7353 & 0 \\ 0 & 0 & 0 & 0.7353 & 0 & 0.847 \\ 0 & 0 & 0 & 0 & 0.847 & 0 \end{bmatrix}. \tag{5.19}$$

The second passband has a centre frequency of 5.22 GHz and a bandwidth of 7.6%. The normalized matrix is

$$m_2 = \begin{bmatrix} 0 & 0.9353 & 0 & 0 & 0 & 0 \\ 0.9353 & 0 & -0.7353 & 0 & 0 & 0 \\ 0 & -0.7353 & 0 & 0.5945 & 0 & 0 \\ 0 & 0 & -0.5945 & 0 & -0.7353 & 0 \\ 0 & 0 & 0 & -0.7353 & 0 & 0.9353 \\ 0 & 0 & 0 & 0 & 0.9353 & 0 \end{bmatrix}. \tag{5.20}$$

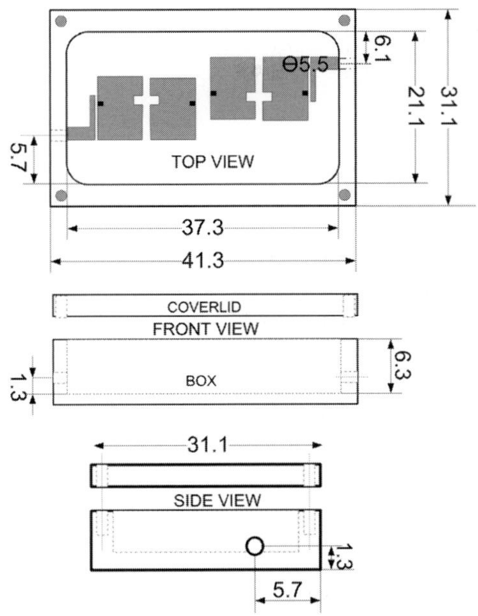

Figure 5.65 Optimized dimensions (in mm) of the metal box with the filter placed inside [10].

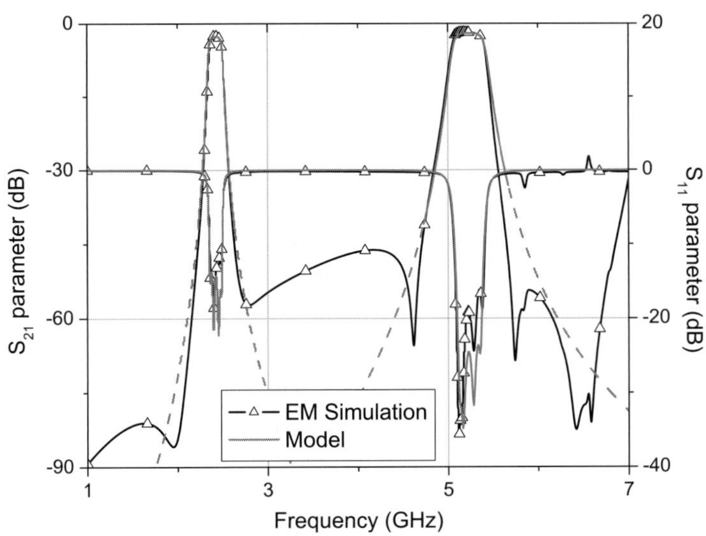

Figure 5.66 Comparison of electromagnetic response and circuit model response [10].

The EM response and the response of the circuit model, Figure 5.66, are in a very good agreement in the passbands. The differences in the stopbands are due to the fact that the circuit model representing coupled resonators is not valid in between the two passbands.

The photograph of the fabricated filter is presented Figure 5.67, while the simulated and measured responses of the filter are compared in Figure 5.68. The size of the filter,

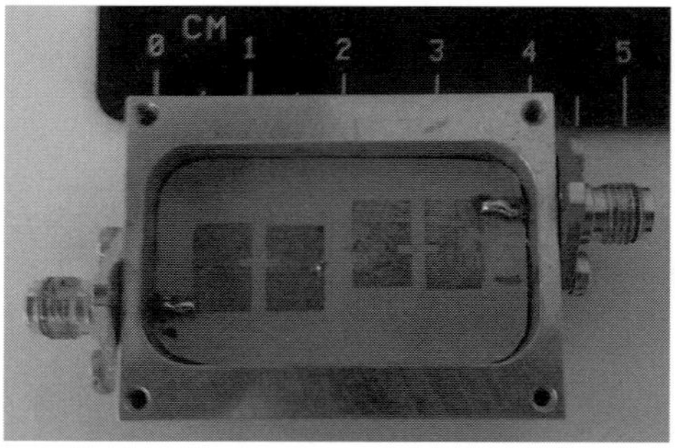

Figure 5.67 Fabricated fourth-order dual-band bandpass filter fitted in a metal box [10].

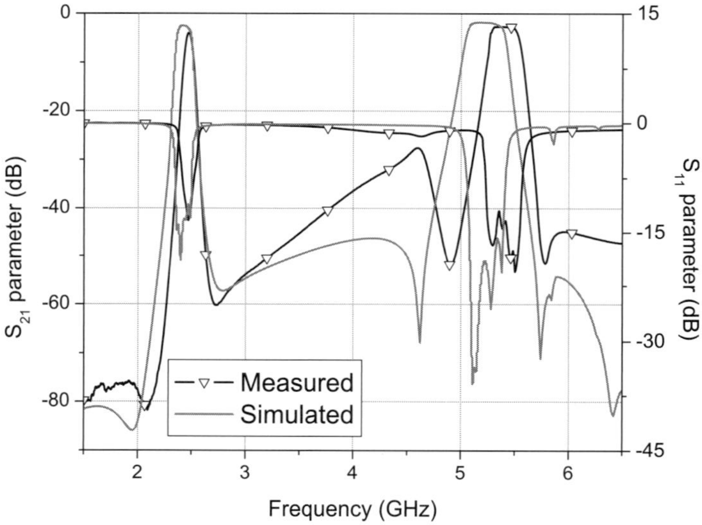

Figure 5.68 Comparisons of the simulated and the measured responses fourth-order dual-band PGPR filter [10].

including the metal box, is 41.5 mm x 31.2 mm x 15.3 mm, i.e. $0.86\lambda_g$ x $0.64\lambda_g$ x $0.315\lambda_g$, where λ_g is the guided wavelength on the Taconic substrate at the first operating frequency of 2.45 GHz.

The measured central frequencies of the passbands are equal to 2.49 GHz and 5.38 GHz, while the corresponding insertion losses are –3.95 dB and—2.68 dB, respectively. Due to the fabrication tolerances of the via and the metal box, the measured response is shifted by 2.5% from the simulated result. The return losses are more than 15 dB in both bands. Also, the rejection in the stopband is somewhat degraded. However, it remains at more than 28 dB and therefore does not diminish the filter's performance.

As shown in [16], different feed schemes and coupling structures can affect the appearance of an additional transmission zero. The same effect can be achieved in the filters based on a grounded patch resonator if the feed scheme and some couplings are changed.

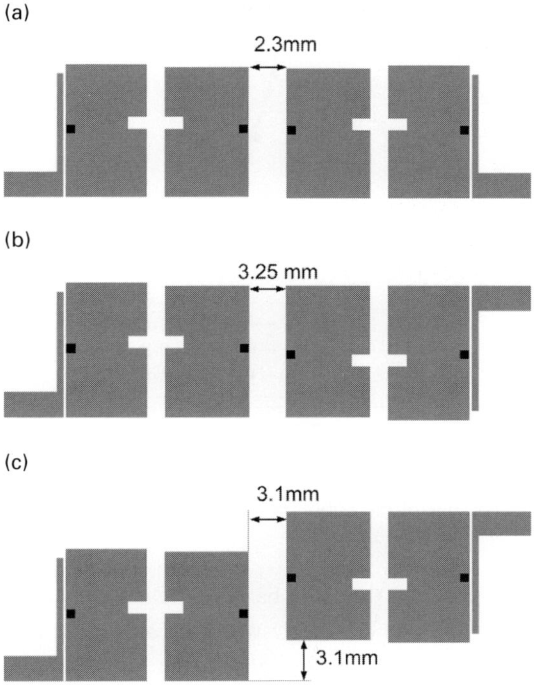

Figure 5.69 Various modifications of the dual-band filter based on the PGPR: (a) Filter 3-I. (b) Filter 3-II. (c) Filter 3-III.

Figure 5.69 shows three different modifications of a fourth-order dual-band filter based on the PGPR. Only coupling between the second and third resonator and feed orientation have been changed. Marked dimensions were changed with regard to the filter proposed in Figure 5.64. Comparison of the simulation results are shown in Figure 5.70. It can be seen that in the case of Filter 3-I the zeros in the transmissions characteristic are not pronounced, while in case of Filter 3-II five transmission zeros are clearly visible: two below the first band, one between the two passbands, and two above the second band. In the Filter 3-III, two zeros are placed at the left side of the first band, two at the right side of the second one, and three between the two passbands.

It is obvious that the same passband characteristic but with a different number and position of transmission zeros can be obtain by changing the position of the feeds and by changing the coupling between non-adjacent resonators.

5.4 Conclusion

In this chapter, the grounded patch resonator (GPR) and its modifications for application in the design of compact single-band and dual-band filters have been presented. A number of examples of single-band bandstop and bandpass filters were designed based on different modifications of the grounded patch resonator using a coupled-resonator approach.

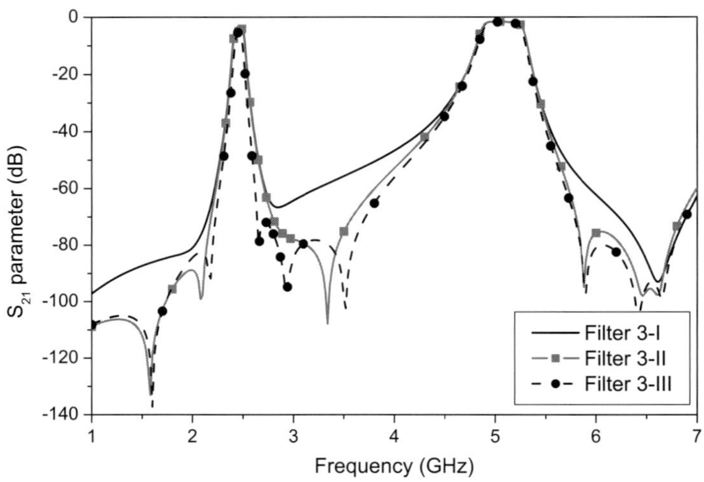

Figure 5.70 Transmission characteristics of the filters proposed in Figure 5.69.

A compact bandstop filter based on the GPR has been realized that exhibits a 35% 10 dB fractional bandwidth centered at 4.36 GHz with a rejection of more than 30 dB and the maximal reflection coefficient around 1 dB in the stopband. The fabricated fourth-order filter has an overall dimension of $0.52\lambda_g$ x $0.13\lambda_g$ with the second stopband at 10 GHz.

Another highly selective bandstop filter of fourth order was designed based on the Hilbert grounded patch resonator embedded in the microstrip. The overall filter dimensions are 12.9 x 3.3 mm, i.e. approximately $0.47\lambda_g$ x $0.12\lambda_g$, where λ_g is the guided wavelength. The filter exhibits a 10 dB fractional bandwidth of 30% at 4.43GHz, with rejection of more than 30 dB.

Design curves forthe GPR have been developed and discussed in detail to facilitate bandpass filter design, and circuit models have been proposed. The highly selective compact filter of third order with Chebyshev bandpass response and the cross-coupled filter based on the on GPR have been designed following the coupling coefficient design theory. The third-order bandpass filter for Wireless LAN applications operates at 2.4 GHz and exhibits a wide stopband region of up to 6.25 GHz, with a more than 60 dB rejection in the stopband. The fabricated cross-coupled filter with an overall size of $0.385\lambda_g$ x $0.34\lambda_g$, is designed with a central frequency of 2.85 GHz and stopband rejection of up to 8 GHz.

Using the proposed PGPR resonator, dual-band bandpass filters with completely independent passband central frequencies and bandwidths have been designed. Design curves for the PGPR and a circuit model have been developed and discussed in detail to facilitate filter design. Finally, two dual-band bandpass filters, one of the second and one of fourth order, have been designed, fabricated, and measured. The filters operate according to the IEEE 802.11b/g/n standard, and they exhibit good performance and small size. The fourth-order dual-band filter operates at 2.49 GHz and 5.38 GHz, and its size is $0.86\lambda_g$ x $0.64\lambda_g$ x $0.315\lambda_g$, where λ_g is the guided wavelength at the first passband.

References

[1] D. Sievenpiper, L. Yhang, R. F. J. Broas, N. Alexopulos, and E. Yablanoviitch, "High-impedance electromagnetic surfaces with a forbidden frequency band," *IEEE Transactions on Microwave Theory and Techniques*, vol. 47, no.11, 1999.

[2] V. Radonić, V. Crnojević-Bengin, and B. Jokanović, "Analysis of metamaterial unit cells based on grounded patch," *Microwave Review*, vol. 12, no. 2, pp. 12–15, 2008.

[3] V. Radonić, V. Crnojević-Bengin, and B. Jokanović, "Novel left-handed unit cell for multi-band filtering applications," *European Microwave Conference EuMC 2008*, Amsterdam, Netherlands, Oct. 2008, pp. 694–697.

[4] V. Radonić, V. Crnojević-Bengin, and B. Jokanović, "Novel unit cell based on the grounded patch for filter applications," *Mediterranean Microwave Symposium MMS 2008*, Damascus, Syria, Oct. 2008.

[5] V. Radonić, N. Janković, and V. Crnojević-Bengin, "Left-handed unit cell based on Hilbert grounded patch for multi-band filtering applications," *9th International Conference on Telecommunication in Modern Satellite, Cable, and Broadcasting Services, TELSIKS 2009*, Niš, Serbia, Oct. 2009.

[6] V. Radonić and V. Crnojević-Bengin, "Super-compact stopband filter based on grounded patch resonator," *Electronics Letters*, vol. 46, no. 2, pp. 146–147, Jan. 2010.

[7] V. Radonić, N. Janković, and V. Crnojević-Bengin, "Super-compact stop band filter based on the grounded Hilbert patch resonator," *4th International Congress on Advanced Electromagnetic Materials in Microwaves and Optics*, Karlsruhe, Germany, Sept. 2010.

[8] V. Radonić and V. Crnojević-Bengin, "Cross-coupled microstrip filter using grounded patch resonators," *European Microwave Conference 2011*, Manchester, UK, Oct. 2011, pp. 763–766.

[9] V. Radonić, R. Geschke, and V. Crnojevic-Bengin, "Bandpass filter using direct-coupled grounded patch resonator for wireless LAN applications," *10th International Conference on Telecommunication in Modern Satellite Cable and Broadcasting Services, TELSIKS 2011*, Nis, Serbia, 2011, pp. 397–400.

[10] V. Radonić, R. Geschke, and V. Crnojević-Bengin, "Dual-band filters with independent passbands based on perturbed grounded patch resonators," *Journal of Electromagnetic Waves and Applications*, vol. 27, no. 16, 2013.

[11] http://www.sonnetsoftware.com

[12] V. Crnojević-Bengin, "Novel compact microstrip resonators with multiple 2-D Hilbert fractal curves,' *Microwave and Optical Technology Letters*, vol. 48, no. 2, pp. 270–273, Feb. 2006.

[13] V. Crnojević-Bengin, V. Radonić, and B. Jokanović, "Fractal geometries of split-ring resonators," *IEEE Transactions on Microwave Theory and Techniques*, vol. 56, no. 10, pp. 2312–2321, Oct. 2008.

[14] P. Jarry and J. Beneat, *Design and Realizations of Miniaturized Fractal Microwave and RF Filters*, John Wiley & Sons, New Jersey, 2009.

[15] H. W. Johnson and M. Graham, *High-Speed Digital Design – A Handbook of Black Magic*, Prentice Hall, 1993.

[16] J.-S. Hong, *Microstrip Filters for RF/Microwave Applications*, John Wiley & Sons, New Jersey, 2011.

[17] M. Kirschning, R. H. Jansen, and N. H. L. Koster, "Measurement and computer-aided modeling of microstrip discontinuities by an improved resonator method," *Proceedings of IEEE MTT-S International Microwave Symposium Digest*, May 1983, pp. 495–497.

[18] J. Garcia-Garcia, J. Bonache, I. Gil, F. Martin, R. Marques, F. Falcone, T. Loperegi, M. A. G. Laso, and M. Sorolla, "Comparison of electromagnetic band gap and split-ring resonator microstrip lines as stop band structures," *Microwave and Optical Technology Letters*, vol. 44, no. 4, pp. 376–379, Feb. 2005.

[19] V. Oznazi and V. B. Ertruk, "A comparative investigation of SRR- and CSRR-based band reject filters: simulations, experiments, and discussion," *Microwave and Optical Technology Letters*, vol. 50, no. 2, pp. 519–523, Feb. 2008.

[20] R. Wu, S. Amari, and U. Rosenberg, "New cross-coupled microstrip band reject filters," *IEEE MTT-S International Microwave Symposium*, vol. 3, pp. 1597–1600, 2004.

[21] G. Mattaei, L. Young, and E. M. T Jones, *Microwave Filters, Impedance-Matching Networks, and Coupling Structures*, Artech House, Norwood, 1980.

[22] R. E. Collin, *Foundations for Microwave Engineering*, 2nd edn, McGraw-Hill, New York, 1992.

6 Fractal-based multi-band microstrip filters

Nikolina Janković, Kiril Zemlyakov, Riana Helena Geschke, Irina Vendik, and Vesna Crnojević-Bengin

6.1 Introduction

Fractal curves are space-filling curves that, theoretically, allow the design of infinite-length lines on a finite substrate area. Due to their space-filling ability, they have great potential for miniaturization of passive microwave circuits.

This chapter is devoted to fractal curves and their application to filter design. It starts by presenting the main characteristics of fractal curves which are of importance to filter design and miniaturization. Special attention will be given to the Hilbert fractal curve, which is very convenient for application in resonators and filters due to its configuration.

Afterwards, two dual-band bandpass filters, one tri-band bandpass filter, and one tri-band bandstop filter based on the Hilbert fractal curve will be presented and discussed.

The two dual-band bandpass filters consist of dual-mode resonators based on the Hilbert fractal curve of second order. The resonator is formed as a serial connection of two Hilbert fractal curves, and owing to its configuration two resonant modes can be independently controlled. Two dual-band filters based on the proposed resonator have been designed, and they are characterized by compact dimensions and good performance.

Tri-band bandpass and bandstop filters are based on the Hilbert-fork resonator, a multi-mode resonator that consists of two dual-mode resonators and supports four resonant modes. The first resonator is based on the Hilbert fractal curve of the second order, whilst the second resonator has the shape of a fork and is positioned to conveniently fill the remaining footprint area of the first resonator. Both resonators are grounded by vias which enables the dual-mode behavior of the resonators. Also, the presence of the vias enables size reduction of the structure.

Using two modes of the Hilbert resonator and one of the fork resonator tri-band bandstop and bandpass filters have been designed which have excellent performance in all tri-bands, small overall size, as well as the possibility of independent control of the passbands.

6.2 Fractal curves and their application in filter design

Besides good in-band and out-of-band performance, another important characteristic of passive circuits is the footprint area they occupy. A conventional approach to miniaturization is based on folding a straight-line resonator in order to fit it in a smaller area.

Fractal curves offer high potential for miniaturization of passive microwave circuits, because, theoretically, fractal curves allow the design of infinite-length lines on a finite substrate area.

The first fractal curve was constructed by Giuseppe Peano in 1890 [1]. Since the constructed curve was a continuous curve that passes through every point of a given region, fractal curves were initially called space-filling curves. Afterwards, it was shown that the dimension of such curves is not an integer number and hence they were named fractal curves.

6.2.1 Fractal generation

Fractal curves are generated in an iterative manner and there are three types of iterations – iteration based on generator, iteration based on formula, and the iterated function system. In the aspect of the application of fractals in microwave circuits, the most interesting iteration is the one based on the generator.

In that case, a fractal curve is generated by successive substitution of one geometrical shape with another that is often a collection of scaled copies of the initial shape. In order to generate an arbitrary fractal, two components are needed – the initiator and the generator. The initiator is a set of linear segments that comprise a starting shape of a fractal, whilst the generator is an arranged collection of scaled copies of the initiator. The fractal curve is generated by applying the following rule: in the first iteration, each segment of the initiator is replaced by a copy of the generator. The dimensions and position of the copy of the generator are changed so that its end points coincide with the end points of the segment which is to be replaced. In the second iteration, the procedure of copying of the generator is repeated for every segment of the curve obtained after the first iteration. A similar procedure is repeated in the following iterations.

Generation of a fractal curve is illustrated in the following example. Let the initiator be a straight line of a finite length which is divided into three segments of equal lengths. The middle segment is replaced with two equal segments, positioned at angles of 60° and 120°. In that manner, a sample (generator) that will be repeated is obtained. The same rule is now applied to each straight segment of the curve – it is divided into three segments, the middle of which is replaced by two segments. This is how the curve in Figure 6.1 is obtained. The curve is known as Koch fractal.

If the initial straight segment has unit length, then the length of Koch curve after the first iteration is 4/3. After the second iteration the length is $(4/3)^2$, after the third $(4/3)^3$, etc. Since $4/3 > 1$, a curve of infinite length is obtained after infinite number of iterations, Figure 6.2.

If the initiator is chosen to be a triangle instead of a straight line, a fractal curve known as the Koch snowflake is obtained, Figure 6.3 and 6.4.

Using the iterated function system (IFS), more complex fractals than those previously presented are generated. The IFS is based on the application of affine transformations of the initiator in a plane. Affine transformations comprise scaling, reflection, rotation, and translation. One, or more, of these operations is applied to the initiator, and after an infinite number of iterations a fractal curve is obtained. It should be noted that the order

Figure 6.1 First three iterations of Koch fractal.

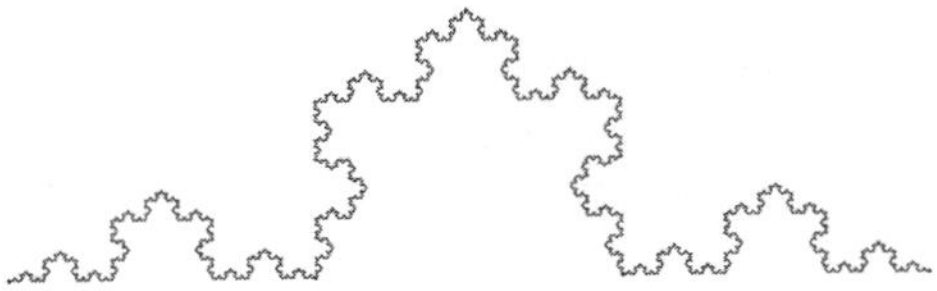

Figure 6.2 Koch fractal.

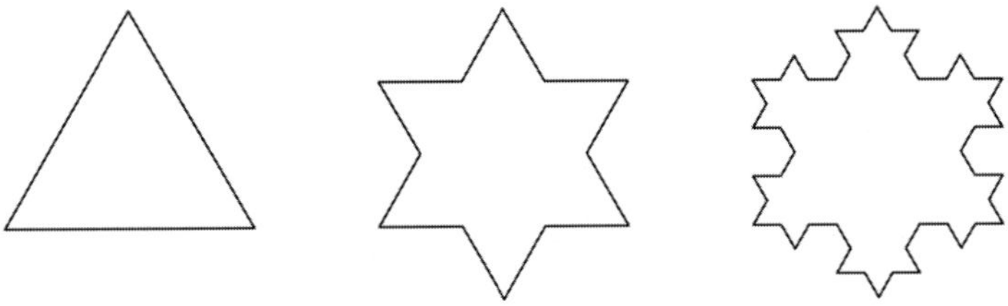

Figure 6.3 First three iterations of Koch snowflake.

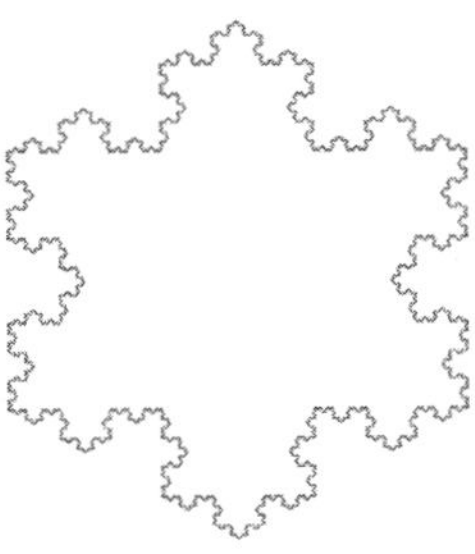

Figure 6.4 Koch snowflake.

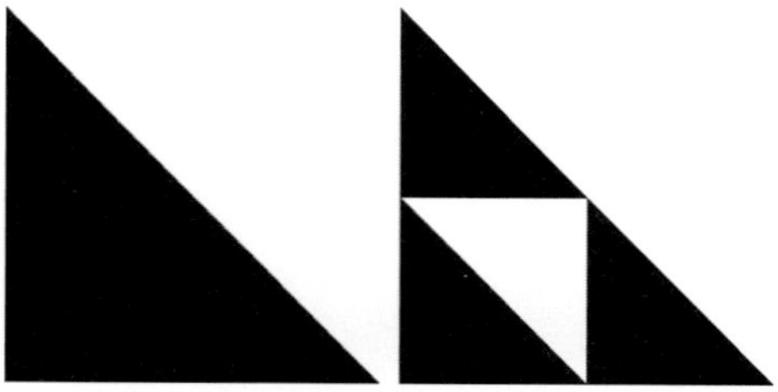

Figure 6.5 Generation of Sierpinski gasket fractal.

Figure 6.6 Sierpinski gasket fractal after ten iterations.

of the operations is very important, i.e. application of operations in different orders gives different fractal curves.

An example of the generation of the Sierpinski gasket fractal is shown in Figure 6.5. In the first iteration, an equilateral triangle is replaced with three smaller triangles. In the next iterations, each triangle is replaced with three smaller triangles, etc. The fractal obtained after ten iterations is shown in Figure 6.6. Mathematically interpreted, the Sierpinski fractal curve is generated using the operations of translation and scaling.

Some fractal curves can be generated only if the IFS method is applied. IFS fractals are most often used in image compression.

Iterations based on formula are the simplest method of fractal generation, yet it enables generation of the most complex fractals. They are based on replacing of an initial number with another obtained using a mathematical formula. In this manner, three types of fractals can be generated – strange attractors, Julia sets, and Mandelbrot sets.

Fractals generated by iterations based on formula are usually applied in modeling of chemical reactions and meteorological phenomena.

6.2.2 Properties of fractal curves

In the previous section, it could be seen that, after each iteration, curves converge to a fractal curve that has many interesting properties. Since fractals are generated by repeating a scaled, translated, and/or rotated generator, they exhibit self-similar behavior, which means that the object is exactly or approximately similar to a part of itself. Self-similarity can be noted in Figure 6.7, which shows the Sierpinski gasket fractal curve and its enlarged part. The whole fractal is comprised of three copies each of which

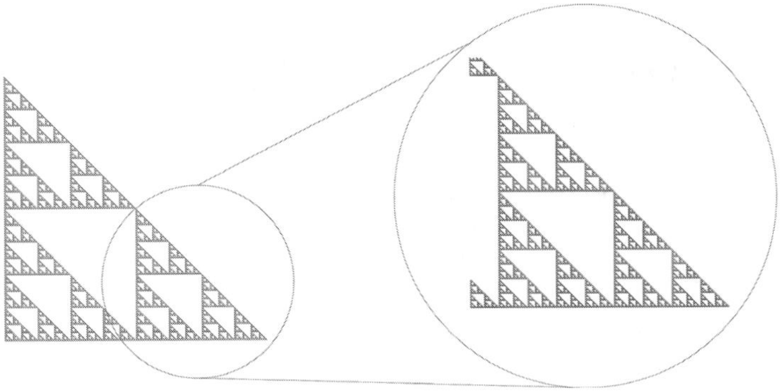

Figure 6.7 Sierpinski gasket fractal curve and its enlarged part.

is twice as small as the fractal. Also, each copy is comprised of three twice as small copies, etc.

Another property of fractals can be also noted – the fact that they are scale-invariant, i.e. a smaller part of a fractal cannot be distinguished from the whole fractal curve.

Apart from self-similarity and scale-invariability, fractals are also characterized by the property of indifferentiability and by their dimension, which is a non-integer number. Namely, conventional (non-fractal) continual curves are differentiable. Since fractal curves consist of an infinite number of infinitely short straight segments, there is no point at which a tangent curve can be positioned. This implies another property of fractal curves – although continual curves, they are also indifferentiable.

The most interesting property of fractal curves is their dimension. In classical geometry, the topological dimension of a curve is 1, because the position of each point can be defined using one value, e.g. distance from the end point of the curve. By analogy, classical planes are two-dimensional because the position of any point can be defined using two values. Integer values in these examples define a topological or Euclidean dimension.

Since fractal curves have infinite length, it is not possible to define the position of a point using only one value, because the point is infinitely distant from any point of a fractal curve. Therefore, the fractal dimension is not equal to 1. On the other hand, the fractal curve belongs to the plane in which it is generated (and whose dimension is 2). It is intuitively clear that the topological dimension of the fractal is higher than 1 and lower than 2, i.e. it is not an integer number.

The dimension of a fractal curve can be understood as a measure of the space-filling ability of the fractal curve. The dimension D can be determined as a logarithmic ratio between the number of self-similar segments obtained from one segment after each iteration k and the number of segments obtained from one segment in each iteration r

$$D = \frac{\log k}{\log r}. \tag{6.1}$$

From the aspect of miniaturization of circuits, the most important criterion in the selection of a fractal curve is its dimension. The higher the fractal dimension, the better the fractal curve fills the given area, therefore achieving higher compactness.

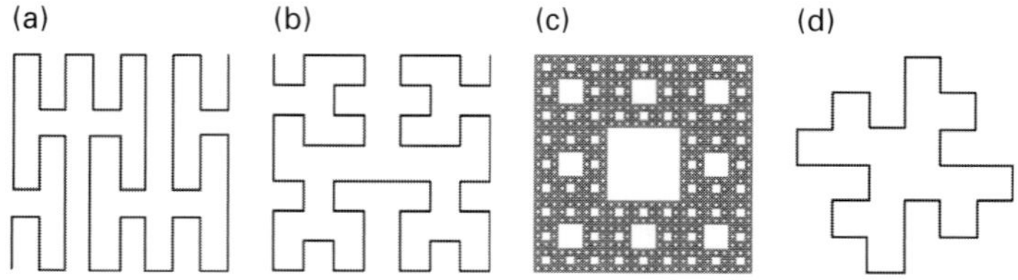

Figure 6.8 (a) Peano fractal. (b) Hilbert fractal. (c) Square Sierpinski fractal. (d) Minkowski fractal.

6.2.3 Application of fractal curves

A great number of fractal curves have been proposed so far and three fractal curves are known that have fractal dimension equal to 2 (i.e. the maximum value) – Peano, Hilbert, and square Sierpinski fractals, Figure 6.8. Apart from them, very widely used are the Minkowski and Koch fractal curves, Figure 6.8 and 6.2.

Fractals find application in various fields, such as fluid dynamics, telecommunication networks, traffic analysis, image compression, microwave electronics, etc. As for their application in microwave circuits, the application is somewhat limited by the fabrication technology. Namely, due to technology limitations, such as minimal line width and the spacing achievable in the fabrication process, fractal curves are not physically realizable. Pre-fractals, fractal curves of finite order, which fill the area only to some extent and whose line segments have finite line width, are used instead.

Different fractal geometries have been utilized in the design of miniaturized antennas [2–7], high impedance surfaces [8–9], left-handed metamaterials [10], RFID tags [11], and filters [12–16].

As for the application in filter design, fractal curves have been used in various filtering configurations. Minkowski, Sierpinski, and Koch fractals are closed curves and as such are primarily used in dual-mode patch resonators. Filters presented in [17–21] are bandpass structures with patch resonators based on Sierpinski, Koch, and Minkowski fractals and they are all characterized by smaller dimensions in comparison to filters based on conventional dual-mode resonators. In addition, in [22] an ultra-wideband filter was presented in which the defected ground structure is based on the Koch fractal.

On the other hand, Peano and Hilbert fractals are open curves that have ends at opposite sides and thus are used in end-coupled resonators. Band-pass and low-pass filters with Hilbert fractal-shaped resonators were proposed in [23–24] and they proved the ability of fractals to provide compactness of the planar filtering structures.

Since the filters that will be presented in this chapter are based on the Hilbert fractal curve, a more detailed explanation on this curve is given below. The first four iterations of the Hilbert curve and an illustration of their generation is shown in Figure 6.9. In order to generate the fractal curve of the second order, three copies of the fractal curve of the first order are made, one of which is rotated 90° clockwise and the other which is rotated 90° counter clockwise. The four same shapes are then connected to form the fractal curve of the second order. The curves of the higher order are generated in the

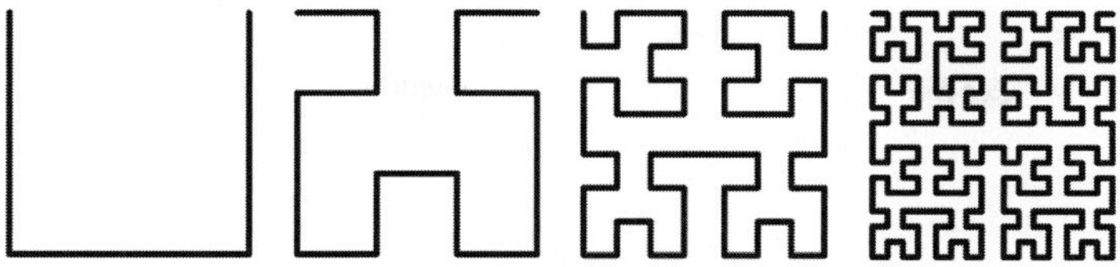

Figure 6.9 First four iteration of Hilbert fractal curve.

Figure 6.10 2-D meander and 2-D Hilbert microstrip resonators with the lines with the same width and spacing.

same manner. After an infinite number of iterations, the Hilbert fractal curve completely fills a square area, which implies that its dimension is equal to two.

An important aspect of the Hilbert fractal is the fact that it is an open curve that has ends on opposite sides, and thus it is very convenient for application in structures such as end-coupled microstrip resonators.

In order to demonstrate the advantages of fractals in comparison to conventional curves, we will compare a resonator based on the Hilbert fractal curve to one based on a non-fractal curve. It is well-known that beside fractals, there are also non-fractal curves that posses the space-filling property. A typical example is a meander. In Figure 6.10, a 2-D meander microstrip resonator is shown which has the same length and occupies the same region as 2-D Hilbert resonator. The lines of the meander and the Hilbert fractal have the same width and spacing. This means that these configurations have the same capacitance towards the ground plane, and thus their resonant frequencies would differ only if their inductances are not equal.

Total inductance per length of the meander, L'_{tot} is equal to the sum of inductances per length of all the straight line segments, L'

$$L'_{tot} = 2nL'. \tag{6.2}$$

where n represents the number of meander turns. Inductance per length of a straight line segment is equal to its self-inductance plus/minus the mutual inductance between the segment and the adjacent straight segments, depending on the direction of the current

$$L' = L'_{self} - 2L'_{12} + L'_{13} - 2L'_{14} + \ldots. \tag{6.3}$$

For the sake of simplicity, we assume that the distance between the adjacent line segments is such that

$$L'_{13} << L'_{12}, \tag{6.4}$$

i.e. that the contribution from all segments except the adjacent ones can be neglected. Now, the inductance per length of the line segment is

$$L' = L'_{self} - 2L'_{12}. \tag{6.5}$$

In the case of the meander, inductance per length of each straight line segment, except for the first and the last one, is equal to the self-inductance per length subtracted by the double value of the mutual inductances per length. Inductance of the line segment is obtained as a product of the inductance per length and the length of the segment a,

$$L' = aL'_{self} - 2aL'_{12}, \tag{6.6}$$

and the total inductance of the meander is

$$L' = 2na(L'_{self} - 2L'_{12}). \tag{6.7}$$

Since the meander resonator has the same length as the Hilbert resonator, their self-inductances are equal. However, due to the particular shape of the Hilbert curve, the size of the individual turns is reduced compared to the meander. Therefore, in the case of the Hilbert resonator, mutual inductance that is subtracted from self-inductance in (6.5) is lower than that in the case of the meander. Furthermore, even line segments that have adjacent segments over minimal distance are not coupled to adjacent lines over their whole length. It can be concluded that due to its specific shape, the Hilbert resonator exhibits lower mutual inductances, and thus smaller decrease in total self-inductance. The Hilbert resonator has greater total inductance than meander and therefore higher potential for miniaturization than any other non-fractal counterpart.

In the following sections, four multi-band filters with resonators based on the Hilbert fractal curve of second order will be presented and discussed.

6.3 Dual-band bandpass filters based on dual-mode Hilbert fractal resonator

Fractal curves find their application predominantly in planar filter structures. In [16] and [25–26], it was shown that filters with resonators based on fractal curves exhibit better performance in comparison to those with conventional resonators, not only in terms of dimensions but also in terms of the quality factor and insertion loss.

As was stressed in the previous section, Peano and Hilbert curves have ends at opposite sides and thus are used in end-coupled resonators. However, the Peano curve has structures that are more complex, and couplings between adjacent segments are stronger than in the case of the Hilbert curve. Therefore, resonators based on the Hilbert curve show better characteristics.

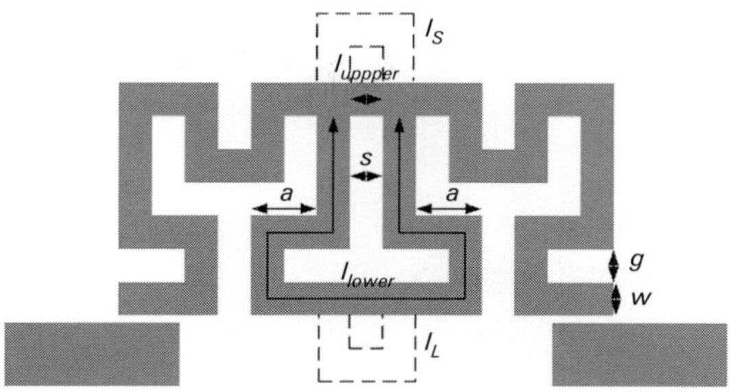

Figure 6.11 Configuration of the dual-mode resonator based on Hilbert fractal curve.

In this section, a dual-mode resonator based on the Hilbert fractal curve of second order is presented [27]. The resonator is formed as a serial connection of two Hilbert fractal curves and owing to its configuration two resonant modes can be independently controlled. Two second-order dual-band filters based on the proposed resonator have been designed, one of which contains edge-coupled resonators and the other containing broadly coupled resonators. They are characterized by compact dimensions and good performance. In the following, the resonator's behavior and filter design will be analyzed and discussed in detail.

6.3.1 Hilbert fractal resonator configuration

The proposed resonator consists of two serially connected Hilbert fractal curves of the second order and its layout is shown in Figure 6.11. The two curves are connected at two points – at the point where the fractal curve is folded and at its end point. The parameter w represents the width of the fractal curve, g the spacing between the segments, s the distance between the two curves, whilst l_{upper} and l_{lower} represents the length of the lines that connect the two curves at the upper and lower point, respectively.

The resonator and filters that will be presented in the following sections are designed on the substrate Rogers RT/Duroid 5880 whose thickness is equal to 1.575 mm, dielectric constant $\varepsilon_r = 2.2$, and loss tangent $\tan\delta = 0.0009$.

It can be seen that there is no main signal path in the resonator, but the signal is rather divided into two paths. Since the proposed resonator is of a symmetrical structure, the even-/odd-mode method can be applied to analyze its behavior. Figure 6.12 shows the even and odd equivalent circuits. In the even circuit, the symmetry plane represents a magnetic wall, whilst in the odd circuit the symmetry plane is an electric wall.

The parameters θ_1, θ_2, and θ_3 represent electrical lengths of the resonator's segments, whilst Z_{ineven} and Z_{inodd} are the even and odd input impedances, respectively. Since the parameters of the Hilbert curve w and g are constant in the whole resonator, all segments have the same characteristic impedance equal to Z_0. For the sake of simplicity, segment bends are not taken into account in the following analysis.

In the case of the even circuit, $Z_{pareven}$ represents the parallel combination of the impedances of the two open stubs. The input impedance of an open stub with the

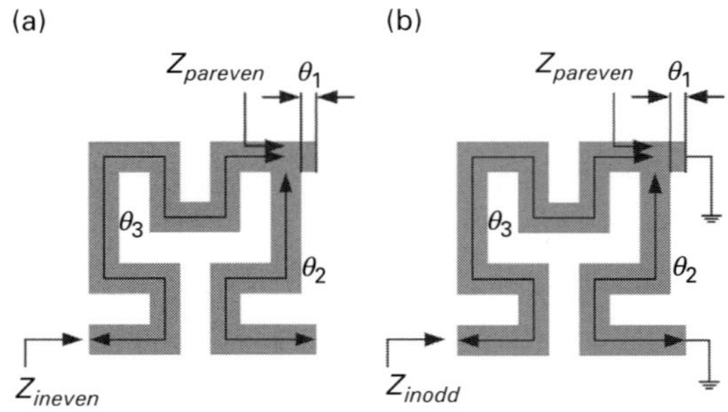

Figure 6.12 Equivalent circuit for: (a) Even mode. (b) Odd mode.

characteristic impedance Z and electrical length θ is equal to $j\tan\theta/Z$. Therefore, $Z_{pareven}$ can be expressed as

$$\frac{1}{Z_{pareven}} = \frac{j\tan\theta_1}{Z_0} + \frac{j\tan\theta_2}{Z_0}. \tag{6.8}$$

The total even input impedance can be expressed as

$$Z_{ineven} = Z_0 \frac{Z_{pareven} + jZ_0 \tan\theta_3}{Z_0 + jZ_{pareven}\tan\theta_3}. \tag{6.9}$$

When the expression (6.8) is inserted into (6.9), we obtain the final expression for Z_{ineven}

$$Z_{ineven} = jZ_0 \frac{(\tan\theta_1 + \tan\theta_2)\tan\theta_3 - 1}{\tan\theta_1 + \tan\theta_2 + \tan\theta_3}. \tag{6.10}$$

The expression for the odd input impedance can be obtained in a similar manner. Namely, Z_{parodd} is a parallel combination of the input impedances of two short-circuited stubs. The input impedance of a short-circuited stub with the characteristic impedance Z and electrical length θ is equal to $1/Zj\tan\theta$. Therefore, Z_{parodd} can be expressed as

$$\frac{1}{Z_{parodd}} = \frac{1}{jZ_0\tan\theta_1} + \frac{1}{jZ_0\tan\theta_2}. \tag{6.11}$$

The total odd input impedance can be expressed as

$$Z_{inodd} = Z_0 \frac{Z_{parodd} + jZ_0\tan\theta_3}{Z_0 + jZ_{parodd}\tan\theta_3}. \tag{6.12}$$

When the expression (6.11) is inserted into (6.12), we obtain the final expression for Z_{inodd}

$$Z_{inodd} = jZ_0 \frac{\tan\theta_1\tan\theta_2 + \tan\theta_3(\tan\theta_1 + \tan\theta_2)}{\tan\theta_1 + \tan\theta_2 - \tan\theta_1\tan\theta_2\tan\theta_3}. \tag{6.13}$$

From the resonant condition $1/Z_{in} = 0$, the even and odd resonant mode equations are derived

$$\tan\theta_1 + \tan\theta_2 + \tan\theta_3 = 0, \tag{6.14}$$

$$\tan\theta_1 + \tan\theta_2 - \tan\theta_1 \tan\theta_2 \tan\theta_3 = 0. \tag{6.15}$$

The previous expressions imply that all electrical lengths θ_1, θ_2, and θ_3 affect both resonances. Also, since the parameter θ_3 has the greatest value, we can note that the resonances are affected predominantly by the segment whose electrical length is θ_3. In other words, the resonances are the most sensitive to the change in the parameter θ_3.

However, the parameters θ_1 and θ_2 also influence the resonances and it will be shown that due to specific configuration of the resonator these parameters can be used for the independent control of the resonant frequencies.

Namely, we can note that $l_{upper} << l_{lower}$, i.e. $\tan\theta_1 << \tan\theta_2$, which means that the function $\tan\theta_1$ changes more slowly than the function $\tan\theta_2$ and thus the even resonance is more influenced by the latter function. Therefore, the even resonance is predominantly determined by the parameters θ_2 and θ_3.

In the case of the odd resonance, it is mainly influenced by the third term on the left side of the expression (6.15), which contains the parameter θ_3. Since θ_3 has the greatest value, $\tan\theta_3$ tends to infinity at a lower frequency in comparison to $\tan\theta_1$ and $\tan\theta_2$, and that frequency is the critical one in meeting the condition (6.15). Namely, the left-hand side in (6.15) is positive before $\tan\theta_3$ changes the sign from positive to negative. In the vicinity of the point where $\tan\theta_3$ changes the sign, the function $\tan\theta_1$+$\tan\theta_2$-$\tan\theta_1 tan\theta_2 tan\theta_3$ crosses zero because the third term has a negative value and the tangent is a monotonically increasing function.

The bandwidth in which the third term tends to infinity is determined by the functions $\tan\theta_1$ and $\tan\theta_2$, but predominantly by $\tan\theta_1$ because $\tan\theta_1 << \tan\theta_2$. The wider the bandwidth is, the lower is the frequency at which the condition is met.

It should be noted that the odd resonance occurs at lower frequencies in comparison to the even one, which is due to the specific configuration of the resonator. The function $\tan\theta_3$ tends to infinity at a lower frequency than the functions $\tan\theta_1$ and $\tan\theta_2$, and in the vicinity of that point the conditions (6.14) and (6.15) are met. Since $\tan\theta_3$ has positive sign in the expression (6.14) and negative sign in the expression (6.15) and since the tangent is a monotonically increasing function, the even resonance always occurs at a higher frequency in comparison to the point at which $\tan\theta_3$ tends to infinity, whilst the odd resonance occurs at a lower frequency.

From the previous analysis, it can be concluded that the two resonances can be independently controlled – the even resonance by the parameter θ_2, and the odd one by the parameter θ_1.

This can also be concluded if the geometrical parameters l_L and l_S are individually varied and the responses of the resonators compared. Figure 6.13 shows how the resonant frequencies of the structure with $w = g = 1.3$ mm are changed if the parameters l_L and l_S are individually varied whilst the other parameter is kept constant. It confirms that the parameter l_L, which corresponds to the electrical length θ_2, affects only the

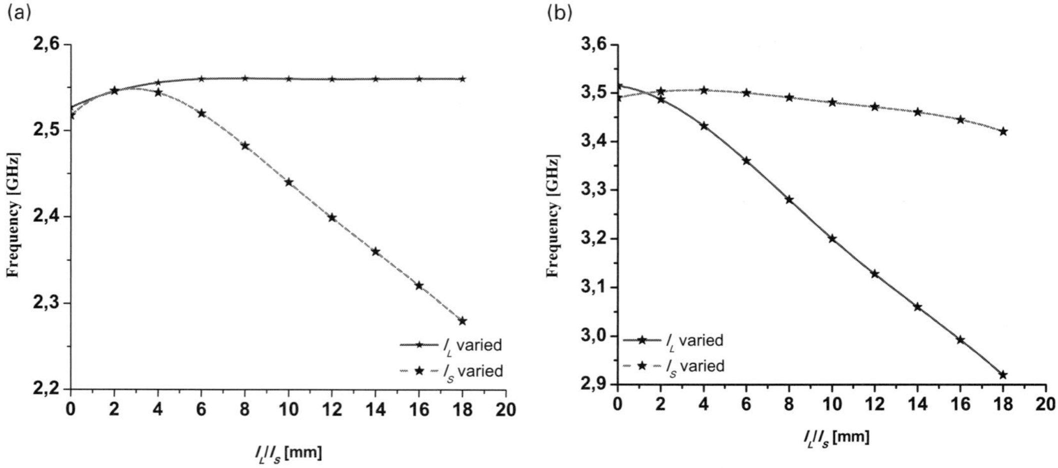

Figure 6.13 Influence of the parameters l_L and l_S on: (a) First resonance. (b) Second resonance.

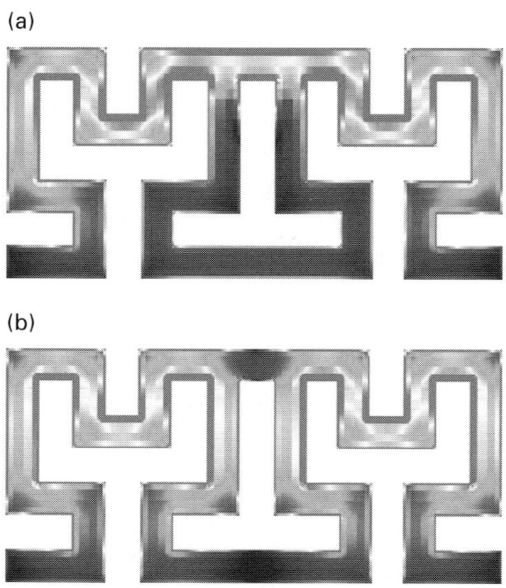

Figure 6.14 Current distribution at: (a) First resonance. (b) Second resonance.

position of the second resonance. On the other hand, the parameter l_S, which corresponds to the electrical length θ_1, strongly affects the position of the first resonance, whilst its influence on the second resonance can be considered as negligible.

Figure 6.14 shows the current distribution at the first and second resonances and reveals that the first resonance is the fundamental odd-moderesonance, whereas the second one is the fundamental even resonance. Since the symmetry plane is an electric wall in the odd circuit, the current is very strong at the midpoints of the resonator. On the other hand, since the symmetry plane represents a magnetic wall in the even circuit, there is no current at the midpoints of the resonator. These figures also confirm that the odd resonance occurs at lower frequencies in comparison to the even one.

Table 6.1 Characteristics of resonators with the same overall dimensions and different ratio *w/g*

w[μm]	100	700	1300	1900	2200
g[μm]	2900	2100	1300	500	100
f_1 [GHz]	1.96	2.24	2.53	3.04	3.44
Q_{u1}	52	97.3	145.1	103.6	83.2
f_2 [GHz]	2.68	3.08	3.52	4.14	4.56
Q_{u2}	57	101.2	153.4	104	75.6

Apart from the two resonances, the proposed resonator provides a transmission zero due to its configuration. Namely, as was stated previously there is no main signal path in the resonator, but the signal is rather divided into two paths. When the signals from the two paths are out-of-phase, a destructive interference occurs, i.e. the resonator exhibits a transmission zero. The transmission zero can also be explained from a mathematical point of view. Namely, the scattering parameter S_{21} is equal to zero when the condition $Z_{ineven} = Z_{inodd}$ is met. In the following sections, the transmission zeros will be analyzed in more detail.

Another degree of freedom in the resonator design represents the possibility to change the ratio of the parameters of the Hilbert fractal curve w and g, i.e. the possibility to change the line width while preserving the overall dimensions of the resonator.

Five resonators with overall dimensions equal to 19.5 x 9.1 mm^2 that differ in the ratio w/g have been analyzed. Table 6.1 shows their geometrical parameters as well as their resonant frequencies and quality factors.

As the line width is decreased, the resonant frequency is shifted towards lower frequencies, i.e. the resonator becomes smaller in terms of guided wavelength. However, the quality factor does not monotonically change with the frequency but it reaches its maximum when $w = g = 1.3$ mm. Also, it should be noted that in the extreme cases when $w = 0.1$ mm and 2.2 mm, the resonator becomes predominantly inductive and capacitive, respectively. Thus, its properties, especially the quality factor, are degraded. Since the resonator, whose line width and spacings are equal, represents the best compromise between performance and resonator dimensions, the structure with $w = g$ is used for the realization of dual-band filters.

From the previous analysis, it can be concluded that the resonator based on serially connected Hilbert fractal curves exhibits dual-mod behavior and that its resonances can be independently controlled by changing certain geometrical parameters. Also, the resonator provides a transmission zero due to its configuration. Thus, such a resonator is a good candidate for dual-band filter design and in the following sections two dual-band filters based on the proposed resonator will be presented and discussed.

6.3.2 Dual-band filter design

Using the proposed resonator two dual-band filters operating at 2.4 and 3.5 GHz that are characterized by good performance, compact dimensions, and independent control of the passbands have been designed.

There are several steps in the design of the filters. The first step is to determine the geometrical parameters of the resonator to obtain resonances at the predefined frequencies. Since the first resonance is the odd one, it is predominantly determined by the parameters θ_1 and θ_3. Therefore, the parameters w and g are determined so that the function $\tan\theta_3$ tends to infinity in the vicinity of the predefined resonant frequency, and then the resonance is finely tuned by the parameter s which corresponds to the parameter θ_1.

The higher resonance is predominantly determined by the parameters θ_2 and θ_3. Since the parameters w and g are determined in the previous step, the higher resonance is finely tuned by the parameter l_{lower} which corresponds to the parameter θ_2.

Following these steps, the dimensions of the resonator that is used in filters operating at 2.4 and 3.5 GHz are determined: $w = g = 1.3$ mm, $s = l_{upper} = 3.2$ mm, $l_{lower} = 23.9$ mm, $a = 1.6$ mm, $l_L = l_S = 0$. As was stated previously, the resonator and the filters are designed on the substrate Rogers RT/Duroid 5880 whose thickness is equal to 1.575 mm, dielectric constant $\varepsilon_r = 2.2$, and loss tangent $\tan\delta = 0.0009$.

Since the filters consist of several resonators which are mutually coupled, once the filter is designed, fine tuning is needed to position the passbands at the desirable frequencies and to obtain good performance.

Both proposed filters consist of two resonators, but they differ in the manner the resonators are coupled. That aspect causes the difference in the filter performance, especially in selectivity. Namely, certain ways of coupling between resonators may provide a transmission zero and thus improve selectivity, and this will be discussed in more detail below.

6.3.2.1 Dual-band filter with edge-coupled resonators

The first proposed filter consists of two resonators that are capacitively coupled through the gap d. Its configuration and response are shown in Figure 6.15. The response is characterized by two passbands and poor selectivity, especially in the stopband region between the two passbands. The transmission zero positioned at the upper side of the second passband occurs because the condition $Z_{ineven} = Z_{inodd}$ is met.

In order to improve the performance of the filter, a transmission zero should be introduced in between the two passbands, which would improve selectivity and allow control of the bandwidths. One way to introduce a transmission zero which does not degrade filter performance and does not excessively increase its dimensions is to extend the feeding lines; namely, at the frequency at which the electrical length of the extension of the feeding line is equal to $\pi/2$ long, a transmission zero occurs.

The extension of the feeding lines changes the coupling between the resonators and the feeding lines which influences the resonant frequencies as well as the position of the transmission zero. Thus, after the configuration of the filter is changed, an additional fine tuning of the geometrical parameters is needed.

Figure 6.16 shows the configuration and the response of the filter with the extended feeding lines. The selectivity of the filter is greatly improved due to a newly introduced transmission zero. It should be noted that there are two transmission zeros between the

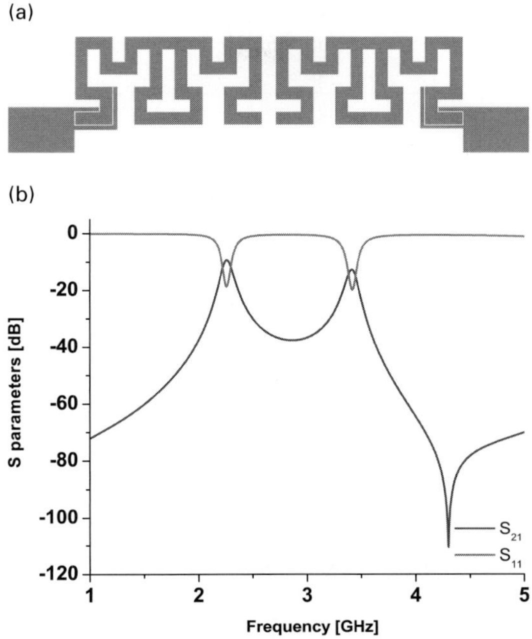

Figure 6.15 Dual-band filter with edge-coupled resonators: (a) Configuration. (b) Response.

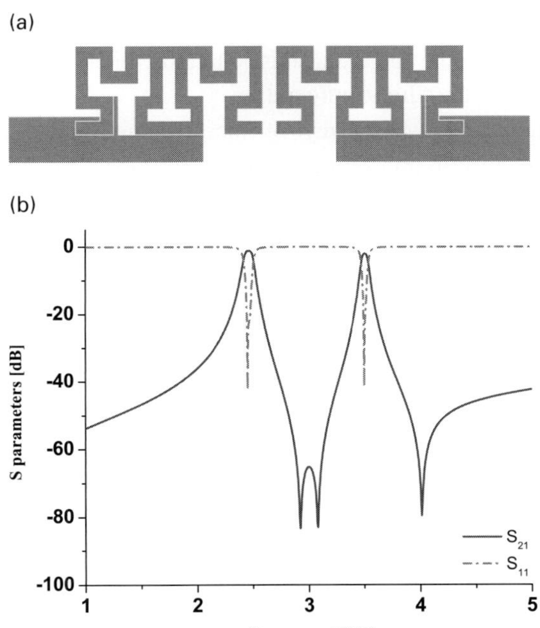

Figure 6.16 Dual-band filter with edge-coupled resonators with extended feeding lines: (a) Configuration. (b) Response.

Table 6.2 Geometrical parameters of the initial and final filter with edge-coupled resonators

parameter [mm]	w	g	s	l_{upper}	l_{lower}	a	l_L	l_S	d
initial configuration	1.3	1.3	3.2	3.2	23.9	1.6	0	0	1.3
final configuration	1.3	1.3	1.3	1.3	22.9	1.5	0	0	1.5

Table 6.3 Characteristics of the final filter with edge-coupled resonators

	f [GHz]	IL [dB]	RL [dB]	3 dB BW [MHz]	3 dB FBW [%]	overall dimesnions [λ_g x λ_g]
first passband	2.45	1.19	22	100	4.17	0.44x0.1
second passband	3.5	2.08	25.8	70	2	

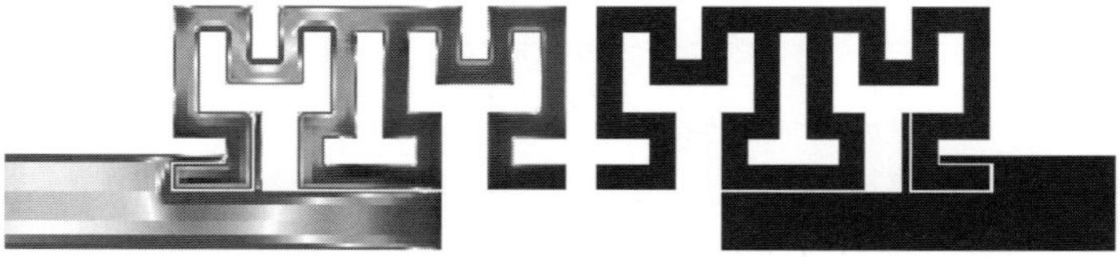

Figure 6.17 Current distribution at the frequency of transmission zero.

passbands which originate from the same transmission zero which has been split into two due to strong couplings in the filter configuration.

The current distribution at the frequency of the transmission zero shown in Figure 6.17 proves that the transmission zero occurs due to the extension of the feeding lines. Namely, there is no current distribution in the second resonator, which implies that there is no signal propagation in the structure, i.e. that the structure exhibits a transmission zero. At the same time, the current distribution is the strongest in the feeding line extension, which means that the extension causes the transmission zero.

As was stated previously, due to the change in the configuration, a fine tuning of geometrical parameters is needed. Table 6.2 shows the parameters of the initial configuration and those of the final configuration. The distance between the two fractal curves s has changed the most, but at the same time the overall dimensions of the filter have decreased. Also, it can be seen that the even five parameters s, l_{lower}, l_{upper}, a, and d have been changed, which may imply that the feeding line extension requires demanding optimization. However, we should note that the parameters s, l_{lower}, l_{upper}, and a, and thus their tuning, are interrelated.

The characteristics of the final filter are summarized in Table 6.3. The filter exhibits low insertion losses, good return losses, and good selectivity, as well as compactness. Another important aspect is the fact that the passbands of the filter can be independently controlled.

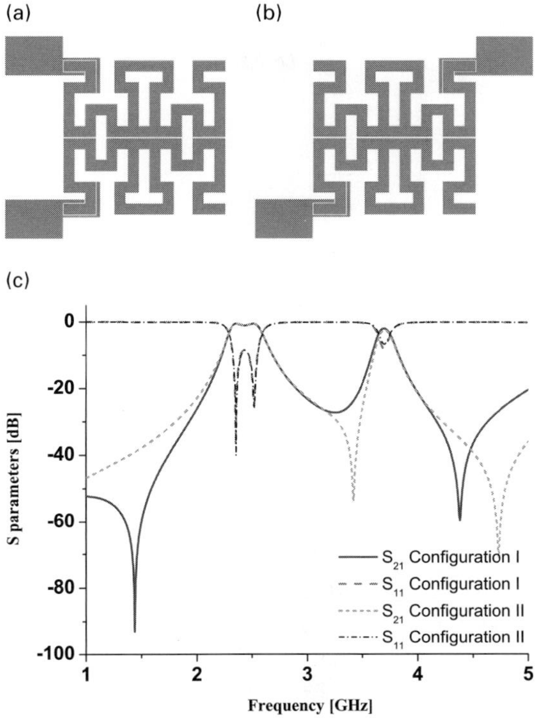

Figure 6.18 Dual-band filter with broadside coupled resonators: (a) Configuration I. (b) Configuration II. (c) Responses of the configurations I and II.

6.3.2.2 Dual-band filter with broadside coupled resonators

The main disadvantage of the filter with edge-coupled resonators is poor selectivity of the first passband. In the following analysis, it will be shown that the selectivity can be significantly improved if the resonators are coupled at the broadside.

Figures 6.18(a) and 6.18(b) show two configurations with broadside coupled resonators which differ in the configuration of the feeding lines, whilst their responses are compared in Figure 6.18(c). The geometrical parameters of the resonators are the same as in the previously presented filter.

The two filters have very similar responses, with two passbands and a transmission zero positioned at the upper side of the second passband. The origin of this transmission zero was discussed in the previous section. Also, both responses are characterized by another transmission zero which is positioned at the lower side of the first passband in the first case, and between the two passbands in the second case.

In Chapter 4, it was shown that specific coupling of transmission lines can provide transmission zeros. In the case of parallel coupling, a transmission zero occurs when the electrical length of the coupled segments is equal to 180°. On the other hand, antiparallel coupling provides a transmission zero at the frequency at which coupled segments are 90° long.

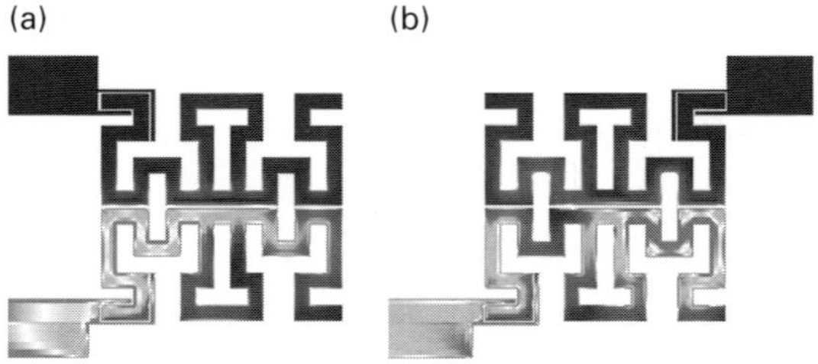

Figure 6.19 Current distribution at the frequency of transmission zero for: (a) Configuration I. (b) Configuration II.

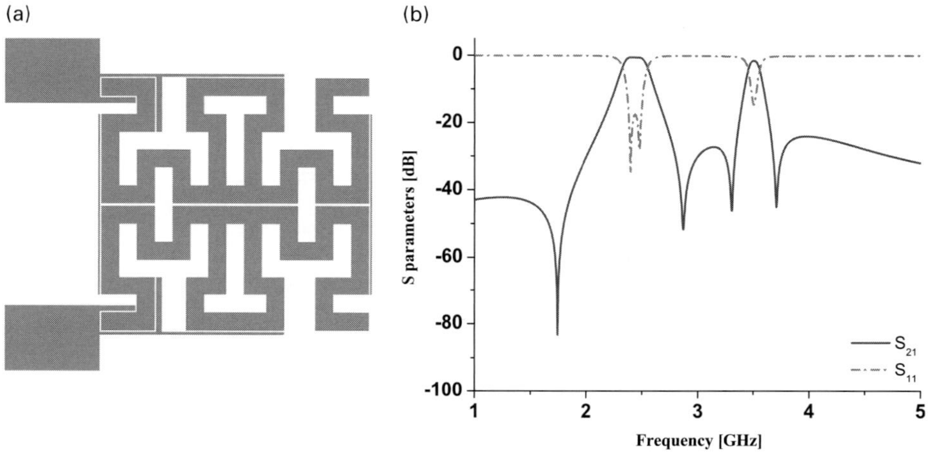

Figure 6.20 Final dual-band filter with broadside coupled resonators: (a) Configuration. (b) Response.

The coupling between the resonators in the two filter configurations can be simplified and treated as an antiparallel and parallel coupling . Namely, although the longer sides of the resonators are loaded with the segments l_{lower}, their coupling can be seen as an antiparallel and parallel coupling. Therefore, in the first case, which corresponds to an antiparallel coupling, the transmission zero occurs at the lower side of the first passband, whilst in the second case, which corresponds to a parallel coupling, the transmission zero occurs in between the passbands. The current distribution, Figure 6.19, shows that at the frequency at which transmission zero occurs only the longer edge of the resonator is "active" whilst the segment l_{lower} is not.

In the previous filtering configuration, it was shown that a transmission zero can be realized if the feeding lines are extended and that it can be positioned between the passbands. A dual-band filter with excellent selectivity can be realized if the configuration with an antiparallel coupling and extended feeding lines is used. The final configuration and the response of such a structure are shown in Figure 6.20. In order to improve the coupling between the resonators, non-resonating nodes are introduced.

Table 6.4 Geometrical parameters of the final filter with broadside coupled resonators

parameter [mm]	w	g	s	l_{upper}	l_{lower}	a	l_L	l_S	d
	1.3	1.3	1.3	1.3	22.9	1.5	0	0	0.2

Table 6.5 Characteristics of the final filter with broadside coupled resonators

	f [GHz]	IL [dB]	RL [dB]	3 dB BW [MHz]	3 dB FBW [%]	overall dimensions [$\lambda_g x \lambda_g$]
first passband	2.45	0.62	17.58	220	9	0.22x0.21
second passband	3.5	1.61	14.68	110	3.14	

Since the filter consists of two resonators and due to the fact that the feeding lines have a complex structure, a fine tuning of the geometrical parameters is needed. Table 6.4 shows the geometrical parameters of the final filter, whilst Table 6.5 shows its characteristics.

It can be seen that the geometrical parameters in the two filters differ only in the parameter d, which in the case of the broadside coupled resonator has been decreased to 0.2 mm.

Similar to the first proposed filter, this filter has good performance in terms of insertion and return loss. At the same time, its selectivity is improved. Additionally, the filter is very compact and allows independent control of the passbands.

6.3.2.3 Fabrication and measurement results

The proposed filters have been fabricated using standard printed circuit technology and they have been realized on the substrate Rogers RT/Duroid 5880 whose thickness is equal to 1.575 mm, dielectric constant $\varepsilon_r = 2.2$, and loss tangent tan$\delta = 0.0009$.

A photograph of the fabricated filter with edge-coupled resonators is shown in Figure 6.21(a), whilst the simulated and measured results are compared in Figure 6.21 (b). It can be seen that the simulated and measured results are in excellent agreement.

Figure 6.22(a) shows a photograph of the filter with broadside coupled resonators and Figure 6.22(b) the simulated and measured results. Similar to the previous filter, there is good agreement between the simulated and measured results.

Table 6.6 and Table 6.7 summarize the characteristics of the simulated and fabricated filters and they confirm the good agreement between them.

6.3.2.4 Comparison to other dual-band filters operating at 2.4 and 3.5 GHz

The proposed filters operate at 2.4 and 3.5 GHz, which represents an advantage in comparison to other dual-band filters. Namely, most dual-band filters published so far are designed to operate at 2.4 and 5.2 GHz, which eases the design procedure and at the

Table 6.6 Characteristics of the simulated and fabricated filter with edge-coupled resonators

	f_1/f_2 [GHz]	IL [dB]	RL [dB]	3 dB BW [MHz]	3 dB FBW [%]
simulated circuit	2.45/3.5	1.19/2.08	22/25.8	100/70	4.17/2
fabricated circuit	2.44/3.49	1/1.6	25/21	101/80	4.1/2.3

Table 6.7 Characteristics of the simulated and fabricated filter with broadside-coupled resonators

	f_1/f_2 [GHz]	IL [dB]	RL [dB]	3 dB BW [MHz]	3 dB FBW [%]
simulated circuit	2.45/3.5	0.62/1.61	17.58/14.68	220/110	9/3.14
fabricated circuit	2.43/3.47	0.38/2.53	18.8/10	213/111	8.7/3.18

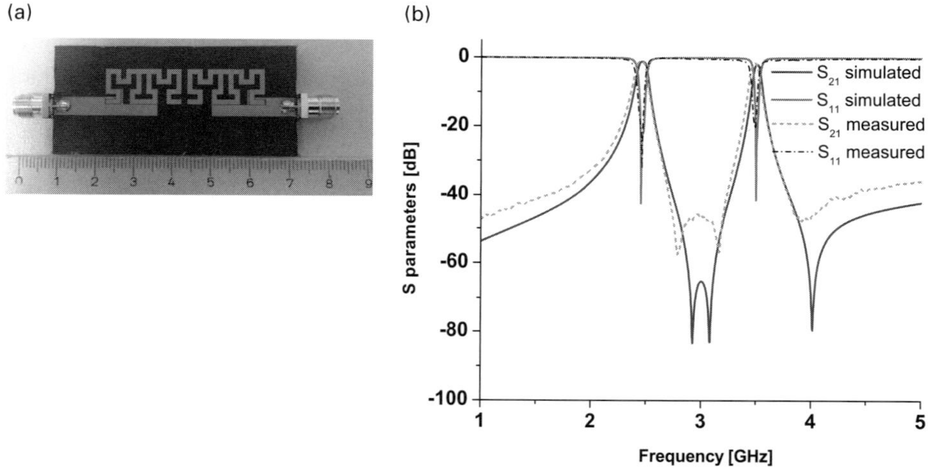

Figure 6.21 Dual-band filter with edge-coupled resonators: (a) Photograph of the fabricated circuit. (b) Simulated and measured responses.

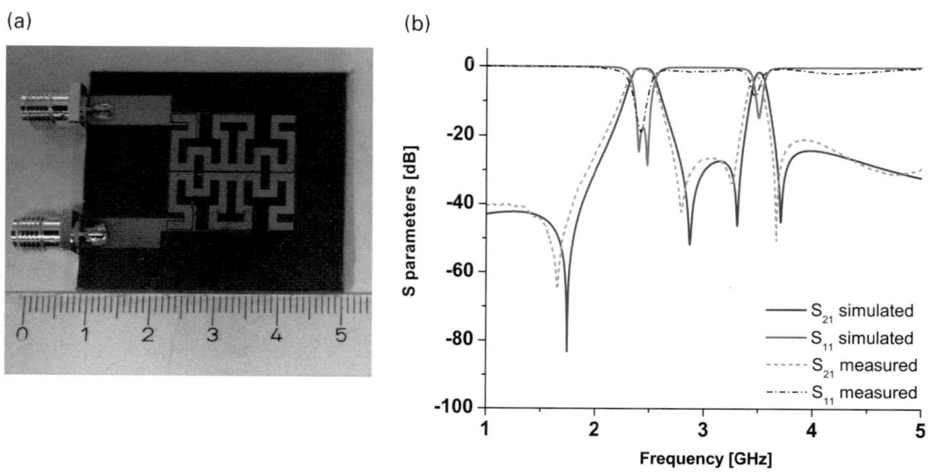

Figure 6.22 Dual-band filter with broadside-coupled resonators: (a) Photograph of the fabricated circuit. (b) Simulated and measured responses.

Table 6.8 Comparison of the characteristics of the proposed filters and other dual-band filters

	f_1/f_2 [GHz]	IL [dB]	RL [dB]	dimensions [λ_gxλ_g]	3 dB FBW [%]	A	B
filter I	2.45/3.5	1/1.6	25/21	0.44x0.1	4.1/2.3	da	planar
filter II	2.45/3.5	0.38/2.53	18.8/10	0.22x0.21	8.7/3.18	da	planar
[28]	2.5/3.5	1.8/2.4	20/15	0.24x0.33	6.3/4.4	ne	planar
[29]	2.35/3.2	1.8/3	25/30	0.18x0.19	3.9/2.8	da	DGS
[30]	2.4/3.3	1.2/1.5	20/20	0.21x0.18	8.1/4.2	ne	DGS+vias

A – possibility of independent control of passbands, B – fabrication technology.

same time reduces the overall filter size. In Table 6.8, the characteristics of the proposed filters are compared to those of other dual-band filters operating at 2.4 and 3.5 GHz.

The proposed filters have good performance in terms of insertion loss and selectivity, and these characteristics are similar to those of other filters. Also, the dimensions are comparable to the dimensions of other filtering circuits. However, the proposed filters allow independent control of the passbands and their fabrication is very simple since the circuits are fully planar and do not require vias or a multi-layer structure.

Therefore, the proposed filters present good dual-band solutions, which unite good performance, compactness, independent control of the passbands, and a simple fabrication process.

6.4 Compact tri-band bandpass and bandstop filters based on Hilbert-fork resonators

In Chapter 2, it was shown that none of tri-band filters published so far exhibits simultaneously good in-band and out-of-band performance, good selectivity of all passbands, and compactness. The structure presented in Chapter 4 has all these properties and its compactness has been achieved by using folded quarter-wavelength resonators, i.e. by using resonators that employ grounding vias.

In this chapter, fractal curves were presented and their great potential for miniaturization of passive microwave circuits discussed. Also, in the previous section it was shown that compactness can also be achieved if resonators have a Hilbert fractal curve shape.

Combining the idea of using grounding vias and Hilbert fractal-shaped resonators, a tri-band filter with excellent performance and very small dimensions can be realized. In this section, tri-band bandpass and bandstop structures based on the Hilbert-fork resonator [31] will be presented.

The Hilbert-fork resonator represents a novel multi-mode structure that consists of two dual-mode resonators. The first resonator is based on a two-dimensional Hilbert fractal curve of the second order, whilst the second resonator has the shape of a fork and is positioned to conveniently fill the remaining footprint area of the first resonator. Both resonators are grounded by vias, which enables the dual-mode behaviour of the resonators, as well as independent control of the resonances. In addition, the presence of the vias enables size reduction of the structure.

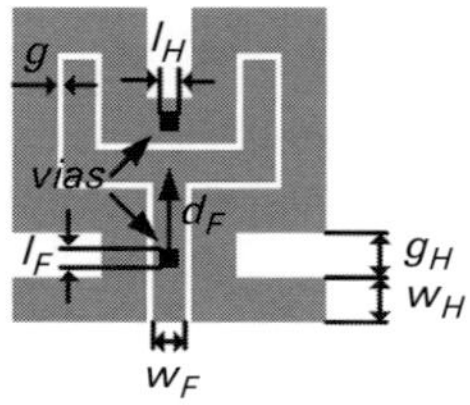

Figure 6.23 Configuration of Hilbert-fork resonator.

Using the two modes of the Hilbert resonator and the one of the fork resonator, tri-band bandstop and bandpass filters have been designed which have excellent performance in all tri-bands, small overall size, as well as the possibility of independent control of the passbands.

The first section is devoted to the analysis of the behaviour of the Hilbert-fork resonator, and it will be shown how its resonant modes are formed and how they can be independently controlled.

In two following sections, a tri-band bandpass and tri-band bandstop filters based on the proposed resonator will be presented and their design procedure discussed. Also, their performance will be compared to those of other published tri-band filters.

Both filters have been fabricated using standard printed circuit board technology and comparison of simulated and measured results will also be presented and discussed.

6.4.1 Hilbert-fork resonator configuration

The configuration of the proposed filter is shown in Figure 6.23. The resonator consists of two dual-mode resonators, one of which is based on the Hilbert fractal curve of the second order, whilst the other has the shape of a fork and conveniently fills the remaining footprint area of the first resonator. In that manner, a significant size reduction has been achieved. Both structures are grounded by vias which are positioned as is shown in the Figure 6.23. As will be shown later, the vias enable dual-mode behaviour, as well as independent control of the resonances.

The width of the fractal curve is denoted by w_H, whilst the spacing between the lines is denoted by g_H. The parameter w_F represents the width of the conductive line of the fork resonator and d_F is the distance of the fork via from the central position of the fork resonator. The dimensions of the Hilbert and fork vias are denoted by l_H and l_F, respectively.

The resonator as well as the tri-band filters are designed on the substrate Rogers RT/Duroid 5880 whose thickness is equal to 1.575 mm, dielectric constant $\varepsilon_r = 2.2$, and loss tangent $\tan\delta = 0.0009$.

Since the structures of both resonators are symmetrical, their behaviour can be analysed using even-/odd-mode analysis. Due to the fact that the resonators are positioned close to each other, they influence the behaviour of each other. For the sake of simplicity, this effect is not taken into account in the analysis of their behaviour.

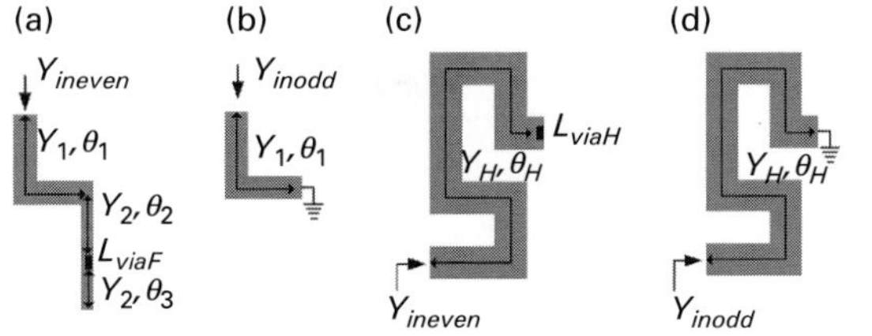

Figure 6.24 Equivalent circuit for: (a) Even mode for fork resonator. (b) Odd mode for fork resonator. (c) Even mode for Hilbert resonator. (d) Odd mode for Hilbert resonator.

However, simplification of the analysis will be compensated in the filter design through optimization of the geometrical parameters.

Figure 6.24 shows the even- and odd-mode equivalent circuits for both resonators. In the even circuit, the symmetry plane represents a magnetic wall, whilst in the odd circuit the symmetry plane is an electric wall. In the case of the fork resonator, the parameters θ_1, θ_2, and θ_3 represent electrical lengths, whilst Y_1 and Y_2 are admittances of the corresponding segments. L_{viaF} is the inductance twice as big as the value of the via inductance, and Y_{ineven} and Y_{inodd} are the even and odd input admittances, respectively. The relations between impedances and admittances are as follows: $Y_1 = 1/Z_1$, $Y_2 = 1/Z_2$, $Y_{ineven} = 1/Z_{ineven}$, $Y_{inodd} = 1/Z_{inodd}$.

The odd input admittance is the admittance of a short-circuited stub and thus it can be expressed as

$$Y_{inodd} = -jY_1 \cot\theta_1. \tag{6.16}$$

The odd resonances occur when the condition $Y_{inodd} = 0$ is met, i.e. when the parameter $\theta_1 = \pi/2 + k\pi$, where k is an integer number.

The even-mode equivalent circuit is somewhat more complex than that of the odd-mode and thus the analysis is more complicated. Namely, the via is not positioned at the very end of the resonator, and to account for the influence of all segments, as well as of the via, *ABCD* parameters are used instead.

Although the segment θ_3 is positioned "after" the via and practically does not influence the resonances, it is included in the calculation so as to obtain a more precise analysis. The *ABCD* matrix of the circuit shown in Figure 6.24(a) is expressed in the following manner

$$\begin{aligned} ABCD = &\begin{bmatrix} \cos\theta_1 & jZ_1\sin\theta_1 \\ jY_1\sin\theta_1 & \cos\theta_1 \end{bmatrix} \cdot \begin{bmatrix} \cos\theta_2 & jZ_2\sin\theta_2 \\ jY_2\sin\theta_2 & \cos\theta_2 \end{bmatrix} \\ &\cdot \begin{bmatrix} 1 & 0 \\ (j\omega L_{viaF})^{-1} & 1 \end{bmatrix} \cdot \begin{bmatrix} \cos\theta_3 & jZ_3\sin\theta_3 \\ jY_3\sin\theta_3 & \cos\theta_3 \end{bmatrix}, \end{aligned} \tag{6.17}$$

and the resonance occurs when the condition $AD = 1$ is met.

In the previous expressions and in Figure 6.24, it can be seen that the odd resonance is determined only by the electrical length θ_1. On the other hand, the even resonance is determined not only by the electrical length θ_1 but also by the electrical length θ_2 and

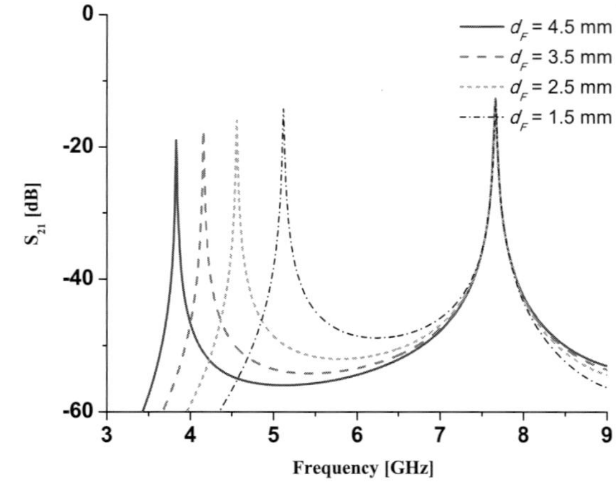

Figure 6.25 Influence of the parameter d_F on the fork resonances.

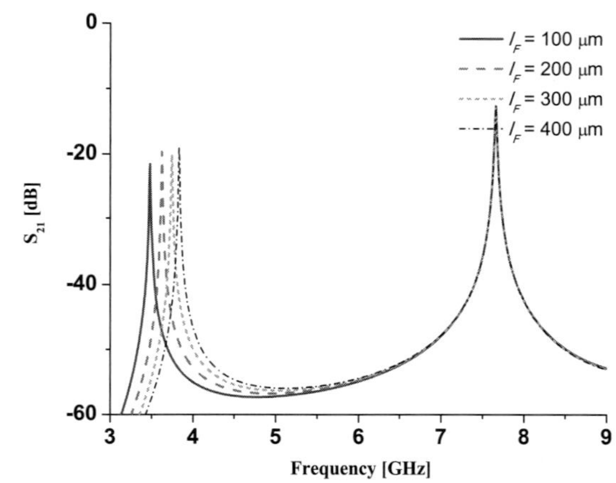

Figure 6.26 Influence of the parameter w_F on the fork resonances.

the via inductance L_{viaF}. It can be concluded that the even resonance always occurs at lower frequencies in comparison to the odd resonance. Namely, the electrical length in the even circuit is always longer than that in the odd circuit since $\theta_1 + \theta_2$ is longer than θ_1. In the case when $\theta_2 = 0$, the fork via inductance shifts the even resonance to the lower frequencies in comparison to the odd mode. More important, the position of the even resonance can be independently controlled from the odd one by variation of the parameters d_H and L_{viaF}.

The responses of the fork resonators that differ in the parameter d_F are shown in Figure 6.25, whilst the responses of the structures that differ in the via inductance are shown in Figure 6.26. In all cases, the parameter w_F is equal to 1.1 mm, the length of the first segment 7.7 mm, and the sum of the lengths of the second and the third segment is equal to 4.6 mm.

The responses confirm that the position of the first resonance can be independently controlled from the second one. Also, they confirm that decrease of the d_F shifts the first

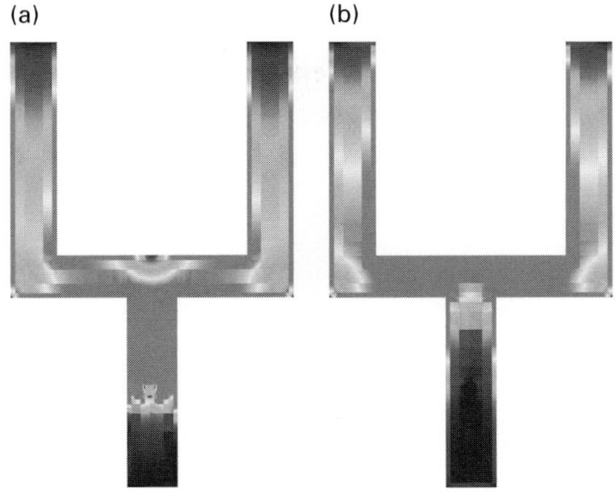

Figure 6.27 Current distribution in the fork resonator at: (a) First resonance. (b) Second resonance.

resonance towards higher frequencies, which implies that the segment θ_3 does not affect the resonator's response.

The current distributions in the fork resonator at the resonant frequencies are shown in Figure 6.27, and they also prove that the even resonance occurs at lower frequencies. Since the symmetry plane is an electric wall in the odd circuit, there is current only in the segment θ_1. On the other hand, since the symmetry plane represents a magnetic wall in the even circuit, there is current in both segments θ_1 and θ_2. At the same time, there is no current in the segment θ_3, which confirms that the segment θ_3 does not affect the resonator's response.

Apart from the resonances, the fork resonator also provides a transmission zero which occurs at the frequency at which the condition $Y_{ineven} = Y_{inodd}$ is met, and this will be discussed in more detail in the following section.

As for the Hilbert resonator, analysis of its behaviour is somewhat simpler. Although the Hilbert resonator is folded, for the sake of simplicity the effect of folding will not be taken into account.

Similar to the odd mode of the fork resonator, the input admittance of the Hilbert resonator is the input admittance of a short-circuited stub and thus it can be expressed as follows

$$Y_{inodd} = -jY_H \cot\theta_H. \tag{6.18}$$

The odd resonances occur when the condition $Y_{inodd} = 0$ is met, i.e. when the parameter $\theta_H = \pi/2 + k\pi$, where k is an integer number.

The even-mode resonance condition can be expressed using *ABCD* parameters of the circuit shown in Figure 6.24

$$ABCD = \begin{bmatrix} \cos\theta_H & jZ_H \sin\theta_H \\ jY_H \sin\theta_H & \cos\theta_H \end{bmatrix} \cdot \begin{bmatrix} 1 & 0 \\ (j\omega L_{viaH})^{-1} & 1 \end{bmatrix}, \tag{6.19}$$

and the resonance occurs when the condition $AD = 1$ is met.

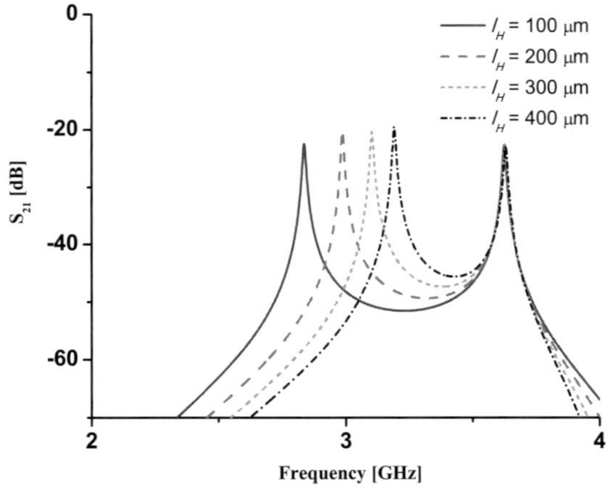

Figure 6.28 Influence of the parameter l_H on the Hilbert resonances.

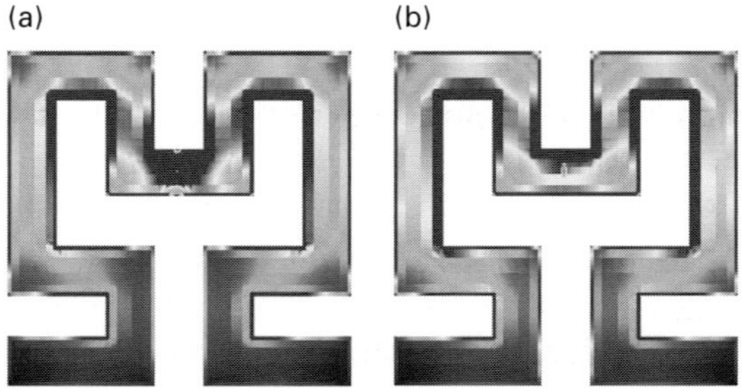

Figure 6.29 Current distribution in the Hilbert resonator at: (a) First resonance. (b) Second resonance.

From the previous two expressions and Figure 6.24, it can be concluded that both resonances are predominantly determined by the electrical length θ_H, but also that the even resonance is determined by the via inductance L_{viaH}. Therefore, the even resonance occurs at lower frequencies than the odd one.

In addition, a change in the via inductance represents a mechanism for independent control of the first resonance and that is illustrated in the Figure 6.28, which shows the responses of the Hilbert resonator with different via inductance L_{viaH}, i.e. with the different via dimension l_H. In other words, it can be seen that the change in l_H causes a change only in the first resonance. The parameters w_H and g_H in the analysed structures are equal to 1.3 mm.

The current distribution in the Hilbert resonator at the resonances are shown in Figure 6.29 and they also imply that the even resonance is positioned at lower frequencies in comparison to the odd one. Since the symmetry plane represents a magnetic wall, the current distribution is very strong at the midpoint of the resonator, especially in the via. However, in the odd circuit the symmetry plane is an electric wall and thus there is no current distribution in the Hilbert via.

In the previous section, it was discussed that an additional degree of freedom in the design of resonators based on the Hilbert fractal curve is the ratio of the parameters w_H and g_H. Also, it was shown that the resonator whose line widths and spacings are equal represents the best compromise between resonator performance and overall size, and thus the Hilbert resonator whose $w_H = g_H$ will be used in the tri-band filter design.

From the previous analysis, it can be seen that the proposed resonator exhibits multimode behaviour and that its resonances can be independently controlled by a change of certain geometrical parameters. Hence, this resonator is a good candidate for multi-band filter design and in the following sections tri-band bandpass and bandstop filters will be presented and their performance discussed.

6.4.2 Tri-band bandpass filter based on the Hilbert-fork resonator

In this section, a tri-band bandpass filter based on the Hilbert-fork resonator that operates at 2.4, 3.5, and 5.2 GHz will be presented. The design procedure and its behaviour will be analysed in detail. Finally, its performance will be compared to the performance of other published tri-band bandpass filters.

6.4.2.1 Configuration and design of a tri-band bandpass filter

In the previous section, it was shown that the Hilbert-fork resonator exhibits four resonant modes. Since the Hilbert resonator has larger overall dimensions than the fork resonator, its modes occur at lower frequencies than those of the fork resonator. Therefore, two resonant modes of the Hilbert resonator and the lower mode of the fork resonator will be used for the design of a tri-band filter.

The filter design procedure consists of several steps. The first step is to determine the geometrical parameters of the resonator which provides resonances at the specified frequencies. The position of the second resonance is determined only by the length of the Hilbert resonator and thus its dimensions are chosen so that the electrical length θ_H is equal to 90° at the specific frequency. Since the second passband is positioned at 3.5 GHz, values of the parameters w_H and g_H on the given substrate are chosen to be 1.3 mm.

The position of the first passband is determined by the length of the Hilbert resonator, as well as by the Hilbert via inductance. Since the parameters w_H and g_H are determined in the previous step, the via dimensions are chosen in a manner that provides the first passband to be positioned at 2.4 GHz. The calculated value of the parameter l_H is 200 μm.

The third passband is provided by the lower resonance of the fork resonator and in this case there is more design freedom since the higher resonant mode of the fork resonator is not used. The dimensions of the fork resonator are mainly dictated by the dimensions of the Hilbert fractal curve because the fork resonator can occupy only the remaining footprint area of the Hilbert resonator. The coupling between the two resonators is chosen to be the strongest one in the given fabrication technology, i.e. the gap g is chosen to be equal to 0.1 mm, which determines the parameter w_F to be equal to 1.1 mm.

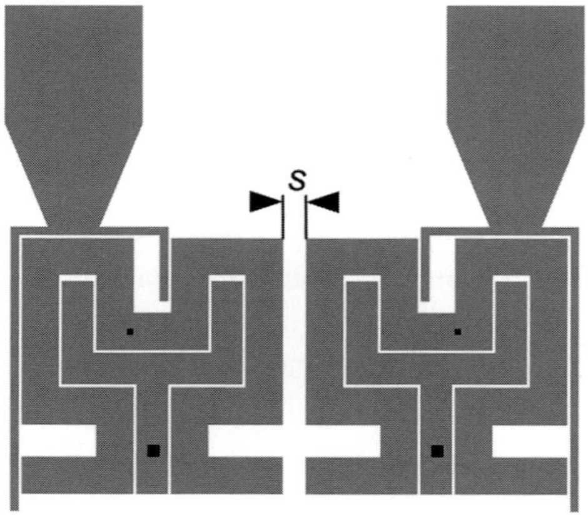

Figure 6.30 Configuration of the tri-band bandpass filter.

In the following step, the parameters d_F and l_F are determined to provide the third passband to be positioned at 5.2 GHz. The fork via dimensions equal to or larger than 400 μm eases the fabrication process and thus l_F is chosen to be 400 μm. This choice is also dictated by the transmission zeros allocation, which will be discussed in more detail below. Finally, the parameter d_F is calculated to be 2.4 mm to obtain the third passband at 5.2 GHz.

Two Hilbert-fork resonators coupled through the gap s are used to design the tri-band bandpass filter. As was stated previously, there is a coupling between the Hilbert and fork resonators which affects the properties of both structures. Also, the filter consists of two coupled resonators and after the filter is formed, a fine tuning of certain geometrical parameters is needed to obtain specified filter characteristics.

Figure 6.30 shows the configuration of the tri-band bandpass filter in which certain geometrical parameters have been tuned to obtain the prescribed filter response. It can be seen that the via in the Hilbert resonator is not centrally positioned, but is shifted by 0.7 mm. In this manner, the position of the first passband is tuned. At the same time, this does not influence the second passband since the shift of 0.7 mm is very small in comparison to the overall length of the Hilbert resonator, and thus it does not significantly degrade the resonator symmetry.

In addition, the fork via has been shifted, i.e. the parameter d_F has been increased to 2.6 mm, in order to tune the third passband.

The bandwidth and insertion loss are controlled by the gap s as well as by the external quality factor which is determined by the configuration of the feeding lines and the position where the resonators are coupled to them. Also, the feeding lines have a role in the formation of the transmission zeros.

The response of the proposed tri-band filter is shown in Figure 6.31, whilst its final dimensions and characteristics are given in Table 6.9 and Table 6.10, respectively. The proposed filter is characterized by small insertion losses and good return losses. Its overall size makes it the most compact tri-band bandpass filter proposed so far, which

Table 6.9 Geometrical parameters of the tri-band bandpass filter

parameter [mm]	w_H	g_H	l_H	l_F	w_F	d_F	g	s
	1.3	1.3	0.2	0.4	1.1	2.6	0.1	0.8

Table 6.10 Characteristics of the tri-band bandpass filter

	f [GHz]	IL [dB]	RL [dB]	3 dB BW [MHz]	3 dB FBW [%]	overall dimensions [$\lambda_g x \lambda_g$]
first passband	2.4	1.5	15	100	4.2	
second passband	3.5	2.54	13	100	2.9	0.21x0.11
third passband	5.2	1.16	20	240	4.5	

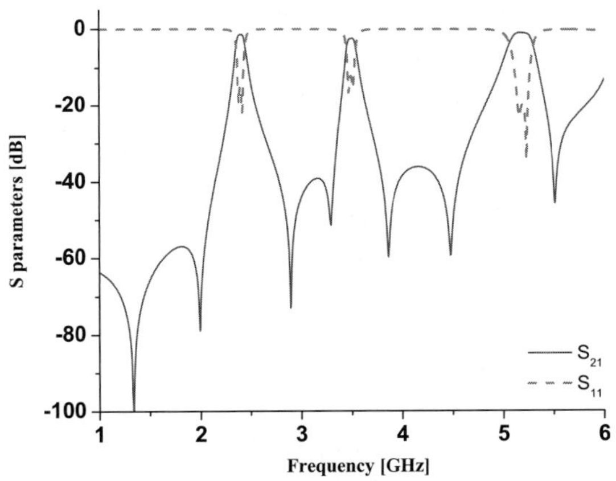

Figure 6.31 Response of the tri-band bandpass filter.

will be discussed in more detail below. Also, it exhibits excellent selectivity provided by seven transmission zeros.

Similarly to the structure in the previous section, the presence of the vias in the structure causes transmission zeros. In this structure, even five transmission zeros are due to the presence of the vias. As was shown, vias act as mutually coupled inductors and in that way another signal path is formed which causes a transmission zero to occur when it is out-of-phase with the main path signal. The position of such transmission zeros is predominantly determined by the via inductances and coupling strength. The higher via inductance and strongest coupling between them shifts the transmission zero to the lower frequencies.

Two lowest transmission zeros occur due to the vias in the Hilbert resonators. The third transmission zero is due to the presence of the other signal path in the filter. Namely, due to the folding of the Hilbert resonator, there is a coupling between the resonator's end segments, which provides another signal path in the filter. When the signals that propagate through the main and secondary signal paths are out-of-phase, a transmission zero occurs.

The vias in the fork resonators causes the fourth transmission zero. Since the Hilbert vias have greater inductance than those of the fork resonators, it is clear why the Hilbert vias cause transmission zeros at lower frequencies. The via dimensions in the fork resonator are chosen to be 400 μm which provides a transmission zero between the first and the second passband and eases the fabrication process.

In addition, a cross-coupling between the Hilbert and fork vias exists and gives the fifth and sixth transmission zeros. The seventh transmission zero occurs when the condition $Y_{ineven} = Y_{inodd}$ in the fork resonator is met.

From the previous analysis, it can be concluded that the proposed resonator enables the design of a very compact filter with excellent performance which exhibits excellent selectivity due to its configuration and various couplings. Also, the proposed filter is characterized by the fact that its passbands can be independently controlled.

6.4.2.2 Fabrication and measurement results

The proposed filter has been fabricated using standard printed circuit board technology and it has been realized on the substrate Rogers RT/Duroid 5880 whose thickness is equal to 1.575 mm, dielectric constant $\varepsilon_r = 2.2$, and loss tangent $\tan\delta = 0.0009$. Figures 6.32(a) and 6.32(b) show a photograph of the fabricated filter and Figure 6.32(c) shows a comparison between the responses of the simulated and measured circuit.

The proposed filter comprises four vias and thus its fabrication is relatively demanding since the positions of two passbands and five transmission zeros depend on the vias and their inductances. However, Figure 6.32(c) shows that there is good agreement between measured and simulated results.

A comparison between the characteristics of the simulated and measured circuit given in Table 6.11 also confirms the good agreement between the simulated and measured results.

6.4.2.3 Comparison to other tri-band bandpass filters

The proposed tri-band filter has been designed to operate at 2.4, 3.5, and 5.2 GHz. As was previously stated, most tri-band filters proposed so far have been designed to operate at the same frequencies. In Table 6.12, the characteristics of the proposed filter are compared to those of other tri-band filters that represent the best tri-band bandpass structures published so far.

The proposed tri-band filter exhibits small insertion losses in all three passbands, 1.5, 2.54, and 1.16 dB, good return losses, excellent selectivity owing to seven transmission zeros and very compact dimensions of $0.22\lambda_g$ x $0.11\lambda_g$.

In comparison to other configurations, the filter has similar properties in terms of insertion and return losses. However, the proposed filter has the most compact dimensions of all published tri-band filters, as well as exceedingly good selectivity. What is more, the proposed filter enables independent control of the passbands.

Table 6.11 Characteristics of the simulated and fabricated tri-band bandpass filter

	$f_1/f_2/f_3$ [GHz]	IL [dB]	RL [dB]	3 dB BW [MHz]	3 dB FBW [%]
simulated circuit	2.4/3.5/5.2	1.5/2.54/1.16	15/13/20	100/100/240	4.2/2.9/4.5
fabricated circuit	2.36/3.51/5.14	2.3/3/1.9	18/13/14	110/120/270	4.6/3.43/5.25

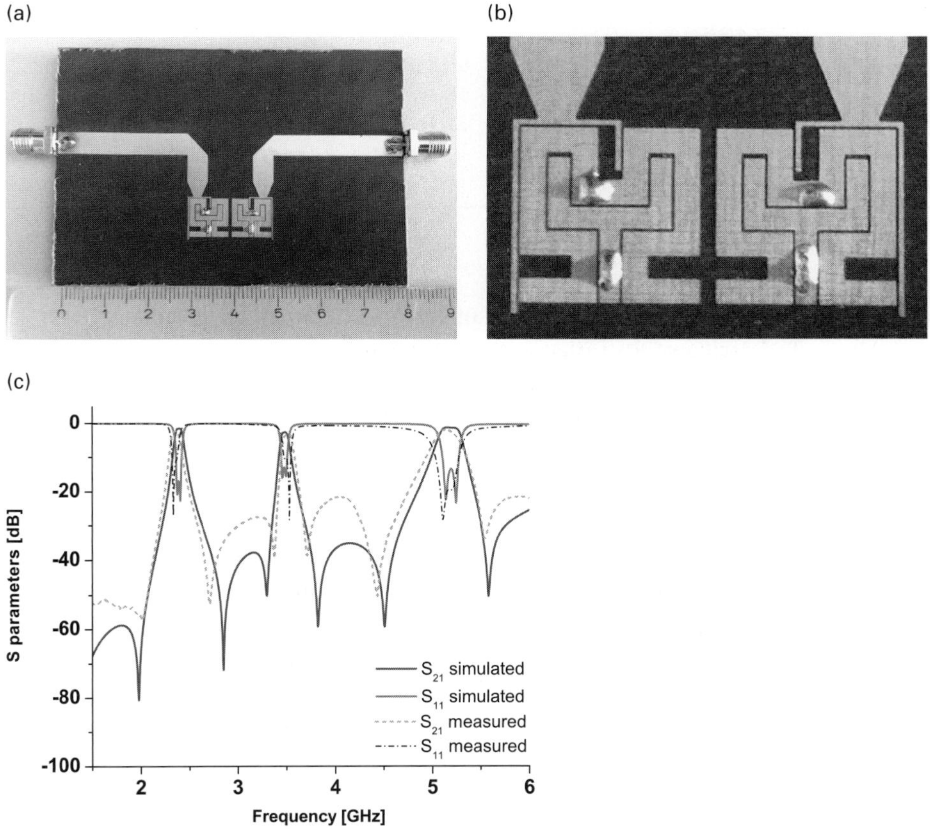

Figure 6.32 Tri-band bandpass filter: (a) Photograph of the fabricated circuit. (b) Enlarged detail of the photograph (a). (c) Simulated and measured responses.

To conclude, the structure presented in this section represents the most compact planar tri-band filter with excellent performance and the possibility of independent control of the passbands.

6.4.3 Tri-band bandstop filter based on Hilbert-fork resonator

In this section, it will be shown how the proposed resonator can be used in the design of a tri-band bandstop filter. The proposed bandstop filter has been designed to operate at 2.4, 3.5, and 5.2 GHz, and its behaviour and characteristics will be discussed in detail below.

Table 6.12 Comparison of the characteristics of the proposed filter and other tri-band bandpass filters

	$f_1/f_2/f_3$ [GHz]	IL [dB]	RL [dB]	transmission zeros	dimensions [λ_g x λ_g]	3 dB FBW [%]
proposed filter	2.4/3.5/5.2	1.5/2.54/1.16	15/13/20	7	0.21x0.11	4.2/2.9/4.5
[32]	2.4/4.8/7.2	1.7/0.9/0.7	15/23/18	5	0.37x0.36	5.3/6.3/8.7
[33]	1.5/2.4/3.5	0.77/1.51/1.8	31/29/28	6	0.35x0.25	12.5/8/6
[34]	2.4/3.5/5.25	1.2/1.1/1.5	16.5/18/14.5	6	0.27x0.21	5/3.7/4.2
[35]	2.4/3.5/5.8	0.8/1.6/1.8	18/16/21	6	0.37x0.297	7.59/5.9/3.7
[36]	1.57/2.4/3.5	1.6/1.5/2.3	9/18.9/13.5	5	0.39x0.35	5.2/3.8/4.6
[37]	2.4/3.5/5.25	2/2.4/1.7	18/16/13	5	0.27x0.25	2.5/1.7/5
[38]	1.8/2.5/3	0.9/1.6/0.8	21/16/17	5	0.27x0.22	5.55/4.17/6.67
[39]	2.4/3.5/5.25	1.9/1.42/1.51	14.3/15/16.8	4	0.205x0.117	6.2/12.2/11.8
[40]	2.4/3.5/5.7	1/1.2/1.5	18.5/25/20	5	0.26x0.22	9.2/6.3/4.2

6.4.3.1 Configuration and design of tri-band bandstop filter

The design of the tri-band bandstop filter is very similar to the design of the bandpass filter. In the first step, the geometrical parameters of the resonator are determined to obtain the resonances at the prescribed frequencies. Since the bandstop filter operates at the same frequencies as the bandpass one, the initial Hilbert-fork resonator is the same as the one presented in the previous section.

However, the coupling mechanism between the resonators and the conductive line in the bandstop filter is different from that in the bandpass filter and thus the parameter optimization result is different. In other words, the optimization procedure in the case of the bandstop filter has resulted in resonators different from those in the bandpass filter.

Namely, in the case of the bandstop filter the parameter d_F is equal to 2.3 mm, which is a slight discrepancy from the parameter value in the bandpass filter.

The configuration of the final tri-band bandstop filter is shown in Figure 6.33. The filter consists of two resonators coupled to the microstrip line through a gap of 0.1 mm. The distance between the resonators is equal to 3.3 mm.

The filter response is shown in Figure 6.34 and its characteristics are given in Table 6.13. The filter is characterized by return losses of 17.23, 24.6, and 24.48 dB, and very compact dimensions of $0.22\lambda_g$ x $0.1\lambda_g$.

So far, there has not been published a tri-band bandstop filter that operates at 2.4, 3.5, and 5.2 GHz. What is more, there have been published only several tri-band bandstop filters [41–43]. However, their performance and overall dimensions are far from those of the proposed bandstop filter.

6.4.3.2 Fabrication and measurement results

Similarly to the bandpass filter, the proposed bandstop filter has been fabricated using standard printed circuit board technology and has been realized on the substrate Rogers RT/Duroid 5880 whose thickness is equal to 1.575 mm, dielectric constant $\varepsilon_r = 2.2$, and loss tangent $\tan\delta = 0.0009$. Figure 6.35(a) shows a photograph of the fabricated filter

Table 6.13 Characteristics of the tri-band bandstop filter

	f [GHz]	IL [dB]	RL [dB]	10 dB BW [MHz]	10 dB FBW [%]	overall dimensions [$\lambda_g x\ \lambda_g$]
first passband	2.4	17.23	0.79	80	3.33	
second passband	3.5	24.6	1.36	85	2.43	0.22x0.1
third passband	5.2	24.84	1.58	130	2.5	

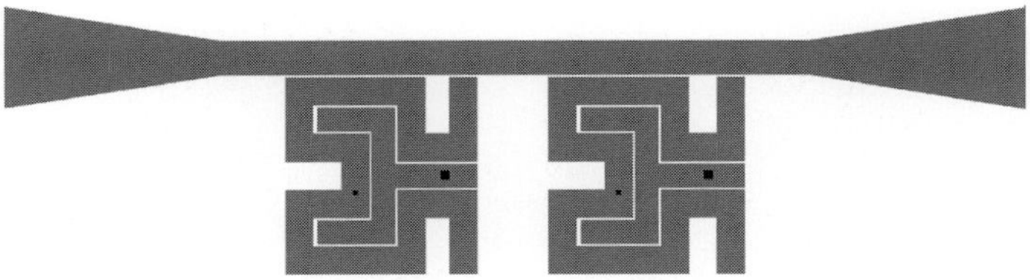

Figure 6.33 Configuration of the tri-band bandstop filter.

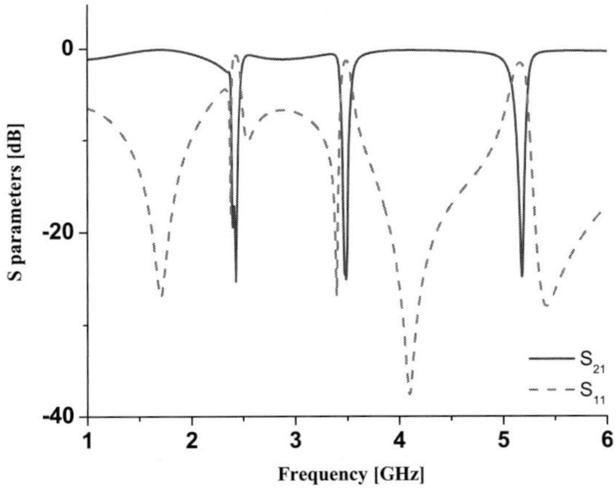

Figure 6.34 Response of the tri-band bandstop filter.

and Figure 6.35(c) shows a comparison between the responses of the simulated and measured circuit.

Similar to the bandpass filter, the bandstop filter comprises four vias and thus its fabrication is demanding since the positions and properties of the two stopbands depend on the vias and their inductances. However, Figure 6.35(c) shows that there is good agreement between measured and simulated results except for the very small discrepancy between the positions of the first and third stopbands which is due to the fabrication tolerance of the vias.

The comparison between the characteristics of the simulated and measured circuits given in Table 6.14 also confirms the good agreement between the simulated and measured results.

Table 6.14 Characteristics of the simulated and fabricated tri-band bandstop filter

	$f_1/f_2/f_3$ [GHz]	IL [dB]	RL [dB]	10 dB BW [MHz]	10 dB FBW [%]
simulated circuit	2.4/3.5/5.2	17.23/24.6/24.48	0.79/1.36/1.58	80/85/130	3.33/2.43/2.5
fabricated circuit	2.36/3.48/5.19	14.4/26.3/34.6	2/2.9/3.7	85/85/145	3.6/2.43/2.79

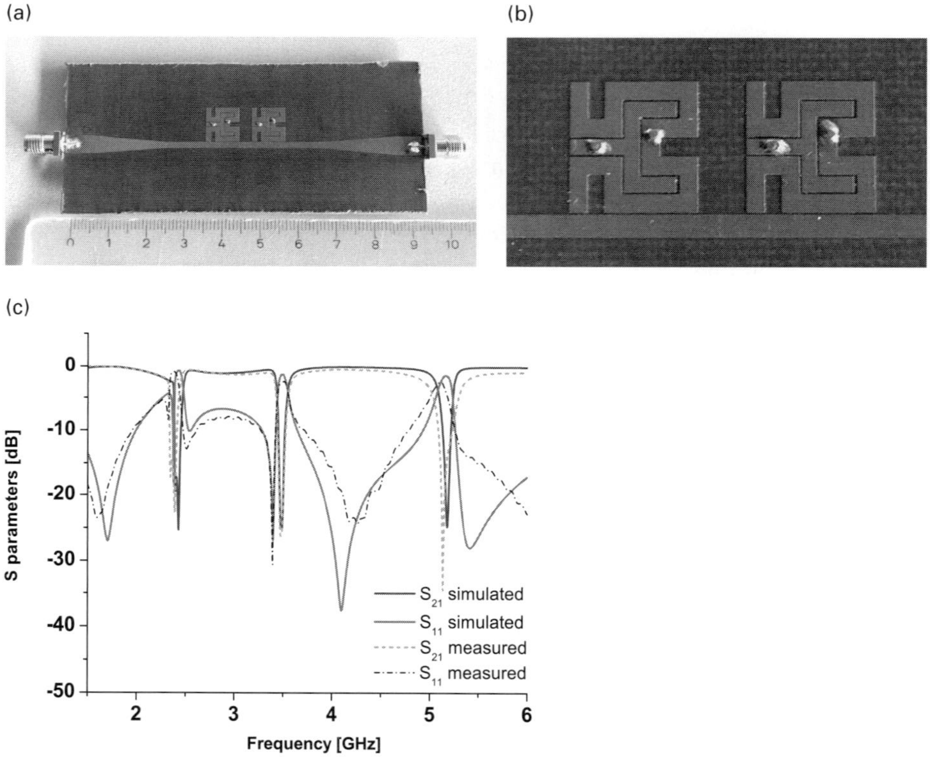

Figure 6.35 Tri-band bandstop filter: (a) Photograph of the fabricated circuit. (b) Enlarged detail of the photograph (a). (c) Simulated and measured responses.

6.5 Conclusion

In this chapter, the applicability of fractal curves to microwave filters has been presented and discussed. The main characteristics of fractal curves were analysed and special attention was given to the Hilbert fractal curve, which is very convenient for application in resonators and filters due to its configuration.

Several multi-band filters based on the Hilbert fractal curve were presented. Firstly, a dual-mode resonator based on a two-dimensional Hilbert fractal curve of the second order was presented. The proposed resonator consists of two serially connected Hilbert fractal curves and there is no main signal path in the resonator, but the signal is rather divided into two paths. Using even-/odd-mode analysis, it was shown that the odd resonance occurs at a lower frequency in comparison to the even resonance and that the resonator provides a transmission zero.

More important, it was shown that the two resonances can be independently controlled due to the specific configuration of the resonator. This property makes the resonator a good candidate for dual-band filter design, and thus two dual-band filters operating at 2.4 and 3.5 GHz were proposed.

The first filter consists of two edge-coupled dual-mode resonators. By extension of the feeding lines, a transmission zero was introduced which significantly improved selectivity of the filter. The configuration is characterized by low insertion loss, good return loss, and compactness.

The second configuration consists of two resonators that are coupled at the broad side. It was shown that such a configuration acts as an antiparallel structure and thus provides a transmission zero below the first passband. In that way, a dual-band filter with excellent selectivity, small dimensions, and good performance was realized.

The proposed filters were fabricated using standard printed circuit technology, and there was good agreement between the simulated and measured results.

The filters were compared to other published filters operating at 2.4 and 3.5 GHz. Insertion loss and overall dimensions are comparable to those of other filters. Nevertheless, the proposed filters are characterized by the fact that they allow independent control of the passbands, as well as by the simple fabrication process that does not require multi-layer structure or vias.

Afterwards, the Hilbert-fork multi-mode resonator that consists of two dual-mode resonators was presented. One of the resonators is based on the two-dimensional Hilbert fractal curve of the second order, whilst the other resonator has the shape of a fork and is positioned to fill the remaining footprint area of the Hilbert resonator. Both resonators are grounded by vias, which enables dual-mode behaviour and structure size reduction.

Using the proposed resonators, tri-band bandpass and bandstop filters operating at 2.4, 3.5, and 5.2 GHz have been realized. Both resonators comprise two Hilbert-fork resonators. A detailed design procedure of the resonator and filters was given.

Owing to the configuration and various couplings in the structure, the proposed bandpass filter has very good performance and very compact dimensions. In comparison to other tri-band bandpass configurations, the proposed structure is the most compact one and has significantly better selectivity, as well as the possibility to independently control the passbands.

The proposed tri-band bandstop filter is also characterized by good performance and represents the first tri-band bandstop filter operating at 2.4, 3.5, and 5.2 GHz and is one of the few tri-band bandstop filters proposed so far.

Both filters have been fabricated using standard printed circuit technology and the comparison between the simulated and measured results show good agreement.

References

[1] G. Peano, "Sur une courbe qui remplit toute une aire plaine," *Mathematische Annalen*, 36, 1890.

[2] K. J. Vinoy, K. A. Jose, V. K. Varadan, and V. V. Varadan, "Hilbert curve fractal antenna: a small resonant antenna for VHF/UHF applications," *Microwave and Optical Technology Letters*, vol. 29, pp. 215–219, 2001.

[3] J. Anguera, C. Puente, E. Martı´nez, and E. Rozan, "The fractal Hilbert monopole: a two-dimensional wire," *Microwave and Optical TechnologyLetters*, vol. 36, pp. 102–104, 2003.

[4] J. Zhu, A. Hoorfar, and N. Engheta, "Bandwidth, cross-polarization, and feed-point characteristics of matched Hilbert antennas," *IEEE Antennas Wireless Propagation Letters*, vol. 2, pp. 2–5, 2003.

[5] J. Zhu, A. Hoorfar, and N. Engheta, "Peano antennas," *IEEE Antennas Wireless Propagation Letters*, vol. 3, pp. 71–74, 2004.

[6] W. Chen, G. Wang, and C. Zhang, "Small-size microstrip patch antennas combining Koch and Sierpinski fractal shapes," *Antennas and Wireless Propagation Letters*, vol. 7, pp. 738–741, July 2008.

[7] L. Yousefi and O. M. Ramahi, "Miniaturised antennas using artificial magnetic materials," *Electronics Letters*, vol. 46, pp. 816–817, June 2010.

[8] J. McVay, N. Engheta, and A. Hoorfar, "High-impedance metamaterial surfaces using Hilbert-curve inclusions," *IEEE Microwave and Wireless Components Letters*, vol. 14, no. 3, pp. 130–132, March 2004.

[9] J. McVay, A. Hoorfar, and N. Engheta, "Peano high-impedance surfaces," *Radio Science,* 40, RS6S03, 2005.

[10] J. McVay, N. Engheta, and A. Hoorfar, "Numerical study and parameter estimation for double-negative metamaterials with Hilbert-curve inclusions," *Proceedings of 2005 IEEE Antennas Propagation Society International Symposium*, Washington, DC, vol. 2B, pp. 328–331, July 2005.

[11] J. McVay, A. Hoorfar, and N. Engheta, "Space-filling curve RFID tags," *Proceedings of IEEE Radio and Wireless Symposium*, San Diego, CA, pp. 199–202, Jan. 2006.

[12] N. Janković, V. Radonić, and V. Crnojević-Bengin, "Novel bandpass filters based on grounded Hilbert fractal resonators," *3rd International Congress on Advanced Electromagnetic Materials in Microwaves and Optics*, London, Aug. 2009.

[13] W. Chen and G. Wang, "Effective design of novel compact fractal-shaped microstrip coupled line bandpass filters for suppression of the second harmonic," *Microwave and Wireless Component Letters*, vol. 19, pp. 74–76, Feb. 2009.

[14] D. Oloumi, A. Kordzadeh, and A. Neyestanak, "Size reduction and bandwidth enhancement of a waveguide bandpass filter using fractal-shaped irises," *Antennas and Wireless Propagation Letters*, vol. 8, pp. 1214–1217, Oct. 2009.

[15] H. Xu, G. Wang, and C. Zhang, "Fractal-shaped UWB bandpass filter based on composite right/left handed transmission line," *Electronics Letters*, vol. 46, pp. 285–287, Feb. 2010.

[16] P. Jarry and J. Beneat, *Design and Realization of Miniaturized Fractal RF and Microwave Filters*, John Wiley, New Jersey, 2009.

[17] M.-H. Weng, D.-S. Lee, R.-Y. Yang, H.-W. Wu, and C.-L. Liu, "A Sierpinski fractal-based dual-mode bandpass filter," *Microwave and Optical Technology Letters*, vol. 50, no. 9, pp. 2287–2289, Sept. 2008.

[18] M.-H. Weng, L.-S. Jang, and W.-Y Chen, "A Sierpinski-based resonator applied for low loss and miniaturized bandpass filters," *Microwave and Optical Technology Letters*, vol. 51, no. 2, pp. 411–413, Feb. 2009.

[19] F. Tong and H. Liu, "Fractal-shaped microstrip dual-mode bandpass filter with asymmetrical sinuous spurlines," *Microwave and Optical Technology Letters*, vol. 51, no. 3, pp. 745–747, March 2009.

[20] J.-C. Liu, S.-H. Chiu, C.-P. Kuei, and B.-H. Zeng, "A novel Minkowski-island-based fractal patch for dual-mode and miniaturization band-pass filters," *Microwave and Optical Technology Letters*, vol. 53, no. 3, pp. 594–597, March 2011.

[21] K.-D. Yeh, H.-H. Liu, and J.-C. Liu, "An equivalent circuit model for the wide-band bandpass filter with the modified Minkowski-island-based fractal patch," *International Journal of RF and Microwave Computer-Aided Engineering*, vol. 24, no. 2, pp. 170–176, March 2014.

[22] J. An, G.-M. Wang, W.-D. Zeng, and L.-X. Ma, "Composite right/left-handed transmission line based on Koch fractal shape slot in the ground UWB filter application," *Microwave and Optical Technology Letters*, vol. 51, no. 9, pp. 2160–2163, Sept. 2009.

[23] J.-C. Liu, P.-C. Lu, C.-H. Chien, J.-M. Chang, J.-Y. Chen, and H.-J. Lin, "Cross-coupled Hilbert spiral resonator for bandpass filter design," *Microwave and Optical Technology Letters,*vol. 49, no. 8, pp. 1890–1894, Aug. 2007.

[24] K. Zemlyakov and V. Crnojevic-Bengin, "Planar low-pass filters based on Hilbert fractal," *Microwave and Optical Technology Letters,* vol. 54, no. 11, pp. 2577–2581, Nov. 2012.

[25] V. Crnojević-Bengin, "Novel compact microstrip resonators with multiple 2-D Hilbert fractal curves," *Microwave and Optical Technology Letters,*vol. 48, no. 2, pp. 270–273, Feb. 2006.

[26] V. Crnojević-Bengin, V. Radonić, and B. Jokanović, "Fractal geometries of complementary split-ring resonators," *IEEE Transactions on Microwave Theory and Techniques,* vol. 56, no. 10, pp. 2312–2321, Oct. 2008.

[27] V. Crnojević-Bengin, K. Zemlyakov, N. Janković, and I. Vendik, "Dual-band bandpass filters based on dual-mode Hilbert fractal resonator," *Microwave and Optical Technology Letters,* vol. 55, no. 7, pp 1440–1443, July 2013.

[28] C. Karpuz and A. Gorur, "Dual-mode dual-band microstrip filters," *Proceedings of European Microwave Conference*, 2009, pp. 105–108.

[29] B. Wu, C. Liang, Q. Li, and P. Qin, "Novel dual-band filter incorporating defected SIR and microstrip SIR," *IEEE Microwave and Wireless Components Letters*, vol. 18, no. 6, pp. 392–394, June 2008.

[30] L. Ren and H. Huang, "Dual-band bandpass filter based on dual-plane microstrip/interdigital DGS slot structure," *IET Electronics Letters*, vol. 45, no. 21, pp. 1077–1079, Oct. 2009.

[31] N. Janković, R. Geschke, and V. Crnojević-Bengin, "Compact tri-band bandpass and bandstop filters based on Hilbert-fork resonators," *IEEE Microwave and Wireless Components Letters*, vol. 23, no. 6, pp. 282–284, June 2013.

[32] S. Luo, L. Zhu, and S. Sun, "Compact dual-mode triple-band bandpass filters using three pairs of degenerate modes in a ring resonator," *IEEE Transactions on Microwave Theory and Techniques,* vol. 59, no. 5, pp. 1222–1229, May 2011.

[33] C. Hsu, C. Lee, and Y. Hsieh, "Tri-band bandpass filter with sharp passband skirts designed using tri-section SIRs," *IEEE Microwave and Wireless Components Letters*, vol. 18, no. 1, pp. 19–21, Jan. 2008.

[34] M. Doan, W. Che, and W. Feng, "Tri-band bandpass filter using square ring short stub loaded resonators," *IET Electronics Letters*, vol. 48, no. 2, pp. 106–107, Jan. 2012.

[35] Q. Yin, L. Wu, L. Zhou, and W. Yin, "A tri-band filter using tri-mode stub-loaded resonators (SLRs)," *Proceedings of IEEE Electrical Design of Advanced Packaging & Systems Symposium*, 2010, pp. 1–4.

[36] W. Chen, M. Weng, and S. Chang, "A new tri-band bandpass filter based on stub-loaded step-impedance resonator," *IEEE Microwave and Wireless Components Letters*, vol. 22, no. 4, pp. 179–181, April 2012.

[37] F. Chen and Q. Chu, "Design of compact tri-band bandpass filters using assembled resonators," *IEEE Transactions on Microwave Theory and Techniques,* vol. 57, no. 1, pp. 165–171, Jan. 2009.

[38] X. Zhang, Q. Xue, and B. Hu, "Planar tri-band bandpass filter with compact size," *IEEE Microwave and Wireless Components Letters*, vol. 20, no. 5, pp. 262–264, May 2010.

[39] Q. Chu, X. Wu, and F. Chen, "Novel compact tri-band bandpass filter with controllable bandwidths," *Microwave and Optical Technology Letters,* vol. 21, no. 12, pp. 655–657, Dec. 2011.

[40] M. Weng, H. Wu, K. Shu, J. Chen, R. Yang, and Y. Su, "A novel triple-band bandpass filter using multilayer-based substrates for WiMAX," *Proceedings of European Microwave Conference*, 2007, pp. 325–328.

[41] K. Singh, K. Ngachenchaiah, D. Bhatnagar, and S. Pal, "Wideband, compact microstrip band stop filter for triband operations," *Proceedings of International Conference on Recent Advances in Microwave Theory and Applications*, 2008, pp. 96–98.

[42] Z. Wang, F. Nasri, and C.-W. Park, "Compact tri-band notched UWB bandpass filter based on interdigital hairpin finger structure," *Proceedings of IEEE 12th Annual Wireless and Microwave Technology Conference*, 2011, pp. 1–4.

[43] J.-K. Xiao and W.-J. Zhu, "New defected microstrip structure bandstop filter," *Proceedings of Progress In Electromagnetics Research Symposium*, 2011, pp. 1471–1474.

7 Multi-band microstrip filters based on near-zero metamaterials

Vesna Crnojević-Bengin, Norbert Cselyuszka, Nikolina Janković, and Riana Helena Geschke

7.1 Introduction

In Chapter 2, four typical approaches to the design of multi-band filters were identified, based on (i) classical filter design, (ii) introduction of transmission zeros in the response of a wideband single-band filter, (iii) use of multi-mode resonators that support several resonant modes in one structure, and (iv) superposition, where two or more filters are combined to obtain a multi-band response. Various state-of-the-art multi-band filters designed using the above approaches have been shown in consecutive Chapters 3 to 6.

Although allowing various degrees of design freedom in terms of independent control of passbands, and exhibiting different relative weaknesses and strengths, all of these approaches have one thing in common: resonators which form filters are regarded as electronic components, and then suitably coupled to result in a desired filtering response.

In this chapter, we propose an alternative view to the design of electrical circuits, multi-band filters in particular, where the resonator is not considered as a component but rather as a piece of material tailored to exhibit specified filtering properties. Namely, we show how artificial electromagnetic materials, also called metamaterials, can be used to design multi-band filters with novel properties – small size, low losses, and reduced group delay. To achieve such behaviour, we will focus on a specific sub-class of metamaterials, called near-zero metamaterials, which have recently been shown to support propagation of electromagnetic waves with the zero propagation constant, giving rise to various interesting phenomena.

In the first part of this chapter, a short introduction to metamaterials is given, followed by a more detailed description of the near-zero metamaterials presented so far in the literature. In the remaining part of this chapter, it is shown how near-zero propagation can be achieved in quasi-TEM circuits, which is very much needed from the point of view of design of microstrip circuits. Near-zero propagation in quasi-TEM circuits is finally used to propose an entirely new class of near-zero bandpass filters, characterized by significantly reduced variation of group delay in comparison to conventional solutions. In the last part of this chapter, multi-band microstrip filters based on near-zero metamaterials are presented, characterized by independent control of the passbands.

7.2 A short introduction to metamaterials

Propagation of an electromagnetic wave in a material is characterized by two material parameters: its dielectric permittivity ε and magnetic permeability μ. Alternatively, another pair of parameters can be used to describe a material from the electromagnetic point of view: its refractive index n and characteristic impedance Z. In any case, a relatively narrow range of values of ε and μ (n and Z) can be found in naturally occurring materials, thus limiting their practical application. For example, relative permittivity ε_r of existing materials varies between 1 and several hundreds, and no natural material exhibits either extremely low (close to zero) or extremely high permittivity. However, a material with extremely low ε_r would allow the design of printed narrow-beam antennas with only one source element, whilst a high-permittivity material would be desirable as a dielectric substrate for the design of miniature planar passive devices, since the guided wavelength in such structures would be reduced by a factor of $\sqrt{\varepsilon_r}$.

In recent years, artificial electromagnetic materials, often called metamaterials, have been proposed as a solution to overcome the problem of the limited characteristics of natural materials. Metamaterials are artificial structures that are comprised of inclusions (unit cells) whose individual dimensions are much smaller than the wavelength of the propagating electromagnetic wave. Since the phase shift across such a sub-wavelength inclusion is extremely small, the whole structure can be considered homogenous. Consequently, effective permittivity and permeability of a metamaterial can be determined, which primarily depend on the shape and mutual geometrical arrangement of the unit cells. Although metamaterials are made of conventional dielectrics and metals, owing to their internal structure they can exhibit unusual values of effective permittivity and permeability, including very high, very low, or even negative values at a certain frequency range.

Although this might seem counterintuitive on a first glance, we have to remember that conventional permittivity and permeability in naturally occurring homogenous materials are also effective parameters, since they are determined for a piece of a material of certain (relatively large) dimensions and they also depend on the inner structure of a material. For example, although both are comprised of hydrogen and oxygen molecules, water and ice exhibit very different permittivity, due to very different inner structures. In a similar manner, design of metamaterials can be viewed as an attempt to mimic nature and to tailor the effective parameters of a metamaterial by careful engineering its internal parts – the unit cells.

Depending on the values of the effective permittivity and permeability, metamaterials can be divided into several classes: single-negative metamaterials (SNG), where one of the constitutive parameters is negative and the other is positive; double-negative metamaterials (DNG), where both constitutive parameters are negative; double-positive metamaterials (DPS), where conventional naturally occurring materials belong; and near-zero metamaterials (NZ), where one or both parameters are equal to zero. As the propagation constant is defined as $\beta = \omega\sqrt{\mu\varepsilon}$, we can easily conclude that SNG metamaterials will support only evanescent waves, as their propagation constant would be imaginary. DNG metamaterials were introduced in [1], where Veselago demonstrated that the propagation constant in such materials would be real but negative, which is not a characteristic of any material found in nature. The direction of the wave vector in such materials would be opposite to the conventional case, resulting in phase velocity antiparallel to the direction of energy propagation, i.e. antiparallel to the group velocity. DPS materials are of course characterized by a positive and real propagation constant, while, as we will show later, a propagation constant in NZ metamaterials can be equal to zero at a non-zero frequency, which again is not a characteristic of any material found in nature.

Although SNG materials actually exist in nature – noble metals such as silver and gold exhibit negative permittivity at THz frequencies whilst ferrites exhibit negative permeability at high frequencies, they are not readily available at arbitrary (lower) frequencies. Thus, Veselago's ideas were not exploited before the late nineties.

Following the line of reasoning based on the internal structure of good conductors, Pendry *et al.* presented in 1996 an artificial material that exhibits negative effective permittivity at microwave frequencies [2].

Three years later, Pendry proposed a structure that consists of metallic split-ring resonators with sub-wavelength dimensions exposed to an axial magnetic field [3]. The effective permeability of such a structure is resonant, and a narrow frequency band exists above the resonant frequency in which the effective permeability is negative, Figure 7.1.

The first experimental demonstration of a DNG metamaterial was presented in [4], based on the two geometries proposed by Pendry – thin copper wires and square

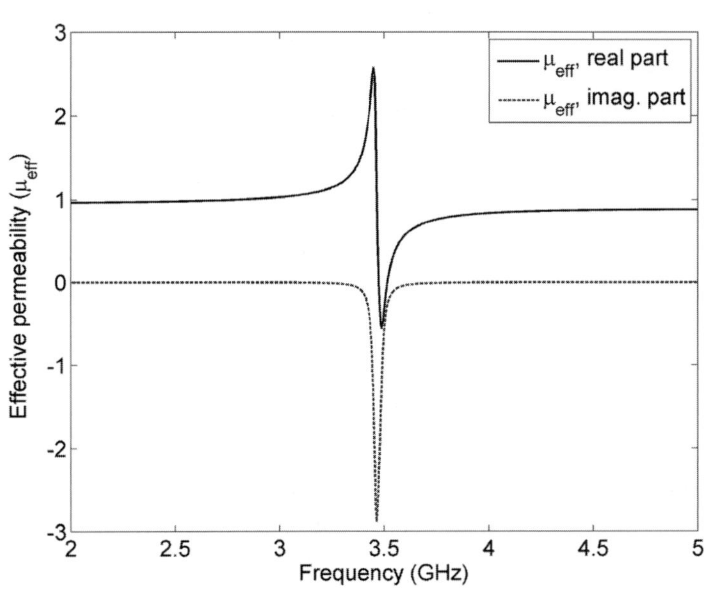

Figure 7.1 Effective permeability of a SRR positioned next to an electrically short 50-Ω microstrip line.

split-ring resonators. It has been shown that negative permittivity and permeability simultaneously occur at the frequency range 10.2–10.8 GHz and the negative refractive index was experimentally demonstrated for the first time. It should be noted here that the passband obtained by simultaneously negative values of effective permittivity and permeability is a fundamentally new type of a passband, with propagation characteristics (value of the propagation constant, phase velocity, etc.) different to any other solution achieved by conventional filtering structures.

In the past decade, significant research efforts have been invested in metamaterials and their application to the realization of devices with small dimensions and/or improved performance. Today, two main approaches to the design of metamaterials can be distinguished: the resonant approach which relies on resonant unit cells such as the split-ring resonator [3, 5], and the transmission-line approach which relies on composite right–left handed (CRLH) lines [6, 7]. Metamaterials have been widely applied in microwave passive circuits, such as antennas [7–11], filters [12–15], and couplers [16–17]. Not only have metamaterials attracted huge attention from the scientific community, they have also entered the commercial market as components of commercial products. One of the first commercial devices to use metamaterial antennas was Netgear 2008 RangeMax wireless routers. The new Rayspan antenna allowed the router to reduce radio frequency interference and boost its range. Also, in 2009 LG Electronics launched LG Chocolate BL40 – the first mobile phone to incorporate metamaterials antenna [18]. As was stated, metamaterials allowed LG to "achieve the dramatically sleek, slim dimensions of the new LG Chocolate and unsurpassed radio-frequency capabilities."

In spite of being very promising for various applications, metamaterials are in principle characterized by two major disadvantages that limit their use in practical engineering systems: significant losses and a narrow operating bandwidth [7]. What is more, the two are not mutually independent. Unlike conventional materials that can be considered dispersionless in a wide frequency range, metamaterials are characterized by a strong dispersion of permittivity and/or permeability. For example, the effective permeability of the split-ring resonator shown in Figure 7.1 follows a typical Lorentz dispersion model. The mathematical description of the Lorentz model, which is also a general model of complex permittivity, is given in the following form

$$\varepsilon = \varepsilon_\infty + \frac{\omega_p^2}{\omega_0^2 - \omega^2 + j\Gamma\omega}, \tag{7.1}$$

where ω is the angular frequency, ε_∞ is permittivity at the infinite frequency, ω_p is the plasma frequency, ω_0 is the resonant frequency, and Γ is the damping factor or collision frequency.

The effective permeability from Figure 7.1 is obviously a resonant phenomenon, due to which losses and the operating bandwidth are mutually dependent: the narrower the bandwidth of the resonance is, the more losses the metamaterial exhibits. Furthermore, losses are an integral part of such resonant phenomena, i.e. they always occur in the frequency range at which the metamaterial exhibits its unusual properties. Also shown in Figure 7.1 is the imaginary part of the effective permeability: in the frequency range at which the real part of the permeability is negative, the imaginary part reaches its maximum value. Therefore, any metamaterial-based device which operates in this frequency region will inevitably exhibit high losses.

On the other hand, metamaterials based on the transmission-line approach obey a simplified Lorentz model, the Drude model, in which the resonant frequency ω_0 is equal to zero. The Drude model also represents a model of permittivity of metals and it is given in the following form

$$\varepsilon = \varepsilon_\infty - \frac{\omega_p^2}{\omega^2 - j\Gamma\omega}. \tag{7.2}$$

It is obvious from (7.2) that at frequencies lower than ω_p, effective permittivity will be negative, but it will not exhibit the resonant nature as was previously the case. A typical example of Drude-type metamaterials are Pendry's first negative permittivity artificial material [2], as well as the composite right–left handed lines [7]. A typical effective permittivity of a Drude-type unit cell is shown in Figure 7.2: although it overcomes the bandwidth-related limitations of the Lorentz-type (resonant) unit cells, large values of the imaginary part of the effective permittivity in the region where the real part is negative will still result in inherently large losses of a final metamaterial-based device.

As a solution to this major drawback, we propose to use metamaterials in a frequency region in which they do not inherently show high losses. In Figure 7.1, it was observed that the imaginary part of the effective permeability is significantly small at the frequency at which the real part of the permeability crosses zero. Metamaterials that exhibit effective permittivity/permeability equal to or close to zero are called

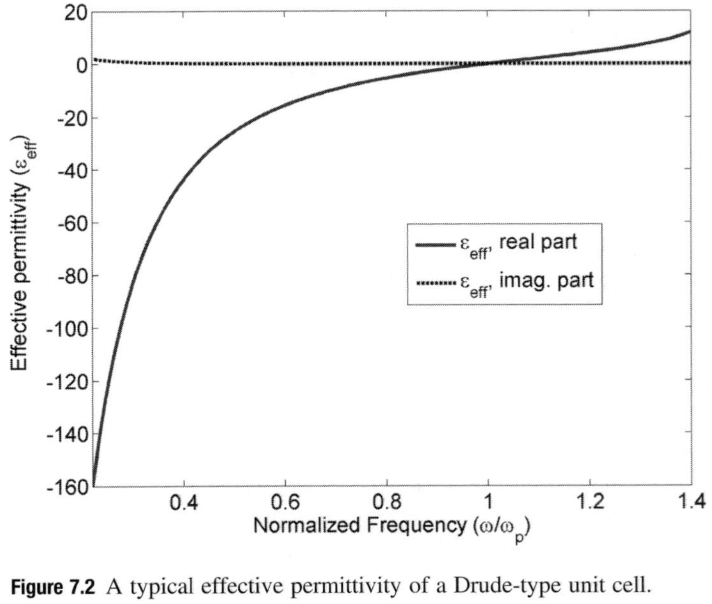

Figure 7.2 A typical effective permittivity of a Drude-type unit cell.

Table 7.1 Some of the most widely used metamaterial unit cells

Dispersion type:	Lorentz	Drude
Negative parameter:		
Permittivity	CSRR	Waveguide section, Shunt inductor
Permeability	SRR	Series capacitor

Epsilon-Near-Zero (ENZ)/Mu-Near-Zero (MNZ) metamaterials. As will be shown in the remaining part of this chapter, such near-zero metamaterials do allow propagation of electromagnetic waves (although with some very peculiar characteristics) and they can be used to design microwave devices which simultaneously exhibit small size and good performance.

To summarize, some of the most widely used metamaterial unit cells are listed in Table 7.1, where their dispersion type is also indicated. We note here that, to obtain a desired behavior, it is of course possible to combine in a single metamaterial unit cells that exhibit dispersion of different types.

7.3 Near-zero metamaterials

Due to sub-wavelength dimensions of constitutive unit cells, the use of metamaterials is obviously a very good approach to the miniaturization of passive microwave devices. On the other hand, as has been shown in the previous section, relatively high losses are inherent to DNG metamaterials, which limits their applicability. However, the effective parameters shown on Figure 7.1 and Figure 7.2 suggest that the losses could be significantly reduced if metamaterials were used in the frequency region where their effective parameters (permittivity and/or permeability) are close to zero. In this section, we analyze in detail such near-zero (NZ) metamaterials.

Furthermore, although NZ metamaterials are very interesting from the aspect of losses, they exhibit some additional interesting properties. Whilst DNG metamaterials support propagation with a negative propagation constant and typically exhibit phase advance, NZ metamaterials make use of the fact that a non-zero frequency exists at which the propagation constant is equal to zero, thus resulting in propagation with infinite phase velocity. (Although such phenomenon cannot be strictly considered as propagation, but rather as tunneling, we will continue to use the term NZ propagation for the sake of simplicity). Infinite phase velocity results in propagation with the constant phase, irrelevant of the distance traveled by the electromagnetic wave through the NZ medium.

Such specific behavior of NZ metamaterials opens up new perspectives for a number of applications: reflectionless sharp bends and corners in transmission lines, matched antennas, novel dielectric sensors, but also passive filters with new topologies, compact dimensions, and high selectivity. In addition, the use of dispersive media and NZ propagation indicates the possibility of realizing devices with reduced group delay.

Namely, as the group delay is defined as the derivative of the phase of the transmission coefficient, $\tau_d = -\frac{d\varphi_{21}}{d\omega}$, in the case of a propagation along the z-axis it is equal to

$$\tau_d = -\frac{d}{d\omega}(-\beta z) = z\frac{d}{d\omega}\left(\omega\sqrt{\mu(\omega)\varepsilon(\omega)}\right) =$$
$$= z\sqrt{\mu(\omega)\varepsilon(\omega)} + z\omega\sqrt{\varepsilon(\omega)}\frac{d}{d\omega}\left(\sqrt{\mu(\omega)}\right) + z\omega\sqrt{\mu(\omega)}\frac{d}{d\omega}\left(\sqrt{\varepsilon(\omega)}\right).$$

If at a certain frequency both permittivity and permeability are equal to (or close to) zero, the group delay would also be equal to (or close to) zero. Alternatively, group delay (and the variation of the group delay) could be controlled by tailoring the slope of the permittivity/permeability curves so as to produce the desired values of the respective derivatives.

Similarly to the case of DNG propagation, passbands obtained due to NZ propagation are fundamentally different than any other type of passbands due to their unique propagation charac teristics (zero value of the propagation constant, infinite phase velocity, etc.). From that point of view, NZ propagation presents a very interesting basis for the design of novel filters.

In principal, the NZ medium can be obtained by either tailoring effective permittivity or effective permeability to be near zero. First NZ metamaterials that were proposed were near-zero-refractive-index materials [19]. Afterwards, ENZ and MNZ metamaterials were introduced.

In the remaining part of this section, recently published ENZ and MNZ metamaterials will be analyzed and various applications of this interesting phenomenon will be discussed.

7.3.1 Permittivity-near-zero case

The first discussion of ENZ metamaterials and their possible application was given in [20], where it was theoretically shown that the ENZ media can be used to "squeeze and tunnel" electromagnetic energy through very narrow waveguide channels, and thus to enhance the efficiency of some waveguide devices and to reduce the reflection coefficient at waveguide junctions. Engheta *et al.* solved a general two-dimensional problem which consists of two parallel-plate metallic waveguides of heights a_1 and a_2, respectively, connected by a channel filled with an ENZ material with the cross-section area A_p. For the wave to tunnel through the channel, it was shown that it is necessary that $a_1 \approx a_2$ and that $k_0\mu_{r,p}A_p << 1$. There are two possibilities to meet the second condition. The first one is to make the permeability of the ENZ channel $\mu_{r,p}$ to be near zero. However, a more interesting possibility is that the reflection vanishes if the area A_p is electrically small, or if one of the physical dimensions of the ENZ channel is electrically small. In other words, as the ENZ channel becomes increasingly narrower, the reflection decreases even more. At the same time, there is no phase change in the ENZ channel, i.e. the phase velocity tends to infinity, and the energy practically squeezes through the channel. This opens up possibilities to realize waveguide bends, which do not exhibit the typically associated losses and reflections [21], as well as to realize the efficient transmission of energy through regions with sub-wavelength cross-sections.

The same authors extended their theoretical investigation in [22] and discussed that the NZ tunneling phenomenon is different from the seemingly similar Fabry–Perot resonance. Namely, Fabry–Perot resonant frequencies are highly dependent on the length of the channel, and their phase and amplitude distributions strongly vary across the channel at the tunneling frequency. On the other hand, the energy can tunnel through the ENZ channel independently of its geometry and length. In other words, ENZ materials support a zero-order resonance characterized by the absence of phase change, which implies that the electrical length of the channel is zero irrelevant of its physical length.

The phenomenon of energy squeezing using ENZ materials at microwave frequencies was experimentally demonstrated in [23]. Since ENZ materials are not readily available in the microwave region, the ENZ medium was designed using a section of a rectangular waveguide. Namely, it is well-known that effective relative permittivity for the TE_{10} guided mode in a rectangular waveguide has the following form

$$\frac{\varepsilon_{eff}}{\varepsilon_0} = n^2 - \frac{c^2}{4f^2w^2}, \tag{7.3}$$

where w is the waveguide width, n is the relative refractive index of the dielectric filling the waveguide, c is the speed of light in the vacuum, and f is the operating frequency. This is indeed the Drude dispersion as given by (7.3) or as shown in Figure 7.2: the effective permittivity of a rectangular waveguide below the cut-off frequency is negative, whilst it is positive above the cut-off. Therefore, at the cut-off frequency the effective epsilon of a waveguide is equal to zero, and such behaviour resembles the response of a plasmonic material.

Following the line of reasoning presented in [20], the structure consisting of the three waveguide sections with the same width was proposed to demonstrate ENZ propagation. To meet the condition of an electrically small cross-section area A_p of the ENZ channel, the height of the middle section was significantly reduced. At the same time, its cut-off frequency is increased by proper choice of a filling dielectric. The TE_{10} mode that is excited in the outer waveguide cannot propagate through the entire structure as it is below the cut-off of the middle section. However, the middle section acts as an ENZ medium at its cut-off frequency, and at that frequency the launched TE_{10} mode is tunneled.

Another experimental demonstration of electromagnetic tunneling through an ENZ metamaterial at microwave frequencies was given in [24]. The proposed structure consists of two planar waveguides separated by a thin ENZ channel that is realized as a channel in which complementary split-ring resonators (CSRRs) are positioned. The three sections support waves that can be considered as transverse electromagnetic waves. The volume bounded by a CSRR surface behaves identically to a volume containing a resonantly dispersive dielectric (i.e. the Lorentz type), and thus this channel acts as an ENZ medium at the frequency slightly above that at which the CSRR exhibits resonance. Experimental measurement and spatial map of the electric field distribution confirmed ENZ propagation in the proposed structure.

Following the idea presented in [23], a microstrip structure that supports ENZ propagation was proposed in [25]. The structure comprises of two conventional microstrip sections, connected by a microstrip section with significantly reduced substrate height. In order to provide energy squeezing at ENZ frequency and reflection at other frequencies, there is a very abrupt discontinuity between the conventional microstrip sections and the section with the reduced height. Although the dominant mode in a microstrip structure is the quasi-TEM mode, wave propagating in the tunnel is converted into a TE_{10} mode, and thus such a structure cannot truly be said to support NZ propagation of the quasi-TEM mode.

Besides energy tunneling, there are other interesting applications of ENZ materials. For example, carefully designed ENZ media can be used to manipulate the phase front of an electromagnetic wave, thus resulting in unusual imaging, lensing, and radiative effects [26]. On the other hand, since energy tunneling is associated with greater enhancement in the magnitude of the electric field inside the ENZ channel, such a high-intensity field ensures high sensitivity to small variations in the material parameters, which is exploited for material sensing applications [27–28].

7.3.2 Permeability-near-zero case

Although a dual to the ENZ phenomenon, MNZ propagation has not been as widely exploited. The main reason for this may be found in the fact that there is no straightforward MNZ equivalent to the ENZ waveguide, i.e. there is no structure that readily exhibits MNZ behavior at a certain frequency, such as a waveguide that exhibits ENZ behavior around its cut-off. Instead, specific unit cells have to be used to provide near-zero values of the effective permeability at certain frequencies.

Most microwave applications of MNZ metamaterials published so far have been related to antenna design. In [29], a flexible dual-frequency broadside radiating rectangular patch antennas based on complementary planar ENZ or MNZ metamaterials were proposed, whilst in [30] an MNZ substrate was shown to have high potential for enhancement of broadside radiation.

As for filtering applications, following the structure proposed in [25], half-mode filters based on ENZ and MNZ concepts were proposed in [31]. The structure consists of an ENZ tunnel, loaded with an interdigital capacitor that provides dispersive behavior of the effective permeability. The structure can be considered a CRLH unit cell that is intentionally designed as unbalanced and which thus provides two tunneling frequencies (one ENZ and one MNZ). Since such a cell acts as a dual-mode resonator, its applicability was demonstrated by designing two-pole and four-pole bandpass filters with good performance. However, the proposed structure again supports NZ propagation of TE_{10} mode, not quasi-TEM.

It should be noted that, although both ENZ and MNZ metamaterials support propagation with a zero propagation constant, a specific interest exists for the design of MNZ media. Namely, since ENZ structures rely on negative permittivity unit cells, their application in filter design is not straightforward. This is due to the fact that most filters use some kind of series capacitors to couple the resonators of the filter. This capacitance acts as a negative permeability unit cell, and, together with the ENZ unit cell, might result in a DNG rather than in the NZ propagation mode. However, this is not the case for MNZ structures which would support a near-zero mode, even when combined with series capacitors of a filter.

7.4 Near-zero propagation in quasi-TEM circuits

7.4.1 Permittivity-near-zero case

The overview presented in Section 7.3 has illustrated that all ENZ structures published so far are in fact based on the propagation of the TE_{10} mode inside a waveguide. Although the structure presented in [25] is a microstrip structure, which is typically associated with the quasi-TEM mode, it ultimately relies on near-zero permittivity at the cut-off frequency of a substrate-integrated waveguide, and therefore does not support NZ propagation of the quasi-TEM mode. The structure based on complementary split-ring resonators [24] supports transversal electromagnetic waves, but ultimately it also relies on an ENZ waveguide. However, NZ propagation of quasi-TEM waves is of great interest for the design of new types of planar filters.

ENZ propagation in conventional quasi-TEM microstrip circuits has for the first time been shown in [32], where a structure that does not comprise narrow channels or tunnels was used, and no conversion of the propagating quasi-TEM to TE_{10} mode was performed. As has been shown in Section 7.3.1, the condition of an electrically small cross-section area of the channel is crucial in TE_{10} ENZ circuits, since the height of a conventional waveguide is comparable to the wavelength of the propagating signal. In the case of conventional microstrip structures, this height corresponds to the thickness of the dielectric substrate, which is typically significantly smaller than the guided wavelength. Thus, microstrip structures do not require additional reduction of the substrate height, since the condition of small cross-section area is always met.

However, whilst transmission lines with a dominant TE_{10} propagating mode inherently exhibit a high-pass response and near-zero permittivity around the cut-off frequency, quasi-TEM microstrip lines are all-pass structures with constant permittivity. Therefore, a microstrip line needs to be loaded with unit cells which will provide near-zero values of effective permittivity at certain frequencies. The most typical metamaterial unit cell that can be used for this purpose is the complementary split-ring resonators (CSRR) etched into the ground plane of a microstrip line [5]. Effective permittivity of a CSRR positioned under a 50-Ohm microstrip line is shown in Figure 7.3, extracted using the procedure shown in [33]. Overall dimensions of the CSRR were arbitrarily chosen to be 4.6 x 4.75 mm, width of the gaps 0.1 mm, and distance between the gaps 0.1 mm. The circuit was designed on the substrate Rogers TMM10 with the thickness of 1.27 mm, dielectric constant $\varepsilon_r = 9.2$, and loss tangent $\tan\delta = 0.0035$. The transmission coefficient of the CSRR-loaded microstrip is also shown in Figure 7.3 and it reveals that the structure behaves as a notch resonator at $f_z = 2.23$ GHz, whilst the CSRR exhibits near-zero permittivity at $f_0 = 2.49$ GHz. In the next section, it will be shown that this

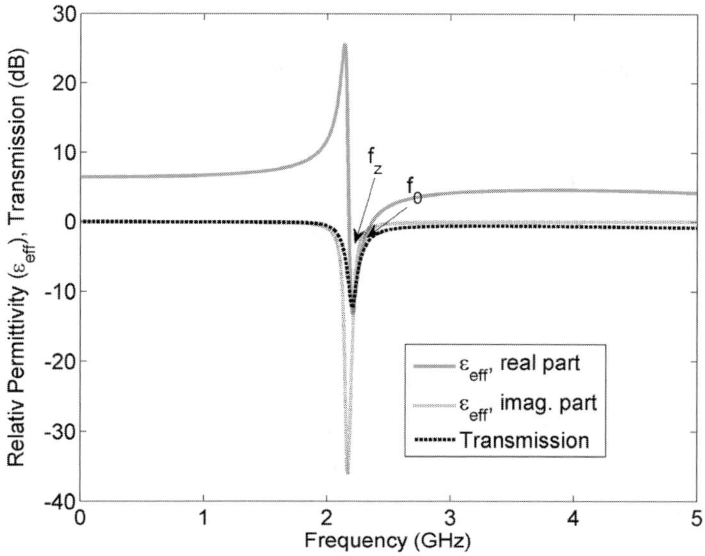

Figure 7.3 Effective permittivity and the transmission coefficient of a CSRR etched in a ground plane of an electrically short 50-Ω microstrip line.

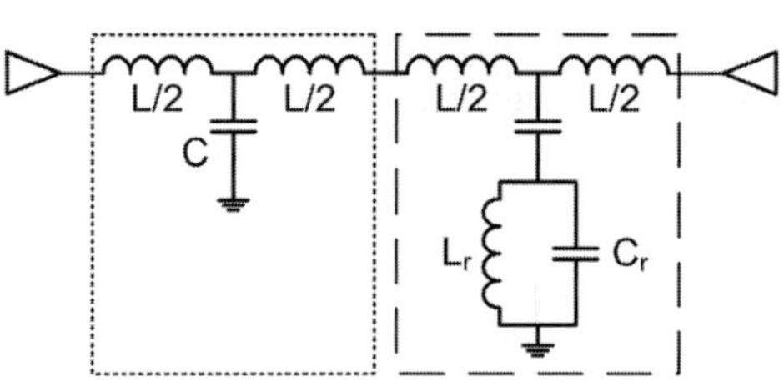

Figure 7.4 Equivalent circuit of a short section of a microstrip line loaded with a CSRR.

frequency is not the only one at which NZ propagation can be supported. The reader is referred to Section 7.4.2 for more details.

The equivalent circuit of the CSRR-loaded microstrip (dashed line) connected to the feed line (dotted line) is shown in Figure 7.4, where L and C denote inductance and capacitance of the 50-Ω microstrip, C_c is the capacitance between the microstrip and the CSRR, and L_r and C_r model the CSRR. When the ring alone is at resonance, i.e. at $f_0 = 1/2\pi\sqrt{L_r C_r}$, it behaves as an open circuit. Effective permittivity of the entire structure is equal to zero and NZ propagation between the ports is possible. Around $f_z = 1/2\pi\sqrt{L_r(C_r + C_c)}$ the structure behaves as a single-negative medium and supports only evanescent waves.

To fully utilize such ENZ behavior, a host medium other than an all-pass 50-Ω microstrip should be used. Namely, to allow impedance matching at NZ frequency, the host microstrip circuit needs to be designed to exhibit a stopband in the region where an ENZ passband is expected to occur. We note here that in the lossless case, ENZ

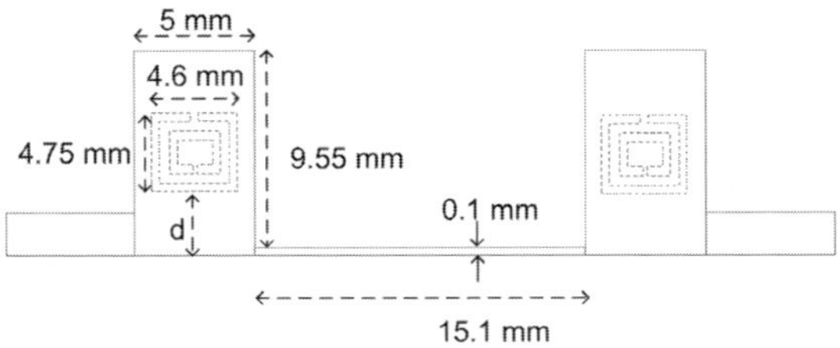

Figure 7.5 Microstrip ENZ structure [32].

metamaterial exhibits infinitely large impedance, since the impedance is inversely proportional to a square root of effective permittivity, which is equal to zero. However, in the real case, the value of the impedance is bounded by losses, so ENZ metamaterial can be matched using practically realizable input microstrip lines.

As a proof-of-concept device, a microstrip ENZ structure shown in Figure 7.5 was proposed in [32]. As a host medium, it uses a simple third-order low-pass filter with open stubs, designed following the standard procedure given in [34]. The capacitive sections of the filter are loaded with one CSRR each, as the axial electric field needed to excite the CSRRs is the strongest there. The dimensions of CSRRs are equal to those given above. The overall length of the structure is 34.4 mm, which corresponds to $0.6\lambda_g$, where λ_g is the guided wavelength at 2 GHz.

The simulated electromagnetic response of the proposed ENZ structure is shown in Figure 7.6 and compared to the response of the same low-pass filter with homogenous ground (i.e. without CSRRs). It can be seen that, due to inclusion of CSSRs, a new passband emerges at the frequency equal to f_0 = 2.27 GHz, determined by the effective permittivity of the CSRR. To validate this, extracted effective permittivity of the CSRR placed at the central position (d = 2.4 mm) under the 9.5 mm wide electrically short capacitive line is shown in the inset of Figure 7.6. It exhibits zero permittivity at exactly 2.27 GHz.

Also shown in Figure 7.6 is the influence of the lateral position of a CSRR, d, which can be used to fine-tune the transmission characteristics of the filter. It should be noted that varying d neither influences the NZ frequency f_0 nor the position of the lower transmission zero f_z, as they are determined by the parameters of the CSRR itself. However, the position of the transmission zero at the upper side of the ENZ passband and thus its bandwidth are strongly affected by the parameter d. For instance, varying d from 0.15 to 2.4 mm changes the fractional bandwidth from 6.7% to 9.3%.

The low-pass filter exhibits a spurious resonance at 3.5 GHz which is not strongly excited in the absence of CSRRs. When CSRRs are introduced into the structure, the resonance is more pronounced and it affects the selectivity of the ENZ passband at its upper side. Nevertheless, this spurious resonance can be successfully supressed by a proper choice of the parameter d, which affects the position of the upper transmission zero. For instance, for d = 2.4 mm, the transmission zero occurs at the frequency of the spurious resonance, which is therefore successfully eliminated.

As was stated previously, a unique property of NZ structures is the fact that they support the propagation of waves with virtually no phase delay, irrelevant of their

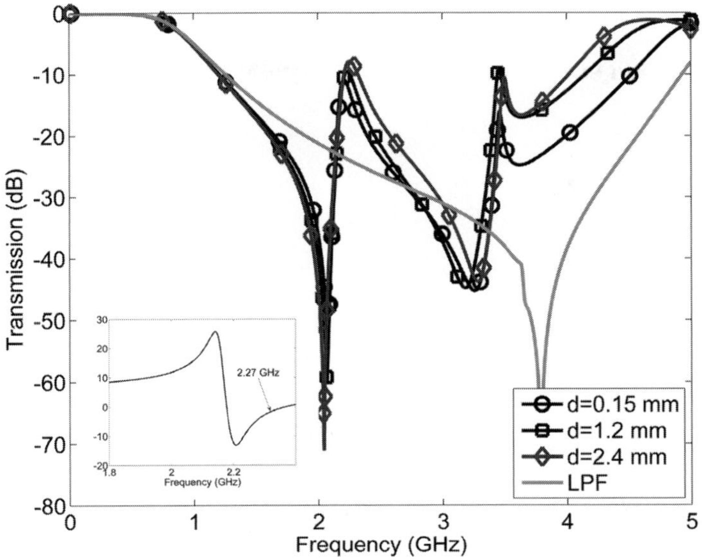

Figure 7.6 Simulated electromagnetic response of the proposed ENZ structure compared to the response of a low-pass filter alone. Extracted effective permittivity is shown in the inset [32].

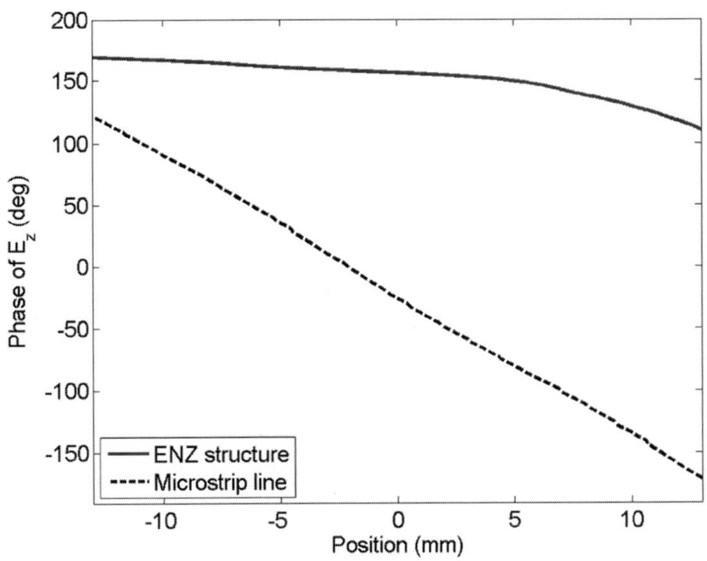

Figure 7.7 Phase distribution along the ENZ structure compared with a 50-Ω microstrip of the same overall length at the central frequency of the ENZ passband.

length. This is validated for the proposed ENZ structure in Figure 7.7, where the phase distribution along the ENZ structure is compared to that of a simple 50-Ω microstrip line of the same physical length. In both cases, the phase has been extracted along a line halfway between the ground plane and the microstrip layer, from the component of the

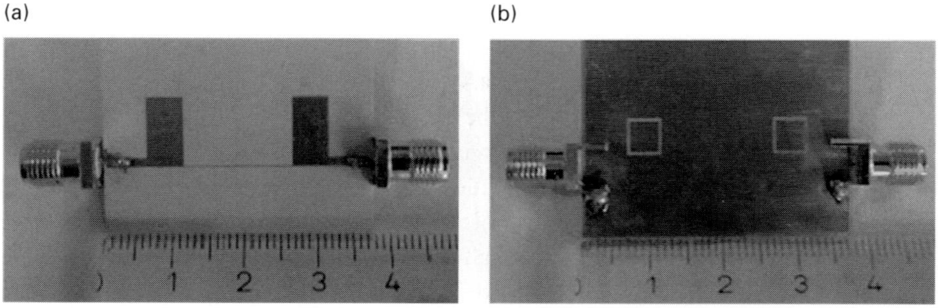

Figure 7.8 Photographs of the fabricated ENZ structure: (a) Top conductive layer. (b) Bottom conductive layer (ground) [32].

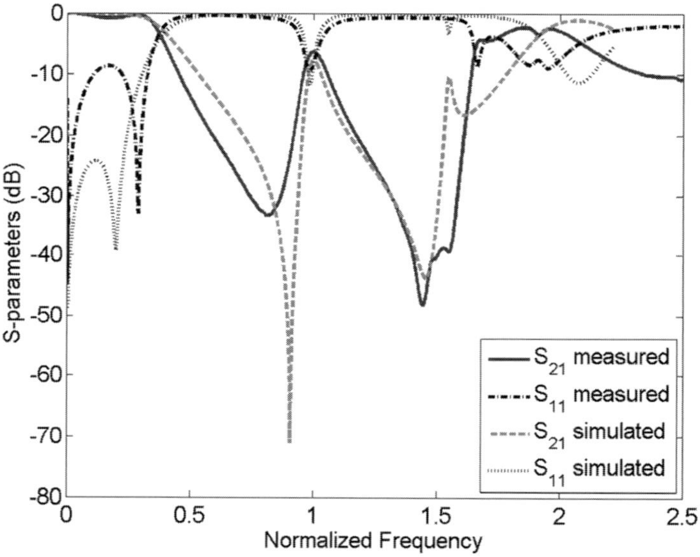

Figure 7.9 Comparison of simulated and measured responses [32].

electrical field normal to the microstrip host and at the ENZ frequency of 2.27 GHz. In Figure 7.7, a zero value of the variable *Position* corresponds to the centre of the whole structure.

Although of the same physical length, the proposed ENZ medium and a 50-Ω microstrip line exhibit very different phase shifts: the phase shift accumulated along the microstrip corresponds well to its length and is equal to 293°, while the phase shift along the ENZ structure equals only 58°, which is consistent with ENZ media.

The proposed ENZ microstrip structure has been fabricated using the conventional printed circuit board technology. Photographs of the fabricated circuit are shown in Figure 7.8, while the simulated and measured responses are compared in Figure 7.9 [32]. Good agreement between the measured and simulated responses can be observed, except for a shift in frequency of approximately 12%. This is explained by the discrepancy between the actual and simulated values of the dielectric constant and the thickness of the substrate, as the manufacturer specifications allow permittivity variations of ± 0.23, as well as variations of substrate thickness.

Finally, we note here that the ENZ structure presented above can also be viewed as a single-band bandpass filter, while the original response of the low-pass filter could be suppressed using some additional circuit elements. Alternatively, a different host structure could be used, which would not exhibit such a low-pass response. For example, the conventional half-wavelength resonator can also be used as a host and loaded with a CSRR [35], provided that the capacitive coupling between the resonator and the feed lines is carefully designed so that the final circuit supports ENZ and not DNG propagation.

7.4.2 Permeability-near-zero case

MNZ propagation in quasi-TEM circuits can be obtained in a manner similar to the one presented above for the ENZ case, with the main difference being the choice of the metamaterial unit cell. To achieve MNZ propagation, a microstrip host needs to be loaded with unit cells which will provide near-zero values of effective permeability at certain frequencies. A widely exploited metamaterial unit cell which provides resonant behavior of the effective permeability is the split-ring resonator (SRR). In the microstrip configuration, SRR is typically positioned next to the host microstrip line so it is properly excited by the axial magnetic field.

Effective permeability and the transmission coefficient of an electrically short section of a 50-Ω microstrip line loaded with one SRR are shown in Figure 7.10. In this example, the SRR has been constructed on a 1.27 mm thick Taconic CER-10 substrate with $\varepsilon_r = 9.2$ and $\tan\delta = 0.0035$, using 0.3 mm lines on 0.25 mm spacing, and with 0.2 mm gaps. When the ring is at resonance, namely at $f_r = 1/2\pi\sqrt{L_r C_r}$, where L_r and C_r model the SRR, it acts as a short circuit and the whole structure behaves as a notch resonator. In other words, around f_r the effective permeability is negative, the structure behaves as a single-negative medium and supports evanescent waves only.

Figure 7.3 and Figure 7.10 fundamentally illustrate the same phenomenon – Lorentz type of dispersion. Apart from the fact that one shows effective permittivity and the other effective permeability, these two figures differ in another important aspect – the amount of losses that have been included in the simulations. More losses have been included in the latter case and one may notice that consequently the extremes of the effective permeability function are not as pronounced as they were in Figure 7.3, and there is a finite slope between the maximum and minimum values. Therefore, in this case two frequencies in the vicinity of f_r may be identified at which MNZ propagation might occur: $f_{01} = 2.39$ GHz and $f_{02} = 2.44$ GHz.

As was discussed above, the main limitation to the wide use of metamaterials is the fact that they are inherently losy. To illustrate this, the imaginary part of the effective permeability is also shown in Figure 7.10. It can be seen that it is very high at f_r; if such a unit cell was used to design a DNG metamaterial, losses would be inherent to the DNG transmission. Let us now analyse two potential NZ frequencies: the imaginary

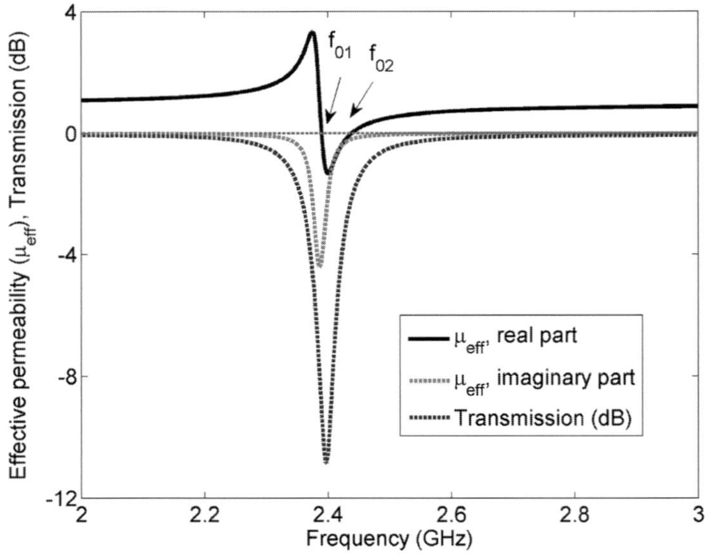

Figure 7.10 Effective permeability and the transmission coefficient of a short section of a 50-Ω microstrip line loaded with one SRR [36].

part is also very high at f_{01} and therefore no propagation occurs at this frequency. At f_{02}, the imaginary part of effective permittivity is smaller, but still non-zero, and consequently the insertion loss is reduced, although it is still not negligible. Therefore, by using a simple microstrip line as a host, MNZ propagation cannot be practically achieved either at f_{01} or at f_{02}.

Similarly to the ENZ case presented in the previous section, a host structure other than an all-pass 50-Ω microstrip line needs to be used. For the proof-of-concept device, let us again use a simple low-pass filter of third order with open stubs (LPF), and load the central section of the LPF with an SRR, designed so that its resonant frequency falls within the stopband of LPF [36]. However, it should be emphasized that the following analysis may be applied to host structures other than LPF.

Let us now analyse two different cases: the first one where LPF consists of two shunt capacitors and one series inductor (denoted here as CLC-LPF), and the second one where the LPF is comprised of two series inductors and one shunt capacitor (LCL-LPF). In both cases, one SRR is placed next to the middle section of the filter, loading the inductive and capacitive section of the filter, respectively. The equivalent electrical circuits of such structures are shown in Figure 7.11. To preserve the symmetry of the circuit, the capacitor of LCL-LPF in Figure 7.11b has been modeled with two capacitors with a capacitance of $C/2$ each.

Simulated lossless responses of both SRR-loaded CLC-LPF and LCL-LPF are compared in Figure 7.12, where also the responses of initial low-pass filters without the SRRs are shown. It can be seen that the inclusion of an SRR in both cases did not influence the original low-pass responses, but that a new transmission peak and transmission zero have occurred in both cases. As expected, the transmission zero

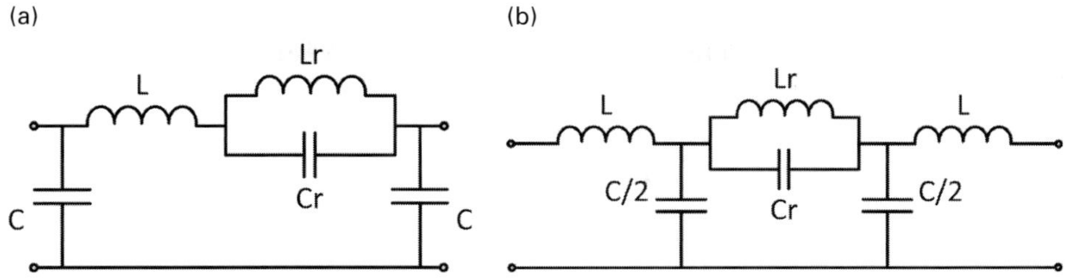

Figure 7.11 The equivalent circuit of the SRR-loaded: (a) CLC-LPF. (b) LCL-LPF [36].

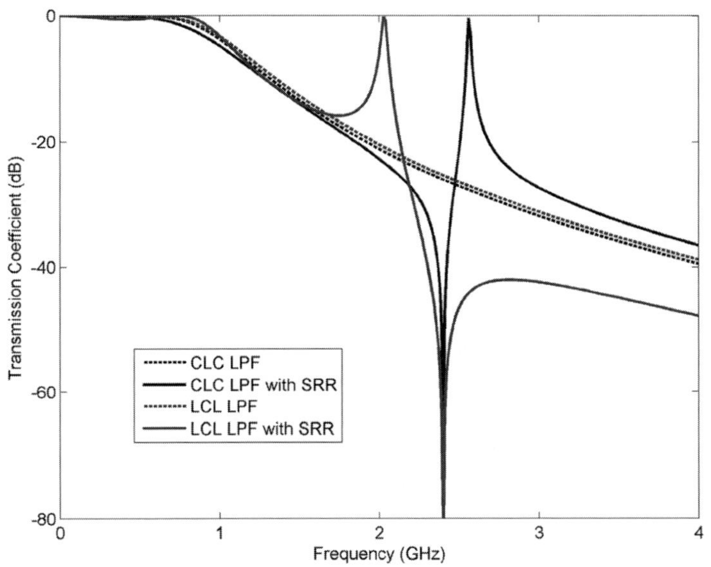

Figure 7.12 Responses of CLC and LCL low-pass filters with and without SRRs coupled to their middle section [36].

always occurs at the same frequency equal to f_r. However, the transmission peak occurs at a frequency higher than f_r for SRR coupled to the inductive section of the host filter, and at a frequency lower than f_r for SRR coupled to the capacitive section of the filter, which might not seem straightforward.

The complex propagation constant $\gamma = \alpha + j\beta$, where α is an attenuation constant and β a phase constant can be determined from the equivalent circuits of Figure 7.11 using the ABCD matrix as $\gamma l = acos(D)$. Let us first note that, in order to determine the effective propagation constants of a circuit, the circuit needs to be considered as homogenous, i.e. its length needs to be significantly smaller than the guided wavelength. Even though the proposed structures might seem electrically long since they are comprised of relatively long low-pass filters, at MNZ frequency they can be regarded as electrically short. This is due to the fact that at the MNZ frequency phase, velocity is infinite and there is no phase change over physically long distances. Therefore, the condition for the electrically short circuit holds even in the case of the proposed, seemingly long, SRR-loaded low-pass filters, and the homogenization process can be applied.

In the case of the SRR-loaded CLC-LPF, D can be obtained from the equivalent circuit of Figure 7.11(a) as

$$D = 1 - \frac{\omega^2 LC}{1 - \omega^2 L_r C_r}\left[1 + \frac{L_r}{L} - \omega^2 L_r C_r\right], \tag{7.4}$$

whilst for the SRR-loaded LCL-LPF, it is equal to

$$D = 1 + \frac{\dfrac{Z_C Z_r}{2Z_C + Z_r} + j\omega L}{\dfrac{Z_C^2}{2Z_C + Z_r}}, \tag{7.5}$$

where $Z_C = 2/j\omega C$ and $Z_r = j\omega L_r/(1\text{-}\omega 2 L_r C_r)$.

The attenuation constant α and phase constant β obtained from the expressions (7.4) and (7.5) are plotted in Figure 7.13. It can be seen that there are two different frequencies, f_{01} and f_{02}, at which the phase constant is equal to zero, and that they are positioned before and after the stopband, respectively. (The stopband is of course characterized by the maximal value of the attenuation constant. The phase constant is not defined in the stopband – in Figure 7.13 this is denoted by the abrupt change of the phase constant from 0 to π rad or vice versa.) It should also be noted that the zero value of the phase constant does not necessarily imply MNZ propagation – this phenomenon

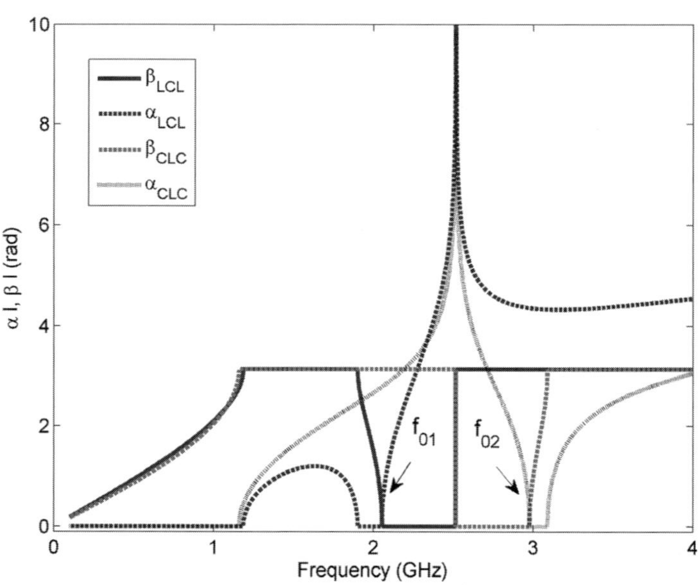

Figure 7.13 Attenuation constant α and phase constant β for the SRR-loaded CLC-LPF and LCL-LPF. The values of circuit elements are chosen as typical for microstrip implementation: $L_{CLC\_LPF} = 5$ nH, $C_{CLC\_LPF} = 5$ pF, $L_{LCL\_LPF} = 6$ nH, $C_{LCL\_LPF} = 6$ pF, $L_r = 2$ nH, $C_r = 2$ pF [36].

will be supported only at the frequencies where the additional condition $D = 1$ is fulfilled, which corresponds to the zero attenuation constant.

By analysing the results presented above, it can be shown that in the case of the SRR-loaded CLC-LPF, the frequency at which MNZ propagation will be supported is f_{02}, which can be obtained from (7.4) by equaling $D = 1$

$$f_{02} = \frac{1}{2\pi\sqrt{L_r C_r}}\sqrt{1+\frac{L_r}{L}}. \tag{7.6}$$

In the case of SRR-loaded LCL-LPF, MNZ propagation will occur at frequency f_{01}, which can be calculated in a similar manner from (7.5)

$$f_{01} = \frac{1}{2\pi\sqrt{\frac{2LL_r}{2L+L_r}\left(C_r+\frac{C}{4}\right)}}. \tag{7.7}$$

MNZ frequencies f_{01} and f_{02} obtained above correspond in fact to the central frequencies of two transmission peaks in Figure 7.12. Equations (7.8) and (7.7) further reveal that MNZ frequency depends not only on the parameters of the unit cell, L_r and C_r, but also on the inductive/capacitive nature of the host, i.e. on L and C.

Moreover, the values of the host inductance and capacitance, L and C, play a very important role in achieving MNZ propagation. Namely, it can easily be noted that the equivalent circuits of low-pass filters shown in Figure 7.11 are identical to those of an electrically short section of an SRR-loaded microstrip line. The only difference is in the values of the parameters L and C – whilst in the case of a low-pass filter L and C are several times larger than L_r and C_r; they are of the same order of magnitude in the case of a 50-Ω microstrip line. Therefore, in order to analyse how the type of host structure affects MNZ propagation, the influence of the parameters L and C needs to be analysed.

It can be seen from equation (7.6) that for a given SRR, i.e. for fixed parameters L_r and C_r, the MNZ frequency of the SRR-loaded CLC-LPF depends only on the parameter L. Furthermore, the MNZ frequency f_{02} is greater than the resonant frequency of the SRR f_r for all values of the host inductance L. Thus, transmission zero will always occur at the lower side of the MNZ transmission peak as was the case in Figure 7.12. Furthermore, if L is increased, f_{02} shifts toward the resonant frequency f_r. On the other hand, if L is decreased to values characteristic of an ordinary microstrip line, f_{02} increases quite rapidly, thus shifting the MNZ frequency out of the range of interest and finally into the range where the second harmonic of the low-pass filter is positioned. For that reason, a simple microstrip line cannot be practically used as a host in MNZ structures, since its inductance L is too low. This is illustrated in Figure 7.14 where the attenuation constant α and phase constant β are plotted for various values of L, for the case of $L_r = 2$ nH and $C_r = 2$ pF. As expected, the resonant frequency f_r is not influenced by varying L, whilst f_{02} increases as L decreases, as denoted by the arrows in Figure 7.14.

A similar analysis can be performed for the case of an SRS-loaded LCL-LPF. As can be seen from equation (7.7), in this case the MNZ frequency depends both on the host inductance L and host capacitance C. Unlike in the case of an CLC-LPF structure in which the transmission zero has always occurred at the lower side of the MNZ

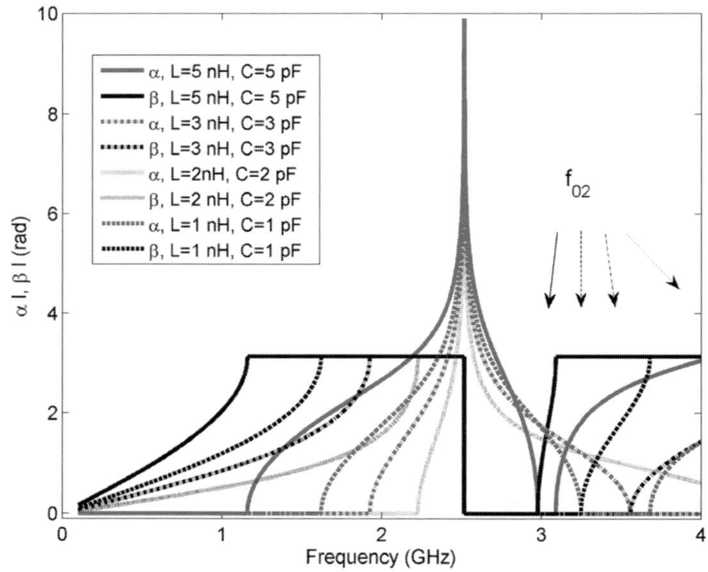

Figure 7.14 Attenuation constant α and phase constant β for various values of the host inductance L in the case of the SRR-loaded CLC-LPF [36].

transmission peak, in the LCL-LPF structure the transmission zero can occur on either side of the transmission peak. For the values of L that correspond to those of a typical low-pass filter, e.g. for L = 5nH and C = 5 pF, f_{01} will be smaller than f_r. This corresponds to a situation in which the MNZ transmission peak occurs below the transmission zero at f_r, as was the case in Figure 7.12. However, if the parameters L and/or C are decreased, the MNZ frequency f_{01} increases and it eventually becomes larger than f_r. This is illustrated in Figure 7.15, where arrows denote the shift of f_{01} with decreasing L and C. As in the previous case, the SRR has been modeled using L_r = 2 nH and C_r = 2 pF. For sufficiently small values of L and C, the MNZ frequency f_{01} behaves as that of the CLC-LPF structure, i.e. both types of low-pass hosts behave as simple microstrip lines not suitable to be used as an MNZ host.

To experimentally verify MNZ propagation of quasi-TEM waves, a one-dimensional MNZ media has been fabricated. It is comprised of a conventional LCL-LPF of third order and both inductive sections have been loaded with one SRR each. Full-wave simulations have shown that changing the rings' position along the inductive sections has negligible influence on the MNZ passband even if extreme displacements are made. Therefore, SRRs are arbitrarily placed in the middle of the inductive sections, as shown in Figure 7.16.

Simulated response of the one-dimensional MNZ media is shown in Figure 7.17 and compared to the response of the host low-pass filter with no SRRs. It can be seen that inclusion of SRRs resulted in a new MNZ passband in the stopband of the host low-pass filter. The MNZ passband is positioned at 2.44 GHz, which corresponds exactly to the frequency at which effective permeability of the SRRs is equal to zero (denoted by f_{02} in Figure 7.10.)

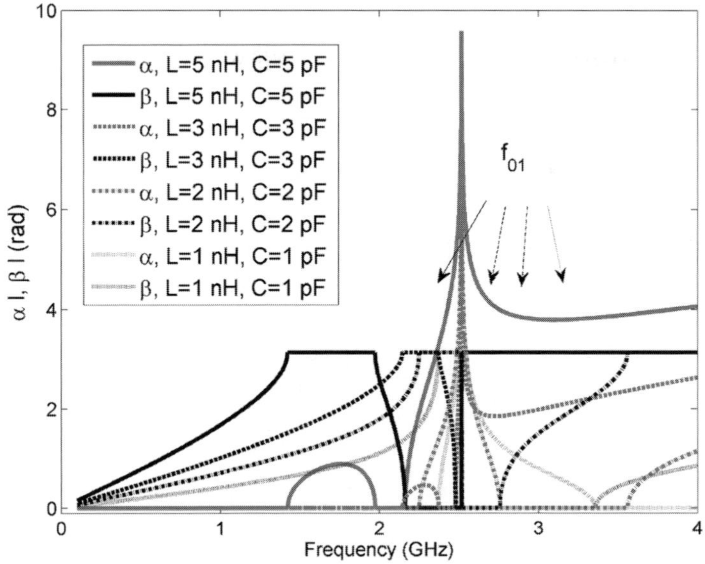

Figure 7.15 Attenuation constant α and phase constant β for various values of the host inductance L and capacitance C in the case of the SRR-loaded LCL-LPF [36].

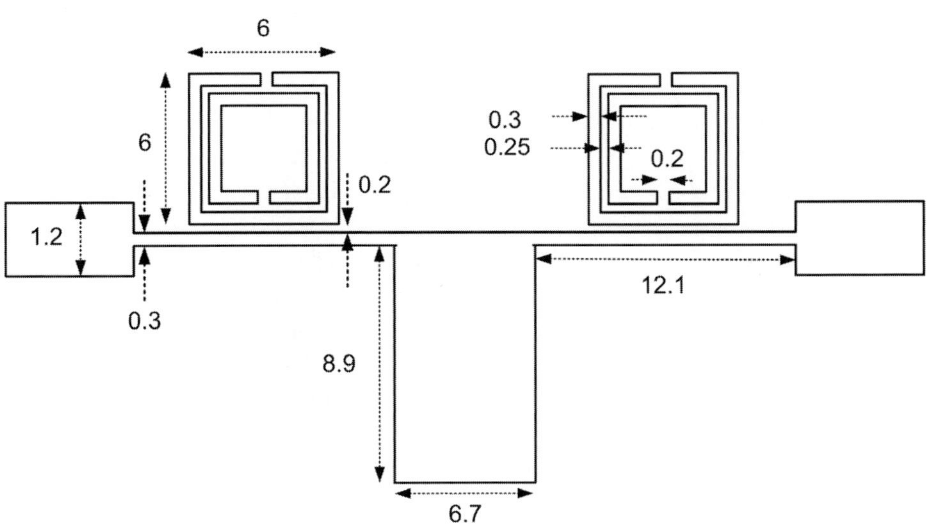

Figure 7.16 One-dimensional MNZ media [36].

The proposed one-dimensional MNZ media has been fabricated using a conventional printed circuit board technology. Comparison between the measured and simulated responses, Figure 7.18, reveals good agreement, except for a shift in the MNZ frequency approximately equal to 5.7%, caused by relatively large variations of the

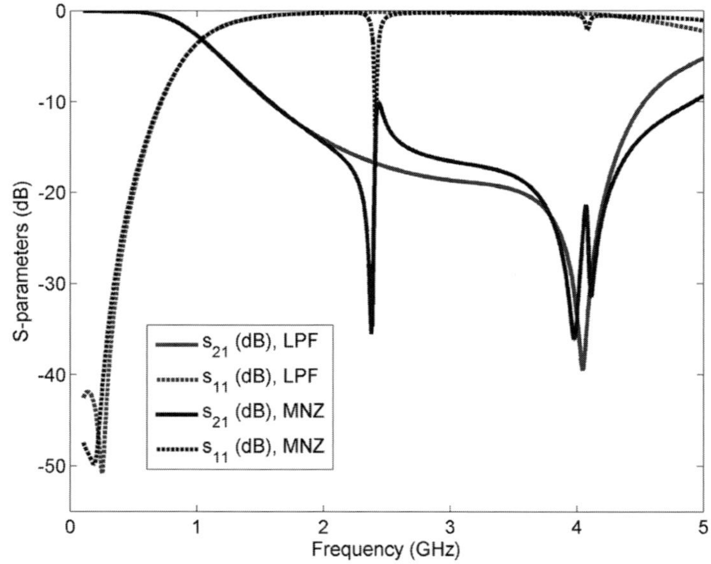

Figure 7.17 Simulated response of one-dimensional MNZ media [36].

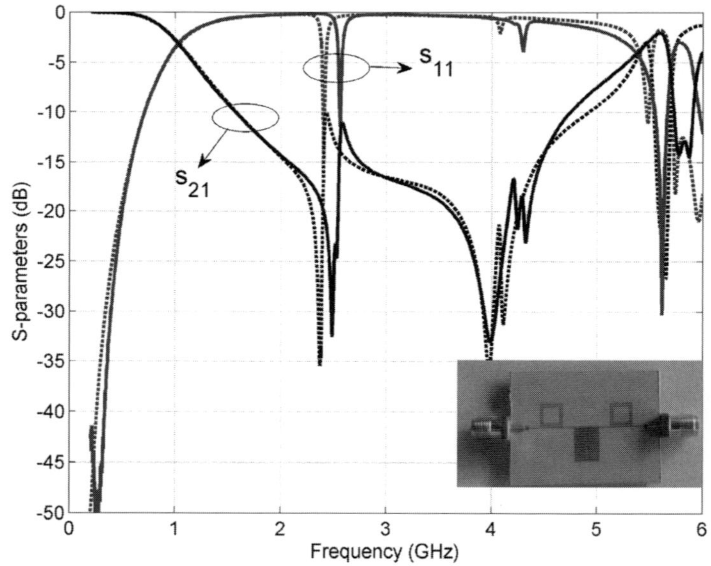

Figure 7.18 Comparison of measured (full lines) and simulated (dashed lines) responses of one-dimensional MNZ media. Photograph of the fabricated structure is shown in the inset [36].

permittivity of the substrate material (± 0.5). Due to the quasi-lumped nature of the low-pass host and its relatively large size, the discrepancy in the permittivity affected the MNZ frequency far more than it influenced the cutoff frequency of the low-pass host.

7.4.3 Design of bandpass filters based on permeability-near-zero propagation

Using the one-dimensional MNZ media presented above, in this section we design a proof-of-concept bandpass filter that utilizes MNZ propagation of quasi-TEM waves. We intentionally use the same (electrically long) low-pass filter as a host, to demonstrate the intriguing aspects of the phase responses of MNZ filters, which are most pronounced with larger structures. However, we note that MNZ filters with a much smaller footprint might easily be designed, e.g. by replacing the host with some of recently developed miniature low-pass configurations such as [34]. Additionally, the original low-pass response of the host structure, which still exists in the final response, could be suppressed by introducing a stopband at lower frequencies (e.g. by using defect ground structures or other techniques).

The MNZ bandpass filter presented here is designed by combining the responses of both MNZ structures shown above. It has been shown in Section 7.4.2 that MNZ propagation can be supported at two frequencies, f_{01} and f_{02}, depending on the position of the SRR relative to the host (SRR next to the capacitive or inductive section of the host, respectively). By using superposition, i.e. by loading each of the host's sections with an appropriate unit cell, propagation at both MNZ frequencies could be simultaneously supported and then coupled to form a useful passband.

A synthesis procedure of bandpass filters of this type is as follows:

1. design the host structure;
2. design unit cells that produce MNZ transmission at the first MNZ frequency;
3. design unit cells that produce MNZ transmission at the second MNZ frequency;
4. optimize the coupling between the unit cells and the host, as well as between the adjacent unit cells.

In the following example, this procedure has been performed using the third-order LCL-LPF with open stubs as a host. Firstly, two SRRs which load inductive sections of the host were designed. In this manner, a structure similar to the one shown in Figure 7.16 has been obtained. We note from Figure 7.17 that MNZ resonance in such structures is relatively weakly excited and that the bandwidth of the obtained passband is small. In the following step, another SRR is placed next to the capacitive section of the host. Its dimensions are optimized so that f_{01} and f_{02} are sufficiently coupled. At the same time, this third (middle) SRR couples the resonances of the outer two and significantly contributes to the level of transmission between the ports and, therefore, to the bandwidth of the final filter.

The layout of the optimized filter is shown in Figure 7.19 with all the relevant dimensions (the dimensions of the two outer SRRs are kept the same as in Figure 7.16). The measured and simulated responses are compared in Figure 7.20. Apart from a frequency shift or around 4% due to unequal nominal and actual substrate permittivity, the responses are in very good agreement. The measured filter exhibits high selectivity, fractional bandwidth of 6%, and insertion loss of 2.3 dB. Due to a specific design of the MNZ filter, there is a weakly excited reflection zero outside the passband which is an artifact of this type of design.

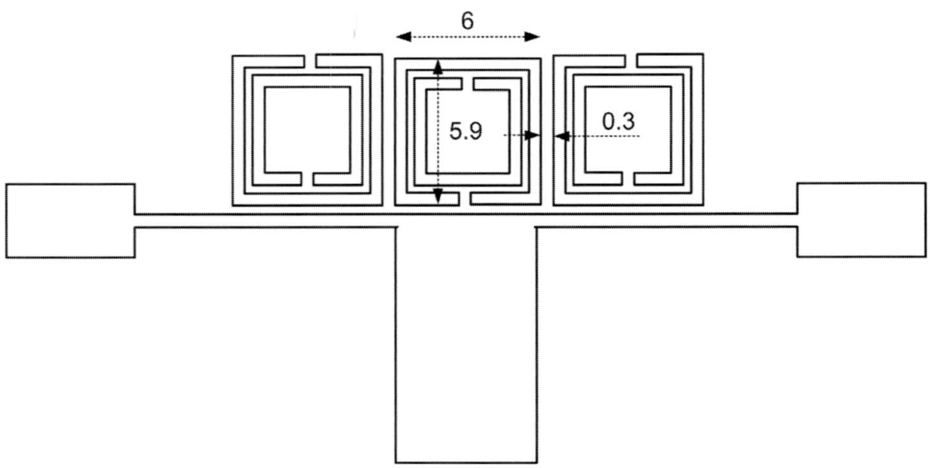

Figure 7.19 Layout of the proposed MNZ filter [36].

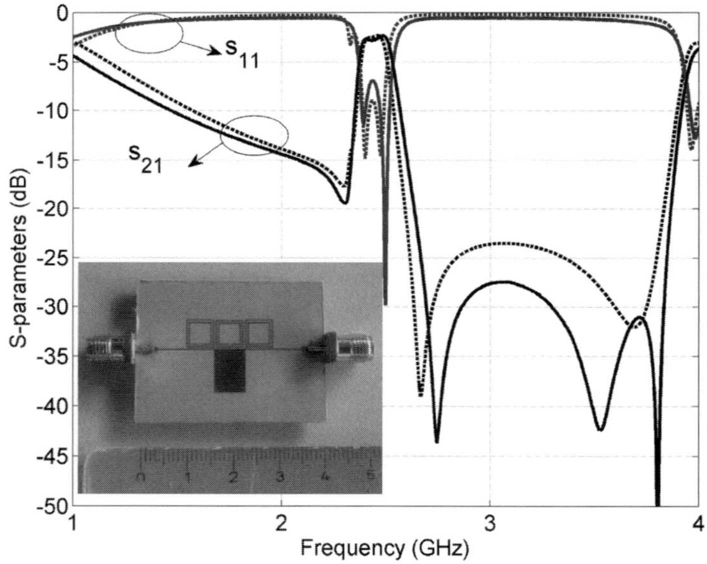

Figure 7.20 Comparison of simulated and measured responses. A photograph of the filter is shown in the inset [36].

An interesting characteristic of the proposed MNZ filter is the fact that, unlike in conventional filters, the spacing between adjacent resonators does not significantly influence the filter's bandwidth. For example, as this spacing is varied from 0.3 mm to 0.5 mm, the fractional bandwidth changes only from 5.33 to 4.9%, while the insertion loss increases by less than 0.1 dB. This low sensitivity to inter-resonator spacing is a special characteristic of MNZ filters, which results from the fact that the filter bandwidth is mainly determined by the slope of the effective permeability function around the zero value.

Low insertion loss of the filter is witness to the fact that NZ metamaterials do not inherently suffer from high losses as DNG media do. The insertion loss obtained by the MNZ filter from Figure 7.19 corresponds to that of a classically synthesized coupled-resonator filter of third order with an unloaded Q of 150, and the same bandwidth. Therefore, MNZ filters exhibit losses on par with microstrip coupled-resonator filters.

To validate the fact that the obtained passband is indeed due to MNZ transmission, we will analyze the phase variation along the structure. To that end, the phase of the transversal component of the magnetic field along a line halfway between the ground and the conductive layer has been extracted at the central frequency of the passband, equal to 2.44 GHz. This has been compared to the phase of a conventional 50-Ω microstrip line of the same length and at the same frequency, Figure 7.21. A phase of approximately 120° has been accumulated along the microstrip line, which precisely corresponds to its electrical length. Although having the same physical length as the microstrip line, the MNZ filter exhibits a total phase shift of only 4.7° where the maximal phase variation over its entire length is only 26.9°. Such behavior – a relatively constant phase over long distances – is the main characteristic of MNZ media.

As the name suggests, NZ propagation occurs not only when the phase constant is equal to zero, but also when it is sufficiently small, i.e. when it is near to zero. To estimate the bandwidth of this phenomenon, i.e. the bandwidth over which the phase remains relatively constant, maximum phase variation accumulated along the MNZ

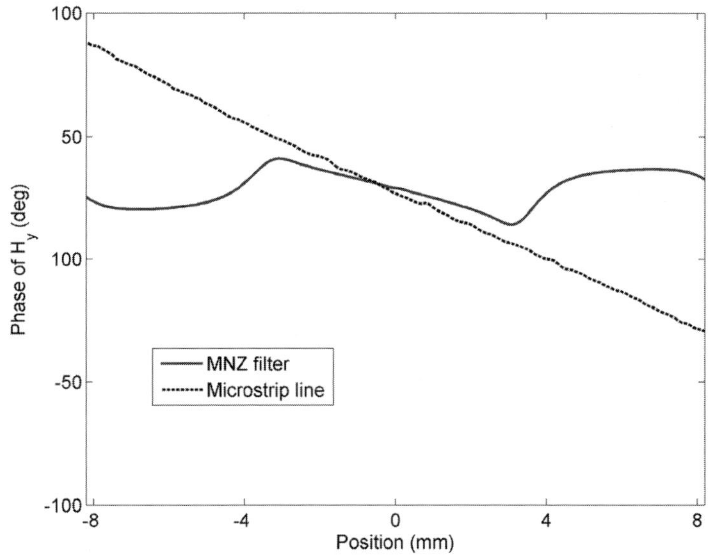

Figure 7.21 Phase distribution along the MNZ filter compared to that of a conventional 50-Ω microstrip line of the same length and at the same frequency. Zero value of the variable *Position* corresponds to the center of the structure [36].

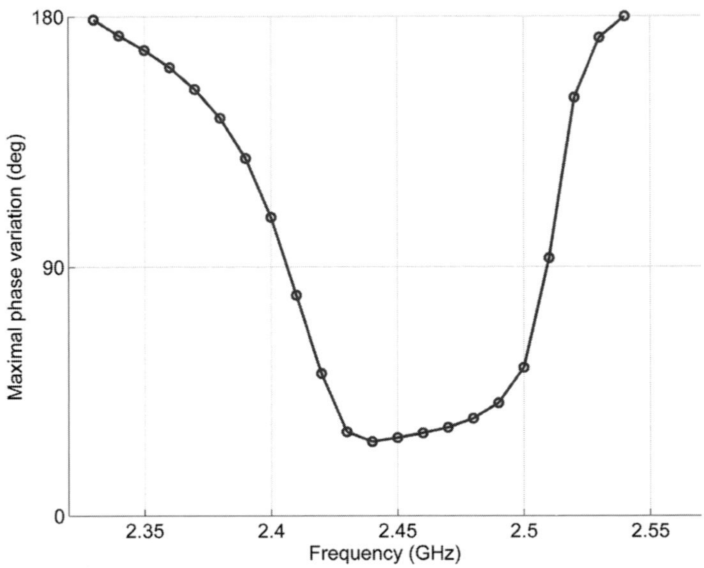

Figure 7.22 Frequency dependence of the maximum phase variation accumulated along the MNZ filter [36].

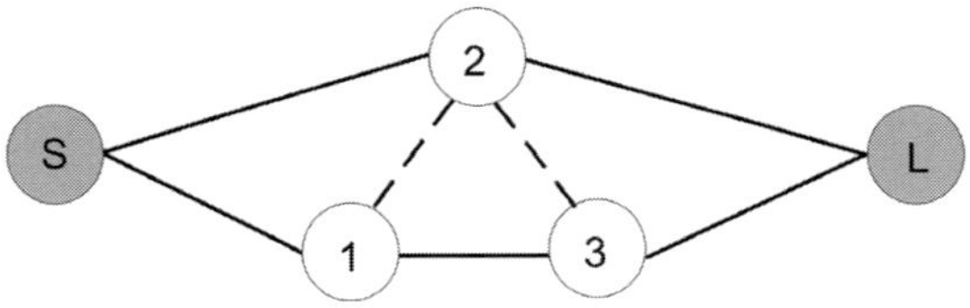

Figure 7.23 Coupling model of the proposed MNZ filter [36].

filter has been extracted at various frequencies in the vicinity of the MNZ frequency, Figure 7.22. As expected, at frequencies far from the MNZ frequency, large phase variations may be observed which correspond to the actual length of the structure. However, in the vicinity of the MNZ, frequency phase variation is reduced significantly. The bandwidth over which this occurs is relatively narrow and approximately equal to 5%. However, it should be noted that this bandwidth is not determined by the topology of the filter but by the metamaterial unit cells used, i.e. by the slope of their effective permeability function around zero. To influence the bandwidth, some unit cell other than the SRR should be used.

The MNZ filter can also be analyzed using the classical coupled-resonator theory. Since there are three resonators in the structure, i.e. three poles in the filter response, a third-order model is needed. Specifically, a modified transversal model is used, Figure 7.23, since the resonators are coupled to a host that exhibits a small phase shift along its length. In this configuration, resonators 1 and 3 are coupled in series and

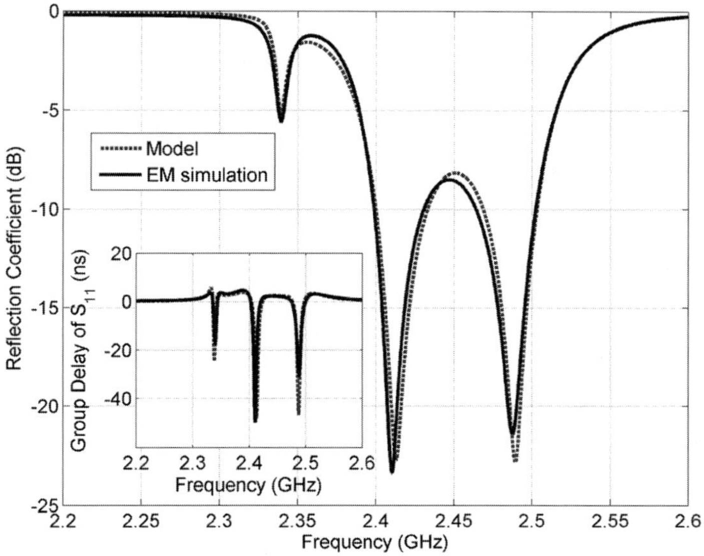

Figure 7.24 Validity of the model from Figure 7.23 is illustrated by good agreement of its response with those of full-wave simulated lossless MNZ filter [36].

placed in parallel with resonator 2 (the middle resonator in Figure 7.19). Smaller couplings also exist between resonators 1 and 2, and 3 and 2. The normalized $N+2$ x $N+2$ coupling matrix for a bandwidth of 5% and a center frequency of 2.44 GHz is

$$M = \begin{bmatrix} 0 & 0.757 & -0.151 & 0 & 0 \\ 0.757 & 0 & -0.042 & 0.848 & 0 \\ -0.151 & -0.042 & -1.863 & -0.143 & -0.151 \\ 0 & 0.848 & -0.143 & 0 & 0.757 \\ 0 & 0 & -0.151 & 0.757 & 0 \end{bmatrix}. \tag{7.11}$$

The matrix describes self-resonant frequencies of 2.44 GHz for resonators 1 and 3 and 2.34 GHz for resonator 2. It should be noted that the couplings $M_{12} = M_{23}$ are relatively small compared to M_{13}, which is due to the fact that the central resonator is in parallel with the series combination of resonators 1 and 3. This is a very different scenario compared to the case of a filter in which the host is removed while the three resonators remain in place. In that case, the three resonators would clearly have been coupled in series and M_{13} would not be significant.

As can be seen from Figure 7.24, the proposed model describes well the couplings between source, load, and the resonators, as the S parameter and group delay responses are well matched to those of the full-wave simulations of the lossless MNZ filter around the passband frequencies. The model also accurately reproduces a weakly excited reflection zero outside the passband, which is a special artifact of this type of configuration. Therefore, the proposed model can be used to perform straightforward design of MNZ bandpass filters and to fine tune the filter's bandwidth in a similar method as used in [37].

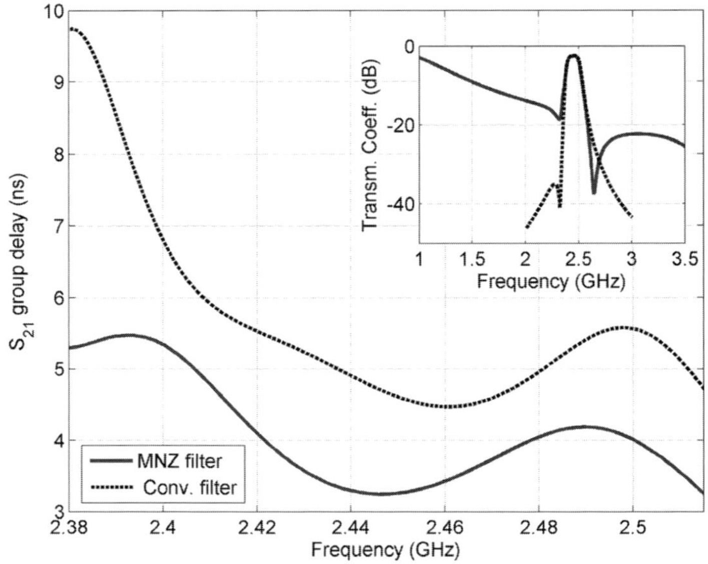

Figure 7.25 Comparison of simulated responses of the proposed MNZ filter and an equivalent coupled-resonator filter [36].

As has been discussed in Section 7.3, it can be expected that NZ propagation is characterized by reduced group delay. To confirm this, group delays of the MNZ filter and of a standard coupled-resonator filter of third order have been compared in Figure 7.25. The standard filter has been synthetized with an unloaded Q-factor of 150, with the center frequency and bandwidth equal to those of the MNZ filter, as well as with one finite transmission zero positioned to match exactly that of the MNZ filter. It can be seen that the group delay of the MNZ filter is significantly reduced compared to the standard prototype. Furthermore, the variation in group delay over the passband is reduced from 5.28 ns for the classical filter to only 2.3 ns for the MNZ filter, i.e. by more than 56%, thus illustrating the unique properties of NZ filters.

7.5 Multi-band filters based on near-zero metamaterials

It has been shown in the previous section how one-dimensional MNZ media can be used to design single-band bandpass filters with good performance and reduced group delay. The proposed idea can be extended to the design of multi-band filters using the superposition principle. Using adequate host structures and adequate unit cells, this concept can be applied to the design of filters with (theoretically) any number of independently controlled passbands. In this section, we will present and analyse in detail a simple dual-band filter realized by superposition of ENZ and MNZ media shown in the previous sections. We note that since NZ propagation is a relatively narrowband phenomenon, properly tuned ENZ and MNZ structures can be combined without altering the operating mode into the unwanted DNG mode.

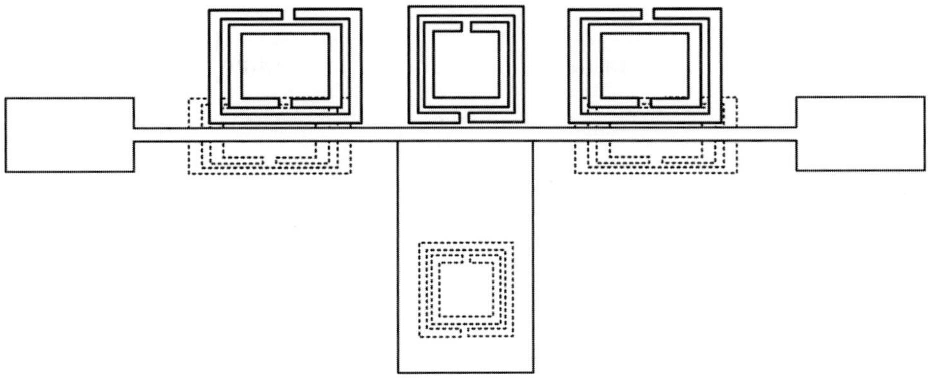

Figure 7.26 Layout of the proposed dual-band near-zero filter.

The dual-band filter proposed here is designed to operate at WLAN frequencies of great practical interest, namely at 2.4 and 3.5 GHz. Its configuration, shown in Figure 7.26, relies on three major segments: the low-pass filter which serves as a host, three complementary split-ring resonators (CSRR) which provide the ENZ passband around 2.4 GHz, and three individually tailored split-ring resonators (SRR) to provide the MNZ passband around 3.5 GHz. We note that either ENZ or MNZ propagation could have been used to obtain the first passband since ENZ and MNZ propagation in quasi-TEM circuits is obtained in a similar manner. As was demonstrated in the previous section, ENZ and MNZ structures differ in the metamaterial unit cell used and their coupling to the host structure – ENZ propagation is obtained using CSRR cells, whilst MNZ propagation is obtained using SRR cells. Furthermore, every unit cell is capable of supporting NZ propagation at two NZ frequencies, and the choice of the actual operating frequency (and the consecutive position the transmission zero) depends on the inductive/capacitive nature of the host's segment to which the unit cell is coupled.

The first step in the design of the near-zero dual-band filter is the design of the host structure. The host needs to exhibit a stopband in the entire frequency range of interest, including both passbands of the final NZ filter. Furthermore, this stopband must not be a consequence of a SNG medium, as in such a case it would couple either with ENZ or with MNZ structure and support DNG propagation.

The final circuit has been designed on the Rogers TMM10 substrate with the thickness equal to 1.27 mm, dielectric constant $\varepsilon_r = 9.2$, and loss tangent $\tan\delta = 0.0035$. The final low-pass host is identical to the one designed in Section 7.4.3, as it exhibits a stopband from 2 to 4 GHz. Namely, since both passbands have to be positioned within the host's stopband region, such a low-pass filter is suitable for the design of a dual-band bandpass filter operating at 2.4 and 3.5 GHz.

In the following steps, ENZ and MNZ single-band filters are independently designed around the identical host, using the approach shown in Section 7.4.3. The layout of the ENZ filter is shown in Figure 7.27. It includes three CSRRs etched in the ground plane – two coupled to the high-impedance sections of the host, whilst the middle one is positioned centrally underneath the low-impedance section so it is exposed to the

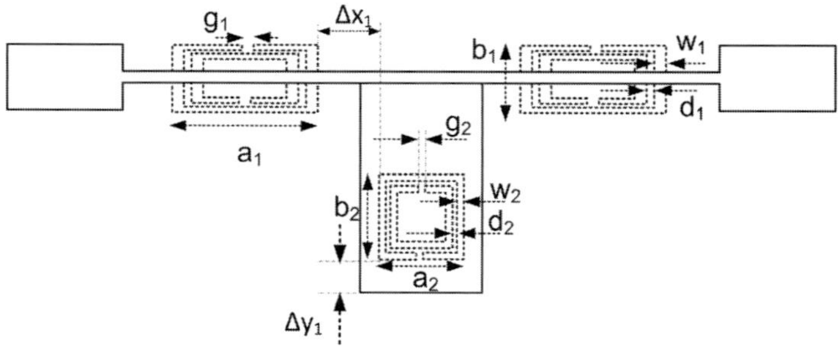

Figure 7.27 Layout of the ENZ single-band bandpass filter. The relevant dimensions are: $a_1 = 5.8$ mm, $b_1 = 2.35$ mm, $g_1 = g_2 = 0.1$ mm, $w_1 = w_2 = 0.1$ mm, $d = d_2 = 0.1$ mm, $\Delta x_1 = 2.3$ mm, $\Delta y_1 = 1.35$ mm, $a_2 = 4$ mm, $b_2 = 4.25$ mm.

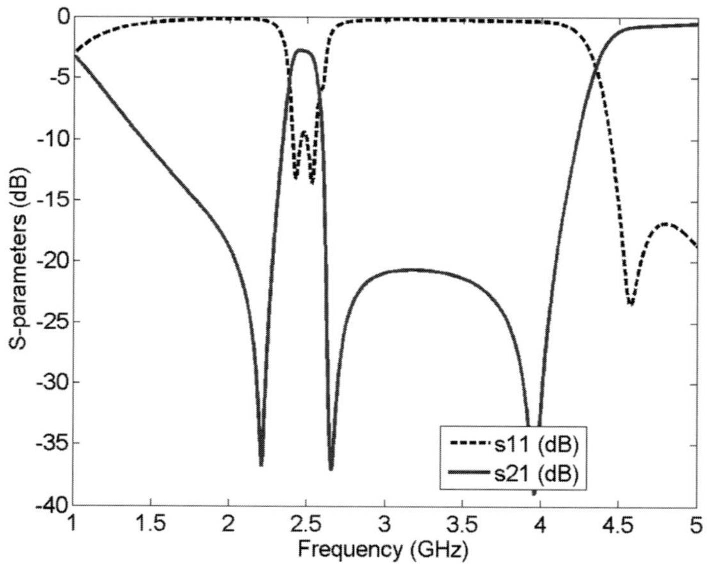

Figure 7.28 Simulated response of the ENZ single-band bandpass filter.

strongest electrical field available. The response of the single-band filter, Figure 7.28, reveals that such a configuration provides good selectivity, insertion loss of 2.6 dB, and a fractional bandwidth of 7.25%.

The layout of the MNZ single-band filter is shown in Figure 7.29 and its response in Figure 7.30. The MNZ single-band filter exhibits good selectivity, insertion loss of 1.96 dB, and fractional bandwidth of 4.8%.

Finally, two single-band filters are combined in one structure. This is straightforward, since ENZ and MNZ structures deploy unit cells positioned in different conductive layers. Furthermore, the fact that both filters use the same host structure reduces the overall size of the dual-band filter. Comparison of the responses of two single-band filters with the dual-band filter obtained by their simple superposition is shown in

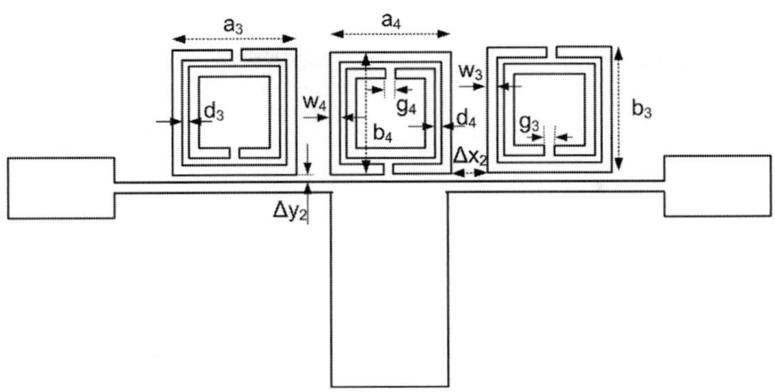

Figure 7.29 Layout of the MNZ single-band bandpass filter. The relevant dimensions are: $a_3 = 5.6$ mm, $b_3 = 3.15$ mm, $g_3 = 0.2$ mm, $d_3 = 0.2$ mm, $w_3 = w_4 = 0.4$ mm, $\Delta x_2 = 1.5$ mm, $\Delta y_2 = 0.1$ mm, $a_4 = 6.15$ mm, $b_4 = 3.2$ mm, $g_4 = 0.15$ mm, $d_4 = 0.15$ mm.

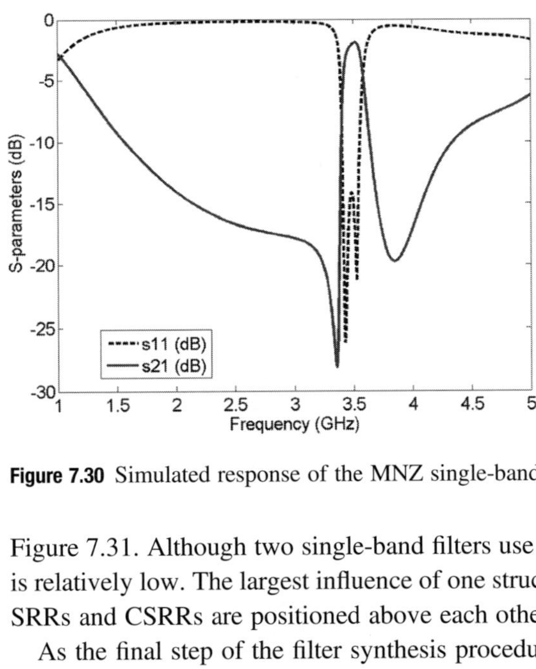

Figure 7.30 Simulated response of the MNZ single-band bandpass filter.

Figure 7.31. Although two single-band filters use the same host, their mutual influence is relatively low. The largest influence of one structure on the other is due to the fact the SRRs and CSRRs are positioned above each other.

As the final step of the filter synthesis procedure, the geometrical parameters of the filter are optimized to meet the specifications. The final dimensions of the dual-band NZ filter are: $a_1 = 6.05$ mm, $b_1 = 2.05$ mm, $g_1 = g_2 = 0.1$ mm, $w_1 = w_2 = 0.1$ mm, $d_1 = d_2 = 0.1$ mm, $\Delta x_1 = 2.2$ mm, $\Delta y_1 = 1.35$ mm, $a_2 = 4$ mm, $b_2 = 4.25$ mm, $a_3 = 5.6$ mm, $b_3 = 3.15$ mm, $g_3 = 0.2$ mm, $d_3 = 0.2$ mm, $w_3 = w_4 = 0.4$ mm, $\Delta x_2 = 1.15$ mm, $\Delta y_2 = 0.1$ mm, $a_4 = 6.15$ mm, $b_4 = 3.2$ mm, $g_4 = 0.15$ mm, $d_4 = 0.15$ mm. It can be seen that only the parameters Δx_1, Δx_2, and the dimensions of CSRR under the inductive sections (a_1 and b_1) have been slightly changed, which implies that the optimization process is not very demanding and that desired characteristics are readily obtained.

The optimized response of the dual-band NZ filter is shown in Figure 7.32. It exhibits passbands at 2.4 and 3.5 GHz, both characterized by good selectivity, low insertion

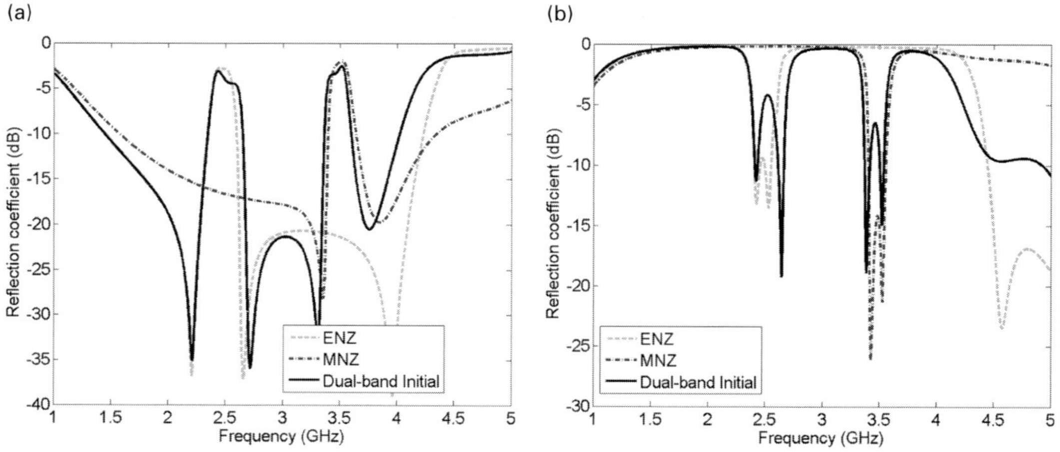

Figure 7.31 Comparison of transmission coefficients (a) and reflection coefficients (b) of two single-band filters with their superposition – the initial dual-band filter.

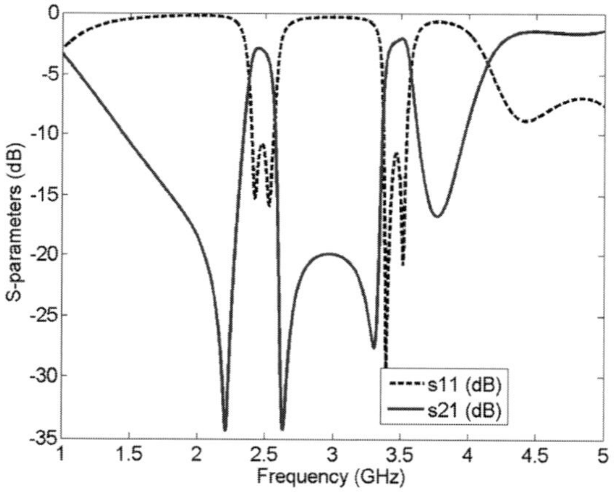

Figure 7.32 The response of the optimized dual-band near-zero filter.

losses of 2.6 and 1.8 dB, return losses of 12 and 13 dB, and fractional bandwidths of 7.2% and 5.5%, respectively.

Finally, to verify that indeed NZ propagation occurs in both passbands, phase distribution along the filter has been extracted at central frequencies of both passbands, using approaches similar to those described in Sections 7.4.1. and 7.4.3. The extracted phase is shown in Figure 7.33 and compared to the phase accumulated along a 50-Ω microstrip line of the same length and at the same frequency. It can be seen that, although the dual-band NZ filter is long in terms of guided wavelength, it exhibits a relatively constant phase in both passbands, which is consistent with NZ media. At the central frequency of the first passband, phase shift of only 7° is accumulated along the filter, compared to 184° accumulated over the conventional microstrip line. A similar

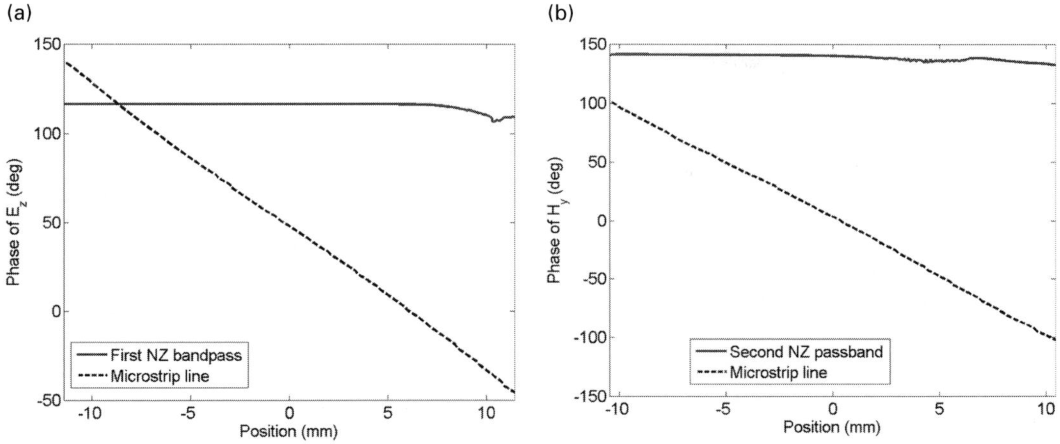

Figure 7.33 Phase distribution along the dual-band NZ filter at the central frequency of the (a) first and (b) second passband.

situation occurs at the central frequency of the second passband, where the phase shift is reduced from 203° accumulated along the microstrip to only 8° in the case of the NZ filter.

7.6 Conclusion

In this chapter, a new approach to the design of multi-band filters has been presented, which relies on the use of so-called near-zero metamaterials. Due to a unique characteristic of this sub-class of metamaterials to support propagation with zero propagation constant at a non-zero frequency, filters with reduced group delay and a reduced variation of group delay can be designed.

The chapter focuses on NZ propagation of quasi-TEM modes, needed for the design of NZ microstrip filters. It examines both the permittivity-near-zero and permeability near-zero propagation of quasi-TEM waves and demonstrates how it can be used to design single-band NZ filters. Finally, this idea is extended to multi-band filters and a dual-band microstrip NZ filter is presented and analyzed. Owing to its flat phase response, such filters are promising candidates for applications where flat group delay is required.

References

[1] V. Veselago, "The electrodynamics of substances with simultaneously negative values of ε and μ," *Soviet Physics Uspekhi*, vol. 92, no. 3, pp. 517—526, April 1967.

[2] J. B. Pendry, A. J. Holden, W. J. Steward, and I. Youngs, "Extremely low frequency plasmons in metallic mesostructures," *Physical Review Letters*, vol. 76, no. 25, pp. 4773–4776, June 1996.

[3] J. B. Pendry, A. J. Holden, D. J. Robbins, and W. J. Stewart, "Magnetism from conductors and enhanced,"*IEEE Transactions on Microwave Theory and Techniques,* vol. 47, no. 11, pp. 2075–2084, Nov. 1999.

[4] R. A. Shelby, D. R. Smith, and S. Schultz, "Experimental verification of a negative index of refraction,"*Science*, vol. 292, pp. 77–79, April 2001.

[5] R. Marques, F. Martin, and M. Sorolla, *Metamaterials with Negative Parameters*, John Wiley & Sons, New Jersey, 2008.

[6] C. Caloz and T. ltoh, "Application of the transmission line theory of left-handed (LH) materials to the realization of a microstrip LH transmission line," *IEEE Antennas and Propagation Society International Symposium, 2002*, vol. 2, pp. 412–415, June 2002.

[7] C. Caloz and T. Itoh, *Electromagnetic Metamaterials: Transmission Line Theory and Microwave Applications*, John Wiley & Sons, New Jersey, 2006.

[8] A. Grbic and G. V. Eleftheriades, "Experimental verification of backward-wave radiation from a negative refractive index metamaterial," *Journal of Applied Physics*, 92, 5930 (2002).

[9] R. Ziolkowski and A. D. Allison, "Application of double negative materials to increase the power radiated by electrically small antennas," *IEEE Transactions on Antennas and Propagation*, vol. 51, no. 10, pp. 2626–2640, Oct. 2013.

[10] D. Sievenpiper, L. Zhang, R. F. J. Broas, N. G. Alexopolous, and E. Yablonovitch, "High-impedance electromagnetic surfaces with a forbidden frequency band," *IEEE Transactions on Microwave Theory and Techniques*, vol. 47, pp. 2059–2074, Nov. 1999.

[11] J. McVay, N. Engheta, and A. Hoorfar, "High impedance metamaterial surfaces using Hilbert-curve inclusions," *IEEE Microwave and Wireless Components Letters*, vol. 14, no. 3, pp. 130–132, March 2004

[12] J. Garcia-Garcia, J. Bonache, I. Gil, F. Martin, M. C. Velazquez-Ahumada, and J. Martel, "Miniaturized microstrip and CPW filters using coupled metamaterial resonators," *IEEE Transactions on Microwave Theory and Techniques,* vol. 54, no. 6, pp. 2628–2635, June 2006.

[13] M. Studniberg and G. V. Eleftheriades, "A dual-band bandpass filter based on generalized negative-refractive-index transmission-lines," *IEEE Microwave and Wireless Components Letters*, vol. 19, no. 1, pp. 18–20, Jan. 2009.

[14] C.-H. Tseng and T. Itoh, "Dual-band bandpass and bandstop filters using composite right/left-handed metamaterial transmission lines," *Proceedings of IEEE MTT-S International Microwave Symposium Digest*, pp. 931–934, June 2006

[15] J. S. Sun, H. Lobato-Morales, J. H. Choi, A. Corona-Chavez, and T. Itoh, "Multistage directional filter based on band-reject filter with isolation improvement using composite right-/left-handed transmission lines," *IEEE Transactions on Microwave Theory and Techniques*, vol. 60, no.12, pp. 3950–3958, Dec. 2012.

[16] C. Caloz, A. Sanada, and T. Itoh, "A novel composite right-/left-handed coupled-line directional coupler with arbitrary coupling level and broad bandwidth," *IEEE Transactions on Microwave Theory and Techniques*, vol. 52, no. 3, pp. 980–992, March 2004.

[17] C. G. M. Ryan and G. V. Eleftheriades, "Design of a printed dual-band coupled-line coupler with generalised negative-refractive index transmission lines," *IET Microwave Antennas and Propagation*, vol. 6, no. 6, pp. 705–712, April 2012.

[18] S. R. Das, "Metamaterials arrive in cellphones," *IEEE Spectrum*, Oct. 2009.

[19] R. W. Ziolkowski, "Propagation in and scattering from a matched metamaterial having a zero index of refraction," *Physical Review E,* vol. 70, no. 4, pp. 046608-046619, Oct. 2004.

[20] M. Silveirinha and N. Engheta, "Tunneling of electromagnetic energy through sub-wavelength channels and bends using epsilon-near-zero (ENZ) materials," *Physical Review Letters*, vol. 97, no. 15, pp. 1574403-1/4, Oct. 2006.

[21] B. Edwards, A. Aluu, M. G. Silveirinha, and N. Engheta,"Reflectionless sharpbends and corners in waveguides using epsilon-near-zero effects," *Journal of Applied Physics,* vol. 105, no. 44, pp. 044905-1/4, Feb. 2009.

[22] M. Silveirinha and N. Engheta, "Theory of supercoupling, squeezing wave energy, and field confinement in narrow channels and tight bends using near zero materials," *Physical Review B*, vol. 76, no. 24, pp. 245109–1/17, Dec. 2007.

[23] B. Edwards, A. Aluu, M. E. Young, M. G. Silveirinha, and N. Engheta, "Experimental verification of epsilon-near-zero metamaterial coupling and energy squeezing using a microwave waveguide," *Physical Review Letters*, vol. 100, no. 3, pp. 033903-1/4, Jan. 2008.

[24] R. Liu, Q. Cheng, T. Hand, J. J. Mock, T. J. Cui, S. Cummer, and D. Smith, "Experimental demonstration of electromagnetic tunneling through an epsilon-near-zero metamaterial at microwave frequencies," *Physical Review Letters*, vol. 100, pp. 023903-1/4, Jan. 2008.

[25] D. V. B. Murthy, A. Corona-Chavez, and J. L. Olvera-Cervantes, "A novel epsilon near zero (ENZ) tunneling circuit using microstrip technology for high integrability applications," *Progress In Electromagnetics Research C*, vol. 15, pp. 65–74, Aug. 2010.

[26] A. Alù, M. G. Silveirinha, A. Salandrino, and N. Engheta, "Epsilon-near-zero (ENZ) metamaterials and electromagnetic sources: tailoring the radiation phase pattern," *Physical Review B*, vol. 75, no. 15, pp. 155410-1/13, April 2007.

[27] A. Alu and N. Engheta, "Dielectric sensing in ε-near zero narrow waveguide channels," *Physical Review B*, vol. 78, no. 4, pp. 045102-1/5, July 2008.

[28] H. Lobato-Morales, D. V. B. Murthy, A. Corona-Chávez, J. L. Olvera-Cervantes, J. Martínez-Brito, and L. G. Guerrero-Ojeda, "Permittivity measurements at microwave frequencies using epsilon-near-zero (ENZ) tunnel structure," *IEEE Transactions on Microwave Theory and Techniques,* vol. 59, no. 7, pp. 1863–1868, July 2011.

[29] J. Xiong, X. Lin, Y. Yu, M. Tang, S. Xiao, and B. Wang, "Novel flexible dual-frequency broadside radiating rectangular patch antennas based on complementary planar ENZ or MNZ metamaterials," *IEEE Transactions on Antennas and Propagation*, vol. 60, no. 8, pp. 3958–3961, Aug. 2012.

[30] P. J. Ferrer, J. Romeu, J. M. Gonzalez-Arbesu, J. Parron, F. Capolino, F. Bilotti, L. Vegni, G. Lovat, and P. Burghignoli, "Broadside radiation enhancement using a spiral resonator MNZ metamaterial substrate," *Proceedings of IEEE International Symposium on Antennas and Propagation Society*, pp. 1–4, 2008.

[31] B. Lopez-Garcia, D. V. B. Murthy, and A. Corona-Chavez, "Half mode microwave filters based on epsilon near zero and mu near zero concepts," *Progress In Electromagnetics Research*, vol. 113, pp. 379–393, 2011.

[32] V. Crnojević-Bengin, N. Janković, and R. H. Geschke, "Epsilon-near zero transmission in quasi-TEM microstrip circuits," *Proceedings of 41st European Microwave Conference*, 2011, pp. 798–801.

[33] Z. Szabó, G. Park, R. Hedge, and E. Li, "A unique extraction of metamaterial parameters based on Kramers–Kronig relationship,“ *IEEE Transactions on Microwave Theory and Techniques,* vol. 58, no. 10, pp. 2646–2653, Oct. 2010.

[34] J. S. Hong and M. J. Lancaster, *Microstrip Filters for RF/Microwave Applications*, John Wiley & Sons, New Jersey, 2011.

[35] V. Crnojevic-Bengin, R.H. Geschke, N. Jankovic, and N. Cselyuszka, "On the nature of transmission in resonant metamaterial transmission lines," *Metamaterials,* The Sixth International Congress on Advanced Electromagnetic Materials in Microwaves and Optics, St. Petersburg, Russia, 17-22 September 2012.

[36] V. Crnojević-Bengin, N. Janković, N. Cselyuszka, and R. Geschke, "Mu-near-zero propagation in quasi-TEM microstrip circuits," *Journal of Electromagnetic Waves and Applications*, vol. 27, no. 17, pp. 2198–2212, Nov. 2013.

[37] D. Cañete-Rebenaque, M. Martínez-Mendoza, J. Pascual-García, and J. S. G. Díaz, "Novel implementations for microstrip resonator filters in transversal and alternative topologies," *IEEE Transactions on Microwave Theory and Techniques,* vol. 59, no. 2, pp. 242–249, Feb. 2011.

8 Miniature microwave filters using multi-layer technologies

Irina Vendik, Dmitry Kholodnyak, Viacheslav Turgaliev, Alexander Rusakov, Shilong Qian, Jia Ni, and Jia-Sheng Hong

8.1 Introduction

Design of miniature, low-cost microwave devices with enhanced functionality has recently been supported by increasing interest in multi-layer structures. Multilayer technology provides 3D flexible design of microwave components, circuits, and subsystems, and it is very promising for the integration of multi-standard functions in a single device. So far, it has been successfully used for the design of various microwave devices that are intended for mass production, such as miniature filters.

In this chapter, we present microwave filters realized in multi-layer architecture. Firstly, in Section 8.2 we present and analyse in detail two widely used multi-layer technologies – low-temperature co-fired ceramics and technology based on liquid crystal polymer material.

Section 8.3 is devoted to single-band multi-layer filters designed using different approaches. Filters based on sections of artificial transmission lines are presented together with a detailed discussion of the properties of artificial transmission lines.

The same methods are applied for the design of high-performance dual-band filters which are presented in Section 8.4. The filters are based on two-mode resonators realized using a combination of artificial transmission line sections and capacitively loaded cavities. Additionally, a very compact dual-band immittance inverter based on artificial transmission lines is presented.

Compact microwave filters, based on capacitively loaded cavities exhibiting a high Q-factor, designed as multi-layer structures are proposed and discussed in Section 8.5. Their performance is analysed in detail.

Conclusions are given in Section 8.6.

8.2 Multilayer technology as a base for miniature filter design

Multilayer technology applied to microwave devices allows bridging 2D planar structures to 3D versions of many microwave designs. Using layers with printed components connected by metallized vias makes it possible to reduce the dimensions of the devices drastically. Moreover the multi-layer technology supports the manufacture of miniature lumped components designed as multi-layer structures of capacitive and inductive elements. This is followed by the practical use of lumped components for the design of many passive microwave devices, including filters, forming a new niche for designers involved in the design of low-cost production.

Multilayer technology can be applied to multi-layer filter design in different ways. Multilayer filters can be composed of coupled-line resonators or circuit elements that are located at different layers without any ground plane inserted between the adjacent layers [1–3]. Also, there are multi-layer filters that utilize aperture coupling realized in the common ground between adjacent layers, whilst another type of multi-layer filter is based on patterning ground to form so-called defected ground structures (DGS) [4–5]. In addition, widely used multi-layer filtering structures employ passive lumped (quasi-lumped) L-C-R-components located in different layers that are connected by metalized via holes.

Recent development of multi-layer microwave filters has been predominantly based on low-temperature co-fired ceramic (LTCC) technology and the technology based on liquid crystal polymer material (LCP). These technologies enable the design of monolithic, 3D, cost-effective microwave circuits and systems. In the following two sections, these technologies will be presented in detail.

8.2.1 Multilayer structures using low-temperature co-fired ceramics (LTCC)

Low-temperature co-fired ceramics are the most commonly used ceramic substrate material for compact RF designs and system-on-package (SoP) structures. In general, an LTCC-based device can be defined as a multi-layer circuit fabricated by laminating green dielectric sheets (terms for unfired tapes; Green Tape^{TM}, Dupont) with printed conductor lines and firing them all together in one step, Figure 8.1 [6–7].

Compared to high-temperature co-fired ceramics (HTCC) technology, the main advantage of the LTCC system is the possibility to use low resistive conductors such as silver and gold instead of tungsten and molybdenum, since the firing temperature for LTCC is around 850°C, i.e. much lower than that of HTCC. Also, unlike thick-film process, where successive lamination and firing steps cause bowing and line degradation, the single-step lamination and firing of LTCC produces a circuit pattern with high-quality line definition. The competitive advantages of LTCC technology are illustrated in Figure 8.2.

The main parameters of ceramic materials used for LTCC implementation are presented in Table 8.1.

Conductive materials are one of the important constitutive materials of LTCC, and they have to meet many requirements in order to find application in a wide range of high-frequency components. The most important requirements for conductive materials are as follows:

- low sheet resistance,
- compatibility with resistive layers,
- ability to create electrical and mechanical fine binding,
- good adhesion to substrate (proof against peeling strength).

The base materials of widely used conductive pastes are Au-Ag, binder of Au, Ag, Pt, and Pd. The Au and Ag pastes have the lowest specific resistance, whilst some Pt-Au

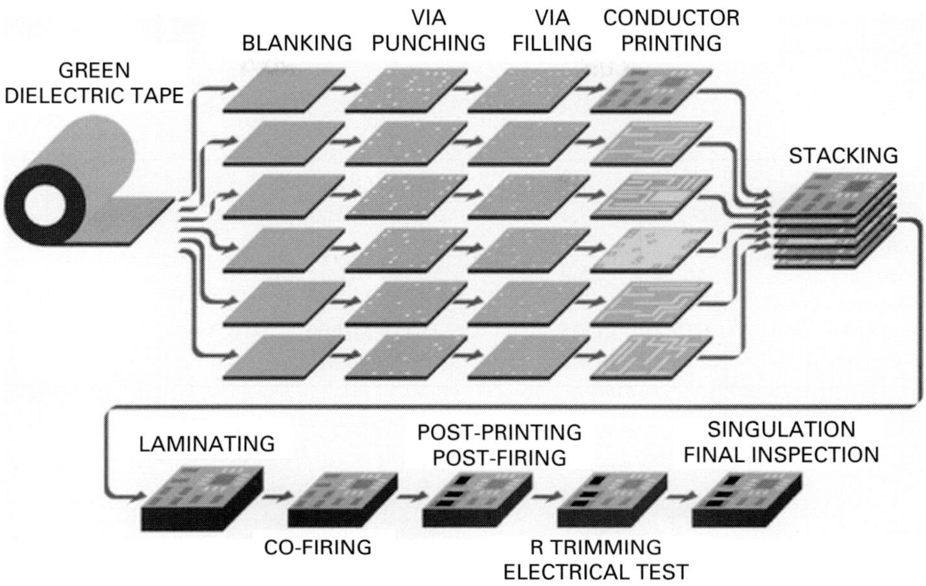

Figure 8.1 LTCC manufacturing process [6].

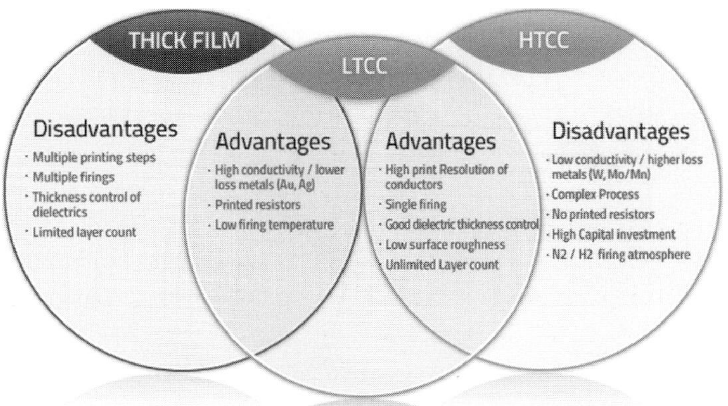

Figure 8.2 Advantages of LTCC technology [8].

pastes have the highest. Since transmission loss has to be reduced in high-speed circuits, it is necessary to use a conductor with low electrical resistance, and thus Cu, Au, Ag, and metals alloyed with them are good candidates for application in LTCC.

As for resistive components, they are realized using resistive paste. Nowadays the most often used resistive paste in LTCC technology is based on RuO_2 since the structure of ruthenium does not change up to the temperature of 1025°C and it is thermodynamically stable up to 1400°C.

On the other hand, capacitors are usually made up of three layers, i.e. two conductive layers separated with one dielectric layer. To realize capacitor values in the required

Table 8.1 Most important parameters of LTCC ceramic materials

LTCC supplier	Product	Dielectric constant ε_r	Q value (1/tan δ)	Thermal Expansion coefficient, ppm/°C	Thermal conductivity, W/m×K
DuPont	951	7.8	300 (3 GHz)	5.8	3.3
www.dupont.com	943	7.4	500 (40 GHz)	6.0	4.4
Ferro	A6-M	5.9	500 (3 GHz)	7	2
www.ferro.com	A6-S	5.9	500 (3 GHz)	8	2
Heraeus	CT700	7.5	450 (1 MHz)	6.7	4.3
www.heraeus.de	CT2000	9.1	1000 (450 MHz)	5.6	-

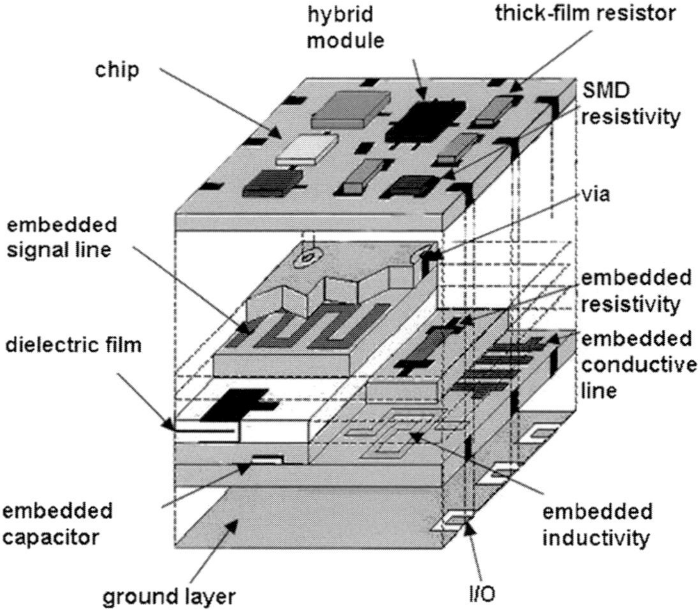

Figure 8.3 Complex LTCC circuit structure [8].

range, different dielectric pastes have to be used. Materials of dielectric pastes are usually $BaTiO_3$, $BaTiO_3$-epoxi, polymer-ceramic, epoxy-glass mixture.

LTCC exhibits a very low electrical loss and thus can be used in multi-layer laminated modules that have densely integrated passive and active devices. Figure 8.3 presents an example of a compact module, where multiple circuitries are handled in a single self-contained, hermetic package by virtue of this technology [8].

Also, the multi-layer characteristic of LTCC technology offers significant benefits in terms of 3D design flexibility and reliability for RF designs. From this aspect, many miniaturized LTCC filters for various wireless applications, which were considered impossible to realize with traditional processes, have been proposed [9–15].

For example, a lumped-element two-pole bandpass filter for Bluetooth application was presented in [14], where a mutual inductive coupling was achieved by overlapping the two inductor strips in the z-direction, resulting in finite transmission zeros. In [15], strong capacitive couplings between the strips positioned at different layers were achieved using LTCC technology, resulting in a high-performance filter with wide bandwidth and reduced size of less than $\lambda_g/12 \times \lambda_g/12 \times h$ (λ_g is the guided wavelength at the operating frequency and h is the substrate thickness).

LTCC technology is also used for the design of miniature microwave filters based on quasi-lumped components. An example of the three-pole bandpass LTCC filter with quasi-lumped elements for Bluetooth and WLAN applications (2400–2500 MHz) is presented in [16–17].

LTCC technology offers a set of benefits:

- a more economical manufacturing process compared with the conventional thick-film technology;
- mass production methods can be readily applied (several processing steps can be automated);
- fabrication techniques are relatively simple and inexpensive;
- tapes of different compositions can be manufactured with desired layer properties;
- electronic circuits can be integrated;
- the possibility of cutting the tape/substrate into different shapes;
- the possibility to bury passive components within the substrate, which reduces the size of circuits (down to about 50% in comparison with the PCB);
- an almost arbitrary number of signal layers;
- the ability to perform at frequencies over 30 GHz;
- high resistance to ambient working temperatures (up to 350°C);
- good thermal conductivity compared to PCB technology;
- compatibility with the equipment of conventional thick-film technology;
- the electrical parameters of conductive materials (gold, silver, etc.) are excellent.

However, LTCC technology has its limits which make it unsuitable for some applications. Namely, the first firing stage of LTCC occurs at 850°C, which might not be acceptable for some fully integrated solutions, such as chips containing active devices. Therefore, they have to be packaged separately, which significantly increases the fabrication cost. Furthermore, LTCC is still expensive compared to some conventional technologies. To realize the full advantages of an integrated SoP system, alternative material technologies need to be explored.

8.2.2 Multilayer liquid crystal polymer (LCP) technology

Liquid crystal polymer (LCP) is a new and promising organic thermoplastic material with unique structural and physical properties. Figure 8.4 presents a typical multi-layer LCP structure. The multilayer structure can be implemented by using a combination of LCP films of different thickness.

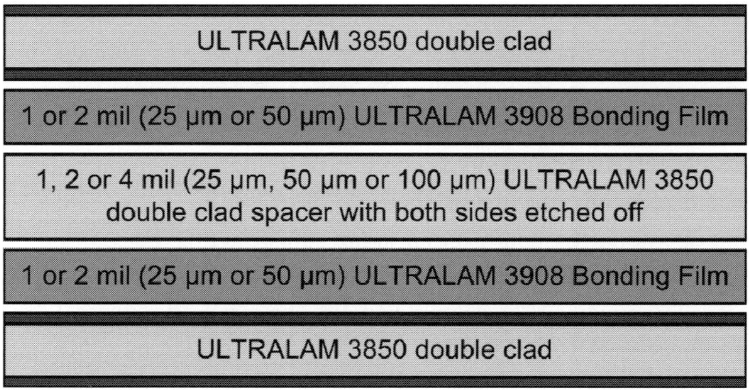

Figure 8.4 Multilayer structure as a combination of LCP films of different thickness.

As can be seen, LCP has two different functional films: one is the core film ULTRALAM 3850 with a melting temperature of 315°C, which is mainly used to support metal circuits and has several available thicknesses, such as 100, 50, and 25 μm; the other one is the bonding film ULTRALAM 3908 with a melting temperature of 280°C, which is utilized as a prepreg layer inserted between different core films for stack and has a typical thickness of 25 or 50 μm. Table 8.2 summarizes the typical characteristics of commercially available LCP materials [18–19].

Generally, LCP demonstrates simultaneous exceptional performance with respect to electrical, thermal, mechanical, and chemical properties, thus it is the perfect match for technically demanding high-frequency, harsh-environment, and direct implantable applications. Being a polymer, the cost of LCP is comparable to that of conventional printed circuit board (PCB) material and is cheaper than LTCC. Active and passive devices can be integrated in compact, vertically integrated RF modules by using homogeneous multi-layer LCP technology at low temperature (about 290°C), which would be more challenging for LTCC technology as stated earlier. Additionally, the low dielectric constant available in LCP can also be used for antenna applications. Moreover, LCP offers a unique combination of properties, such as low cost and excellent electrical and mechanical characteristics, which make it a viable technology for SoP-based systems. Overall, LCP has great potential to act both as a substrate and a package.

The numerous benefits of using LCP as an organic platform include [18]:

- excellent electrical properties up to millimeter-wave frequencies (stable ε_r of 3.16 $\pm$ 0.05 and a low loss tangent from 0.0028 to 0.0045 for $f < 110$ GHz);
- low cost;
- quasi-hermetic (water absorption $< 0.04\%$);
- a low *xy* coefficient of thermal expansion (CTE), which may be engineered to match metals or semiconductors;
- thermally stable electrical characteristics as compared with many other substrates;
- lamination capabilities to generate homogenous multi-layer RF architectures;
- relatively low temperature processing (~290°C);
- flexibility for conformal and/or flex-circuit applications;

Table 8.2 Typical Characteristics of Liquid Crystal Polymer (LCP) Films [19]

	Value		
Property	ULTRALAM 3850	ULTRALAM 3908	Unit
Mechanical Properties			
Dimensional MD	−0.06	< 0.1	%
Stability CMD	−0.03	< 0.1	
Tensile strength	200	216	MPa
Tensile modules	2255	2450	MPa
Density	1.4		gm/cm^3
Thermal Properties			
Coefficient of thermal expansion, CTE			
CTE X (30°C to 150°C)	17	17	ppm/°C
CTE Y (30°C to 150°C)	17	17	ppm/°C
CTE Z (30°C to 150°C)	150	150	ppm/°C
Melting temperature	315	280	°C
Relative thermal index (RTI)			
RTI Mechanical	190	190	°C
RTI Electrical	240	240	°C
Thermal conductivity	0.5		W/m·K
Electrical properties			
Dielectric constant (10 GHz, 23°C)	2.9	2.9	
Dissipation factor (10 GHz, 23°C)	0.0025	0.0025	
Surface resistivity	1.0×10^{10}	1.0×10^{12}	MOhm
Volume resistivity	1.0×10^{12}	2.6×10^{14}	MOhm cm
Dielectric breakdown strength	1378	118	KV/cm
Environment properties			
Chemical resistance	98.7	98.7	%
Water absorption (23°C, 24-h)	0.04	0.04	%
Coefficient of hygroscopic expansion, CHE (60°C)	4	4	ppm/%RH
Flammability	VTM-0	VTM-0	

- naturally non-flammable;
- recyclable.

In the light of these advantageous characteristics, various miniaturized LCP filters have been developed for a wide range of applications. Design methodology for multi-layer coupled line filters can be effectively used [20]. At the same time, a classical approach to the filter design using lumped components is promising for multi-layer structures. For example, in [21] a miniature quasi-lumped-element wideband bandpass filter was presented based on LCP technology (0.5–2 GHz) by cascading low-pass and high-pass filters, Figure 8.5. The multi-layer structure is shown in Figure 8.6.

UWB filters have also been designed using multi-layer LCP technology [22–24]. With multi-layer capability, not only broadside coupling can easily be implemented for designing an ultra-wideband bandpass filter, but also more functionality can be integrated in a compact structure. For millimetre-wave applications, the development of

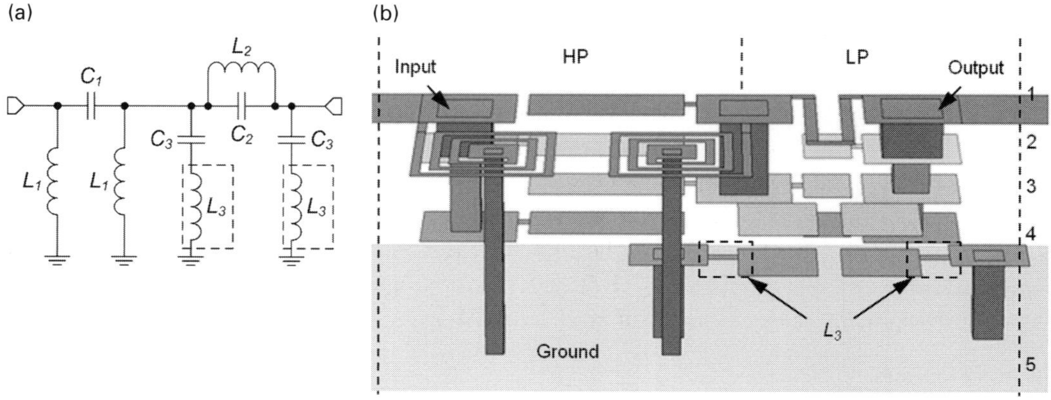

Figure 8.5 UWB filter implemented in LCP package. a) Equivalent diagram. b) Multilayer LCP structure.

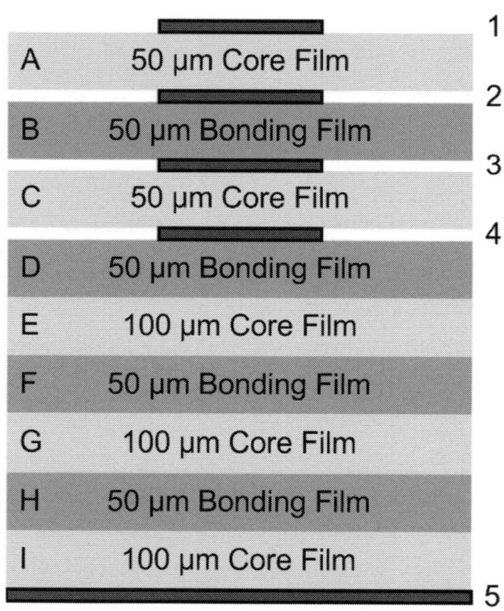

Figure 8.6 Multilayer structure of the filter in Figure 8.5.

integrated passive devices for V-band (50–75 GHz) transceiver applications using LCP technology was reported [25], where filter, matching networks, duplexers, and antenna arrays have been designed, implemented, and characterized.

8.3 Design of miniature microwave filters based on RH/LH transmission lines using multi-layer technology

In this section, we present resonators and filters realized in multi-layer architecture that have been designed using different methods. We present compact bandpass filters designed with quasi-lumped elements embedded in multi-layer packages. We give a thorough discussion on behavior of artificial transmission lines (TL) exhibiting right-handed (RH) and left-handed (LH) properties and propose miniature single-band and dual-band filters based on a combination of such transmission line sections. Among them the following filters are considered: the bandpass filter based on the signal-interference concept; ultra-wideband filter based on a section of composite right/left-handed (CRLH) transmission lines; dual-band filter using dual-mode resonators on RH/LH TLs; tunable dual-band filters on RH/LH TLs. Since all these filters are designed using a combination of RH/LH TLs, for a better understanding we give a brief description of the main properties of artificial RH- and LH-transmission lines.

8.3.1 Characteristics of right/left-handed (R/LH) artificial transmission lines

Different approaches are used for a description of the fundamental electromagnetic properties of metamaterials and the practical realization of these materials as well. Among them, the transmission line (TL) approach gives an efficient design tool for microwave applications providing a correct description of the physical properties of metamaterials [26–29].

A conventional transmission line with the positive phase velocity behaves as a right-handed transmission line (RH TL). An artificial RH TL can be formed as a ladder network of capacitors connected in shunt and inductors connected in series. The unit cell of the RH TL is shown in Figure 8.7(a). The dual transmission line can be designed as a ladder network of inductors connected in shunt and capacitors connected in series as in the unit cell shown in Figure 8.7(b). In this TL, the phase velocity is negative and the line is referred to as the left-handed transmission line (LH TL). A backward wave propagates along the LH TL, which can be considered as a one-dimensional metamaterial.

A more general model of an LH TL is the composite right/left handed (CRLH) structure, which includes RH effects [28]. In practical applications, the influence of RH effects is negligibly small, and many interesting features can be observed when a cascaded connection of RH TL and LH TL is used. The most important feature of the LH and RH TLs is that their dispersion characteristics are described by different equations. That feature can be used for many beneficial applications.

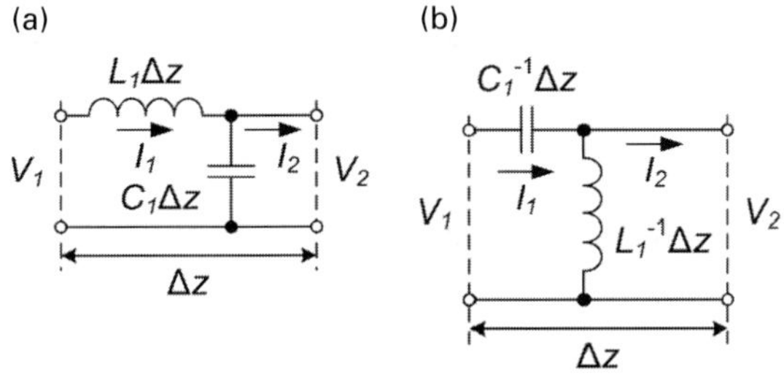

Figure 8.7 (a) Unit cell of the RH TL. (b) Unit cell of the LH TL.

The homogeneous RH TL presented as a cascaded connection of the unit cells is described by the telegraph equations. The voltage and current waves in the line are described as follows

$$V = V_0 e^{i(\omega t - kz)}, \tag{8.1}$$

$$I = I_0 e^{i(\omega t - kz)}, \tag{8.2}$$

with the wave number $k = k_R$ defined as

$$k_R = \omega\sqrt{L_1 C_1} > 0, \tag{8.3}$$

where L_1 and C_1 are the inductance and capacitance per unit length correspondingly. Both phase velocity v_{ph} and group velocity v_g are positive, therefore the RH TL supports the propagation of the forward wave. In line with (8.3), describing the dispersion law, the wave number is proportional to the frequency.

The characteristic impedance is defined as

$$Z_0 = \frac{V}{I} = \sqrt{\frac{L_1}{C_1}}. \tag{8.4}$$

For a homogeneous LH TL formed as a cascaded connection of the unit cells, Figure 8.7(b), the telegraph equations also can be written using the following parameters: the inductance per unit length $L^{-1}{}_1$and capacitance per unit length $C^{-1}{}_1$.

The solutions to the wave equations for voltage and current in this case look as (8.1) and (8.2) with the wave number defined as

$$k_L = -\frac{1}{\omega}\cdot\sqrt{\left(\frac{1}{L}\right)_1\cdot\left(\frac{1}{C}\right)_1} < 0. \tag{8.5}$$

The wave number and the phase velocity are both negative, whereas the group velocity is positive. Hence, the LH TL supports propagation of the backward wave. In line with (8.5), the wave number k_L (propagation constant) is inversely proportional to the frequency.

The characteristic impedance of the LH TL is defined as

$$Z_0 = \sqrt{\frac{\left(\frac{1}{C}\right)_1}{\left(\frac{1}{L}\right)_1}}. \tag{8.6}$$

Equations (8.3)–(8.6) describe the homogeneous infinitely long TL using the reactive parameters per unit length. A section of such a line of length l can be described by the electrical length θ defined as

$$\theta_{R,L} = k_{R,L}\cdot l. \tag{8.7}$$

In accordance with (8.3), the frequency dependence of the electrical length of a section of the RH TL is

$$\theta_R(\omega) = \theta_{0R}\frac{\omega}{\omega_0}, \tag{8.8}$$

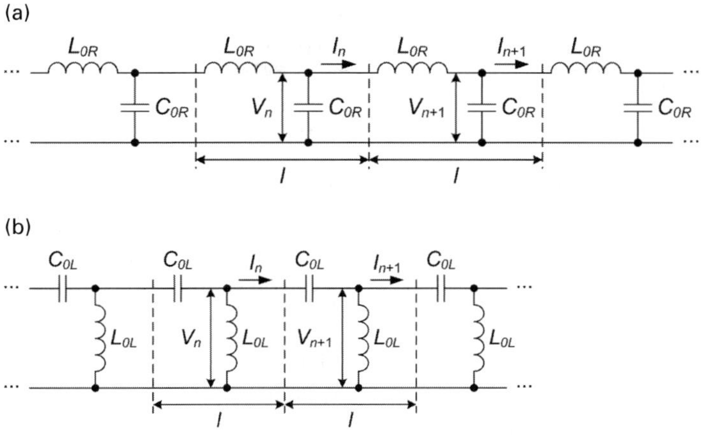

Figure 8.8 (a) Artificial RH TL on lumped elements. (b) Artificial LH TL on lumped elements.

where $\theta_{0R} = k_R l > 0$ is the electrical length at the frequency ω_0. In the case of a section of the LH TL, the frequency dependence of the electrical length is written as

$$\theta_L(\omega) = \theta_{0L}\frac{\omega_0}{\omega}. \tag{8.9}$$

Here $\theta_{0L} = k_L l < 0$ is the electrical length at the frequency ω_0.

In practice, the artificial RH and LH TLs can be composed as a periodical structure containing one inductive and one capacitive component in the unit cell of the length l defined as a real geometric length of the lumped components, Figure 8.8.

Using the translation symmetry for a description of the *I-V* wave propagating along the one-dimensional structure [30–31] and solving the telegraph equations for the RH TL, we obtain after some transformations the following dispersion equation

$$\sin^2\left(\frac{\theta}{2}\right) = \frac{1}{4}\omega^2 L_{0R}C_{0R} = \frac{\omega^2}{\omega_{cR}^2} \tag{8.10}$$

with the cut-off frequency

$$\omega_{cR} = \frac{2}{\sqrt{L_{0R}C_{0R}}}. \tag{8.11}$$

For $\omega > \omega_{cR}$, θ is an imaginary quantity and the wave attenuates: the higher frequency, the greater the attenuation is. In the low-frequency limit ($\omega << \omega_{cR}$)

$$\frac{\omega}{\omega_{cR}} = \pm\sin\left(\frac{\theta}{2}\right) \cong \pm\frac{\theta}{2} \tag{8.12}$$

and

$$k_R = \frac{2\omega}{l\omega_{cR}} = \omega\sqrt{L_{1R}C_{1R}} \tag{8.13}$$

with $L_{1R} = L_{0R}/l$ and $C_{1R} = C_{0R}/l$. The artificial lumped element RH TL behaves at $\omega << \omega_c$ as an infinitely long perfect TL with the linear dispersion law. In a wide frequency range, the artificial periodic RH TL, Figure 8.8(a) is considered as a low-pass lumped element TL.

Now we consider the LH TL shown in Figure 8.8(b). It is important to mention that the real capacitances and inductances in this artificial line are related to the parameters per unit length in Figure 8.7(b) as $C_{0L} = [C_1{}^{-1}l]^{-1}$ and $L_{0L} = [L_1{}^{-1}l]^{-1}$.

The same consideration of the artificial LH TL, Figure 8.8(b), as was used for the RH TL, Figure 8.7(a), leads to the following dispersion equation

$$\sin^2\left(\frac{\theta}{2}\right) = \frac{1}{4}\frac{1}{\omega^2 L_{0L} C_{0L}} = \frac{\omega_{cL}^2}{\omega^2} \tag{8.14}$$

with the cut-off frequency

$$\omega_{cL} = \frac{1}{2\sqrt{L_{0L} C_{0L}}}. \tag{8.15}$$

For $\omega < \omega_{cL}$, θ is imaginary and the wave attenuates: the lower frequency, the more is the attenuation. In the high-frequency limit ($\omega >> \omega_{cL}$)

$$\frac{\omega_{cL}}{\omega} = \pm \sin\left(\frac{\theta}{2}\right) \cong \pm\frac{\theta}{2} \tag{8.16}$$

and

$$k_L = \frac{1}{\omega}\cdot\sqrt{\left(\frac{1}{L}\right)_{1L}\left(\frac{1}{C}\right)_{1L}} \tag{8.17}$$

with $\left(\frac{1}{L}\right)_{1L} = \left(\frac{1}{L_{0L}}\right)\cdot\frac{1}{l}$ and $\left(\frac{1}{C}\right)_{1L} = \left(\frac{1}{C_{0L}}\right)\cdot\frac{1}{l}$. The artificial lumped element LH TL, Figure 8.8(b), behaves as an infinitely long perfect TL with k_L inversely proportional to ω. The artificial periodic LH TL is considered as a high-pass lumped element TL. The dispersion characteristics of the RH and LH TLs are shown in Figure 8.9. In further consideration, the frequency range is limited by the inequalities: $\omega << \omega_{cR}$ for the RH TL and $\omega >> \omega_{cL}$ for the LH TL.

Taking into account the symmetry of the section of a homogeneous TL, it is possible to replace it by the symmetric lumped element T- or Π-circuits [32]. The unit cells

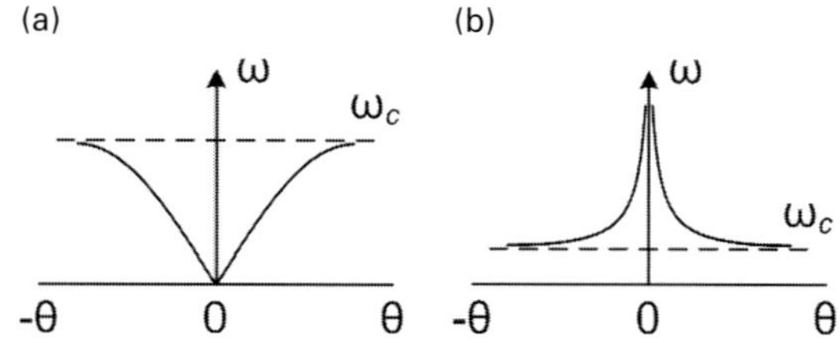

Figure 8.9 (a) Dispersion characteristic of RH lumped element TL. (b) Dispersion characteristic of LH lumped element TL.

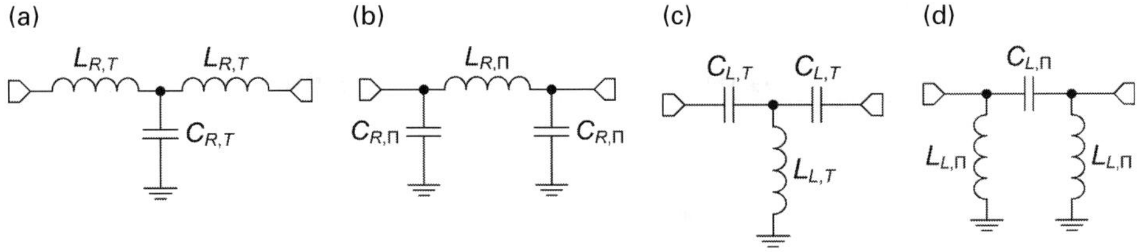

Figure 8.10 Lumped-element equivalent circuits of the RH TL and LH TL sections. (a) T-section of RH TL. (b) Π-section of RH TL. (c) T-section of LH TL. (d) Π-section of LH TL.

presented in Figure 8.10 are used as equivalent circuits of the sections of RH and LH transmission lines. It is supposed that the geometrical length of the lumped element unit cell is equal to zero. The L-C components of the equivalent unit cells designed as T- and Π-circuits can be found by the following equations [32]

$$L_{R,T} = \frac{Z_0 \tan(\theta_{0R}/2)}{\omega_0}, \quad C_{R,T} = \frac{\sin\theta_{0R}}{\omega_0\ Z_0}, \tag{8.18}$$

$$L_{R,\Pi} = \frac{Z_0 \sin\theta_{0R}}{\omega_0}, \quad C_{R,\Pi} = \frac{\tan(\theta_{0R}/2)}{\omega_0\ Z_0} \tag{8.19}$$

for the RH TL and

$$L_{L,T} = \frac{Z_0}{\omega_0\ \sin|\theta_{0L}|}, \quad C_{L,T} = \frac{1}{\omega_0\ Z_0 \tan(|\theta_{0L}|/2)}, \tag{8.20}$$

$$L_{L,\Pi} = \frac{Z_0}{\omega_0\ \tan(|\theta_{0L}|/2)}, \quad C_{L,\Pi} = \frac{1}{\omega_0\ Z_0 \sin|\theta_{0L}|} \tag{8.21}$$

for the LH TL.

At the frequency range close to the chosen frequency ω_0, the characteristics of the T or Π circuits with LC parameters defined by (8.18)–(8.21) are the same as for the corresponding RH/LH TL sections with known Z_0 and θ. This equivalent presentation of the RH/LH TL sections is used in the design of microwave devices.

A generalized approach to the metamaterial TLs was suggested in [28, 33] introducing the concept of a composite right/left-handed (CRLH) transmission line. Figure 8.12(a) shows the equivalent circuit of a unit cell of a transmission line containing series capacitor C_L and the inductance L_R. These reactive components form a series resonance circuit with resonance frequency ω_{se}. The second part of the unit cell is represented by a parallel circuit with capacitor C_R and inductance L_L. This circuit has resonance frequency ω_{sh}. At low frequencies, the properties of the transmission line are determined primarily by parameters L_L and C_L, which provide a negative phase velocity of the wave in the transmission line, that is impart the properties of an LH TL to the line. At higher frequencies, the properties of the transmission line are governed primarily by reactive components L_R and C_R, which provide a positive phase velocity of the wave in the transmission line, that is impart the properties of a RH TL to the line (subscripts R and L refer to the RH- and LH TLs, respectively).

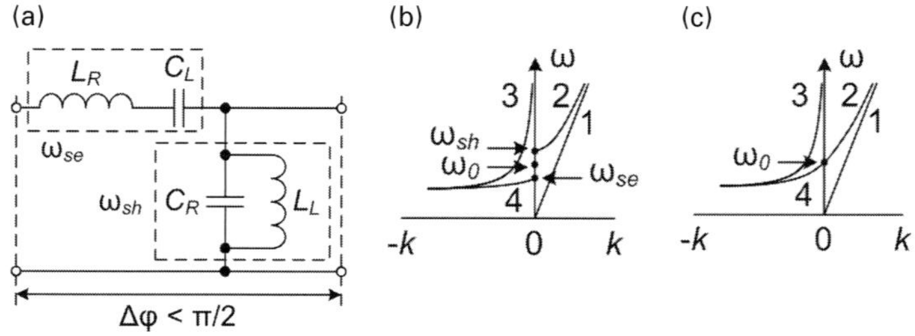

Figure 8.11 (a) Equivalent circuit of a section of CRLH TL. (b) Dispersion diagram of the CRLH TL: unbalanced case. (c) Dispersion diagram of the CRLH TL: balanced case.

The dispersion diagram of the unit cell [28] is presented in Figure 8.11(b) by curves 2 and 4. Curves 1 and 3 correspond to ideal RH TL and LH TL. At higher frequencies, the absolute value of the impedance of C_L tends to zero, while that of the impedance of L_L goes to infinity; accordingly, the cell contains only series inductance L_R and parallel capacitance C_R. The unit cell acts as a low-pass filter with upper cut-off frequency ω_{sh}. Accordingly, at low frequencies, the cell has the properties of a high-pass filter with lower cut-off frequency ω_{se}. In the general case, the frequencies ω_{sh} and ω_{se} are different and therefore there exists a forbidden gap between the positive and negative dispersion regions. However, if these frequencies are the same, the gap disappears, Figure 8.11(c). In this case, the unit cell is said to be "balanced" and the length of a wave propagating in the transmission line at frequency ω_0 tends to infinity.

8.3.2 Design of multi-mode resonators based on a combination of R/LH TL sections

Resonator on a transmission line section has a set of resonant frequencies, which are defined by the following resonance condition:

$$\theta_\Sigma = \pi \cdot n, \tag{8.22}$$

where $n = 0, \pm 1, \pm 2, \pm 3, \ldots$, θ_Σ is the total electrical length of the resonator and n is the resonance number. For the resonator on the natural TL section (RH TL), n is a positive number. In case of a cascaded connection of RH and LH TL sections, the spectrum of the resonant frequencies is formed by positive and negative numbers of responses, including the zero-order resonance [28]. The spectrum of the resonant frequencies is a set of values defined by the parameters of the RH and LH TL sections.

Let us consider the equivalent diagram of the symmetrical structure of the RH-LH-RH resonator, Figure 8.12. The resonator is formed by a cascaded connection of the RH and LH TL sections.

The resonance condition is defined as [34–35]

$$\cos^{-1}\left[\cos\left(2\theta_R\cdot\frac{f_{+n}}{f_0}\right)\cos\left(2\theta_L\cdot\frac{f_0}{f_{+n}}\right)+0.5(Z_{RL}^{-1}-Z_{RL})\sin\left(2\theta_R\cdot\frac{f_{+n}}{f_0}\right)\sin\left(2\theta_L\cdot\frac{f_0}{f_{+n}}\right)2\theta_L\right]=n\pi, \tag{8.23}$$

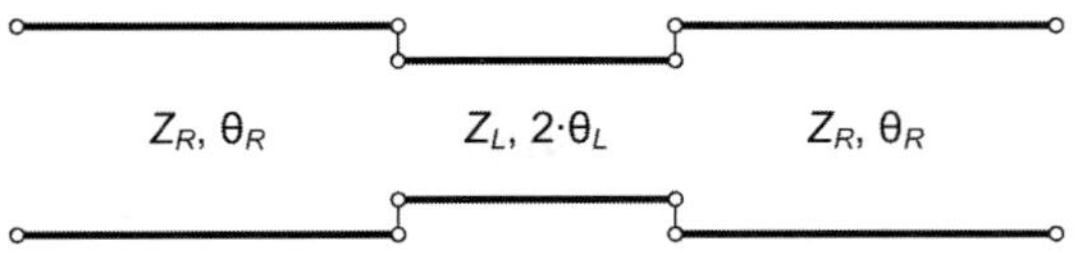

Figure 8.12 Symmetrical structure of the RH-LH-RH (R-L-R) resonator.

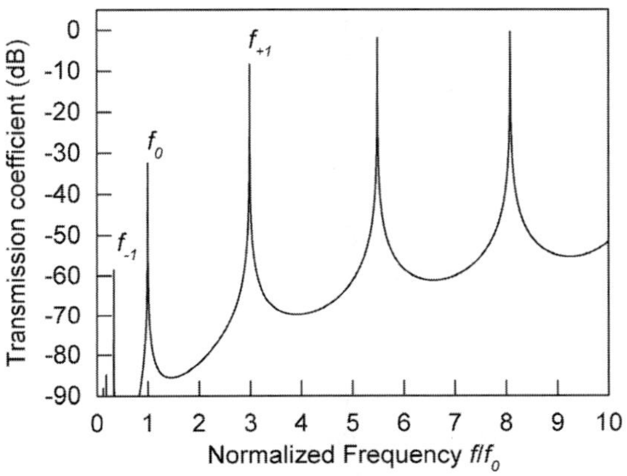

Figure 8.13 Resonant responses of the RH-LH-RH resonator shown in Figure 8.12 with $Z_{RL} = 1$.

$$\cos^{-1}\left[\cos\left(2\theta_R\cdot\frac{f_{-n}}{f_0}\right)\cos\left(2\theta_L\cdot\frac{f_0}{f_{-n}}\right)+0.5(Z_{RL}^{-1}-Z_{RL})\sin\left(2\theta_R\cdot\frac{f_{-n}}{f_0}\right)\sin\left(2\theta_L\cdot\frac{f_0}{f_{-n}}\right)2\theta_L\right]=-n\pi, \tag{8.24}$$

where f_n and f_{-n} are the resonant frequencies for $n > 0$ and $n < 0$ correspondingly, θ_R and θ_L are the electrical lengths of the TL sections at the chosen frequency f_0, and $Z_{RL} = Z_R/Z_L$ is the ratio of the characteristic impedances of the RH and LH TL sections. The resonant frequency spectrum of the resonator is shown in Figure 8.13. The number of resonances is infinite.

The resonator based on a combination of RH/LH TL sections has a unique resonant property, referred to as the zero-order resonance when the phase constant of the dominant mode is zero. The phase and amplitude of standing waves in the resonator are distributed uniformly over the whole structure at the resonant frequency. Since the LH TL does not exist in nature, the practical realization is provided by using artificial TL sections. The resonator is designed as a cascaded connection of T- or Π-lumped element RH and LH TL unit cells shown in Figure 8.10. The resonant frequency depends on the configuration and parameters of the unit cells used in the resonator design.

The zero-order resonator can be designed using the resonance condition defined by the following system of equations

$$\cos^{-1}[\cos 2\theta_R \cos 2\theta_L + 0.5(Z_{RL}^{-1} - Z_{RL}) \sin 2\theta_R \sin 2\theta_L] = 0, \tag{8.25}$$

$$\sin^{-1}\left[\frac{2 \sin \theta_R \cos \theta_L}{(Z_{RL} - Z_{RL}^{-1}) + (Z_{RL} + Z_{RL}^{-1}) \cos \theta_R}\right] - \theta_L = 0. \tag{8.26}$$

In many practical applications, it is necessary to reject higher harmonics. The popular way to suppress the higher harmonics is using a stepped-impedance resonator (SIR). In order to obtain more freedom in the design of a resonator with suppressed harmonic responses, a SIR based on a cascaded connection of the LH TL and RH TL sections is considered [34–36]. In contrast to the ordinary SIR on RH transmission line sections, the advanced structure gives a possibility to change the resonant frequency of higher modes by using different dispersion characteristics of the RH and LH TLs even in the case of the same characteristic impedance of the TL sections. Since the LH TL can be realized as the lumped element equivalent of the LH TL only, such a design makes it possible to decrease the resonator dimensions drastically, just using the RH TL part of the resonator as a distributed TL section.

The ratio of the characteristic impedances $Z_{RL} = Z_R/Z_L$ and the electrical lengths of the transmission line sections θ_R and θ_L are considered as variable parameters of the SIR. For the two-mode resonator, these parameters can be found from equations written for two modes with the resonance frequencies ω_0 (n = 1) ω_1 (n = 2)

$$\tan(\theta_{0L}) = -Z_{RL} \cot(\theta_{0R}), \tag{8.27}$$

$$\tan\left(\theta_{0L}\frac{\omega_0}{\omega_1}\right) = Z_{RL}^{-1} \tan\left(\theta_{0R}\frac{\omega_1}{\omega_0}\right). \tag{8.28}$$

Here θ_{0R} and θ_{0L} are the electrical lengths of RH and LH TL sections at the fundamental resonant frequency ω_0. The similar resonance conditions can be defined for any pair of adjacent resonant frequencies.

In line with (8.27) and (8.28), the shorter the length of the RH TL section, the higher is the resonant frequency ω_1. At the same time, the higher the ratio Z_R/Z_L, the higher is ω_1 and the more effective is the suppression of the higher harmonics. This is illustrated in Figure 8.15 with the results of the simulation of the SIR with different values of the length of the RH TL section (f_0 = 1.75 GHz).

The open-ended lumped element SIR based on a cascaded connection of RH and LH TL sections is shown in Figure 8.15(a). The resonator has the R-L-R structure with T-type of the RH TL section and Π-type of the LH TL section. The characteristic impedance and electrical length of each section is defined by the LC parameters. For the simulation of scattering parameters, the small coupling capacitances C_1 on input and output of the resonator have been introduced. The transmission characteristic with three resonance responses for n = -1, n = 0, and n = +1 is presented in Figure 8.15(b). In case of the SIR designed on lumped elements, the number of resonances is limited and only three resonances exist for the resonator under consideration. In general, the number of resonances depends on the number of L- and C-components and the method of their connections.

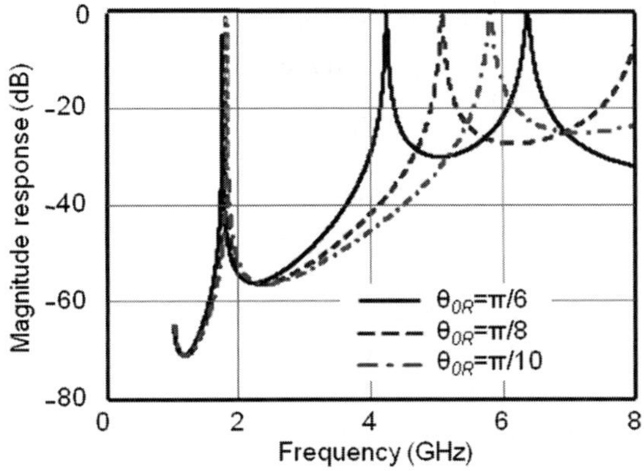

Figure 8.14 The resonant responses for $n = 1$ and $n = 2$ of the dual-mode resonator for different electrical lengths of RH TL sections and $Z_R/Z_L = 2.5$.

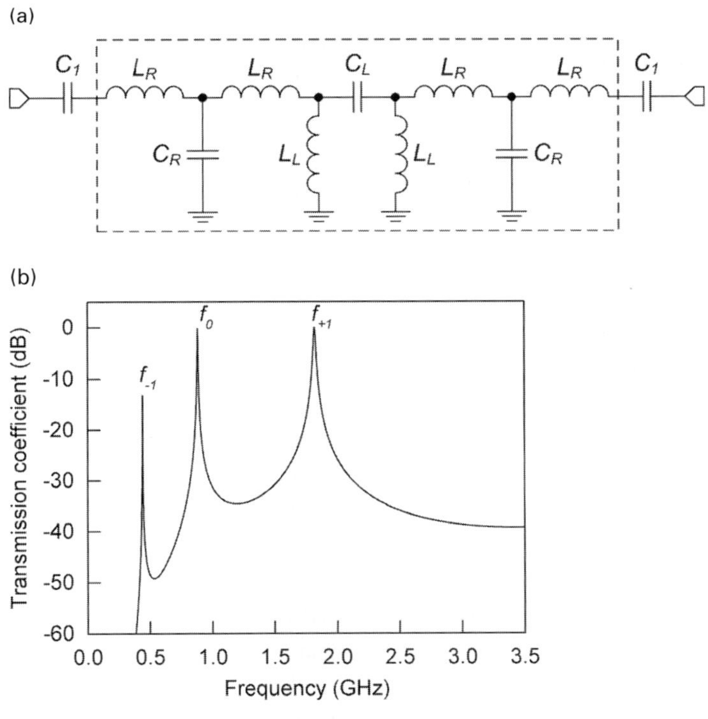

Figure 8.15 a) Equivalent diagram of the resonator on cascaded connection of the RH/LH TL sections (R-L-R-structure). b) Resonant responses of the R-L-R-resonator for $n = -1$, $n = 0$, and $n = +1$.

The same approach is applicable for a variety of possible combinations of TL sections in a resonator: (i) only LH TL sections with $Z_{RL} = 1$ or $Z_{RL} \neq 1$; (ii) only RH TL sections with $Z_{RL} \neq 1$ (SIR); and (iii) cascaded LH and RH TL sections with $Z_{RL} = 1$ or $Z_{RL} \neq 1$.

The most attractive feature of the resonators with a cascaded connection of RH and LH TL sections using lumped components is their ability to provide any desirable number of resonant responses with arbitrary chosen frequencies.

8.3.3 Compact bandpass filter based on transversal signal-interference technique using RH/LH TL sections

Among the various filter topologies and design techniques, the signal-interference technique is one of the popular solutions for microwave planar filter design, since filtering functions with sharp skirt selectivity can be attained in simple networks. The usefulness of signal-interference filters has been proven in the past for moderate-to-broadband bandpass/bandstop specifications with both single- and multi-band characteristics [37–40]. However, as the main drawbacks, these filter designs presented excessive circuit sizes and difficulties in carrying out frequency-asymmetrical filtering actions, as they exclusively consisted of distributed transmission-line segments whose electrical lengths were selected as multiples of $\pi/2$ (e.g. $3\pi/2$ used in [38]) at a chosen design frequency. In addition, in order to broaden stopband bandwidths, these filters usually require more than one transversal filtering section, as well as extra matching lines, leading to increased circuit size as well. Even so, there are always some signals transmitted near the dc frequency, which degraded the out-of-band performance to some extent.

To overcome the aforementioned shortcomings, a novel signal-interference filter using artificial LH and RH TL sections is discussed in this section. As demonstrated in [26–28] [32], and [41], the LH TL supports backward-wave propagation and exhibits negative electrical length. Hence, replacing the conventional TL with long electrical length by a short artificial LH TL in the single-interference filter, the circuit size can be significantly reduced. Additionally, it is noteworthy that owing to the high-pass nature of the LH TL, the proposed filter achieves an advanced characteristic of the frequency-asymmetrical stopband. To further reduce the dimensions of the filter, all the artificial RH and LH TL sections are implemented in the form of multi-layered structures using LCP technology.

Figure 8.16 presents a miniature bandpass filter topology having two transmission zeros placed near both passband edges [42]. As shown, the filter structure is based on the transversal signal-interference concept, but consists of a LH TL and a RH TL sections connected in-parallel. The RH and LH TLs have the characteristic impedances Z_R, Z_L and the electrical lengths θ_R, θ_L, respectively. It is assumed that $\theta_R = 90°$ and $\theta_L = -90°$ at the centre frequency f_0.

Based on the fundamental theory of the artificial transmission lines, as indicated in Figure 8.16(b), the RH TL section is implemented as a cascade of two identical T-networks consisting of the series inductances L_R and the shunt capacitance C_R. The LH TL section is realized as one unit-cell of the Π-network with the shunt inductances L_L and the series capacitance C_L. The relevant circuit parameters of each unit cell can be easily derived using (8.18) and (8.21).

It is well known that during the design process of the signal-interference filters, the electrical lengths and characteristic impedances of the transversal filter lines are the key

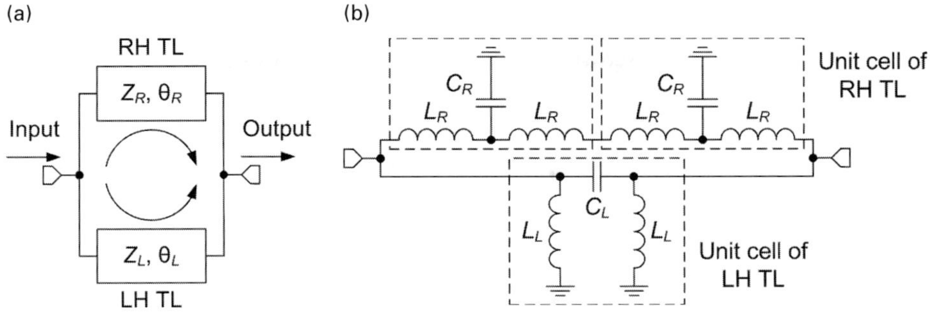

Figure 8.16 (a) The bandpass filter structure based on signal-interference concept. (b) Equivalent circuit of the filter.

parameters in achieving proper amplitude and phase responses. Specifically, the generation of perfect-constructive signal interference at the desired centre frequency is the main condition to be forced. For the structure in Figure 8.16(a), the normalized admittance matrix at the desired centre frequency f_0 can be written as

$$[Y(f_0)] = [Y(f_0)]_R + [Y(f_0)]_L = j\begin{bmatrix} 0 & -y_R + y_L \\ -y_R + y_L & 0 \end{bmatrix}, \tag{8.29}$$

where $y_R = Z_0/Z_R$ and $y_L = Z_0/Z_L$.

In turn, the power reflection coefficient of the proposed filter at the central frequency can be readily obtained by the transformation

$$S_{11}(f_0) = \frac{(1+Y_{11})(1+Y_{22}) + Y_{12}Y_{21}}{(1+Y_{11})(1+Y_{22}) - Y_{12}Y_{21}} = \frac{1-(y_R-y_L)^2}{1+(y_R-y_L)^2}. \tag{8.30}$$

To this end, if a desired value of the reflection coefficient at the central frequency is given as $S_{11}(f_0) \leq k$, the admittances y_R and y_L of the RH TL and LH TL should satisfy the following condition:

$$\frac{1-k}{1+k} \leq (y_R - y_L)^2 \leq \frac{1+k}{1-k}. \tag{8.31}$$

Note, there are two optional solutions obeying (8.31) when the admittance y_R is either smaller or larger than the admittance y_L. For better understanding, let us illustrate this fact by some typical examples summarized in Table 8.3, where we assume $k = 0.316$ corresponding to the desired reflection coefficient value $S_{11}(f_0) \leq 10 \quad dB$. For the given passband specification, once the impedance Z_L is determined at the earlier step of the filter design, then the range of approximate values of the impedance Z_R can be obtained or vice versa. However, it was found that two transmission zeros appear around the passband allowing to improve the filter selectivity only when $Z_R > Z_L$, Figure 8.17.

In addition, the fractional bandwidth is mainly determined by the impedance ratio Z_R/Z_L: the broader bandwidth is obtained as the higher impedance ratio is chosen. For example, as implied in Figure 8.18, the fractional bandwidth of around 14% is provided

Table 8.3 Typical examples of filter RH TL and LH TL characteristic impedance values

$k = 0.316$ (S_{11} <= – 10 dB)		
LH TL impedance	**RH TL impedance**	
	$Z_R > Z_L$ ($y_R < y_L$)	$Z_R < Z_L$ (($y_R > y_L$))
Z_L = 50 Ohm (y_L = 1)	Z_R > 178 Ohm	29 Ohm > Z_R > 21 Ohm
Z_L = 100 Ohm (y_L = 0.5)	None	40 Ohm > Z_R > 26.5 Ohm
Z_L = 25 Ohm (y_L = 2)	82 Ohm > Z_R > 39 Ohm	18 Ohm > Z_R >14 Ohm
Z_L = 10 Ohm (y_L = 5)	14 Ohm > Z_R > 11 Ohm	9 Ohm > Z_R > 8 Ohm

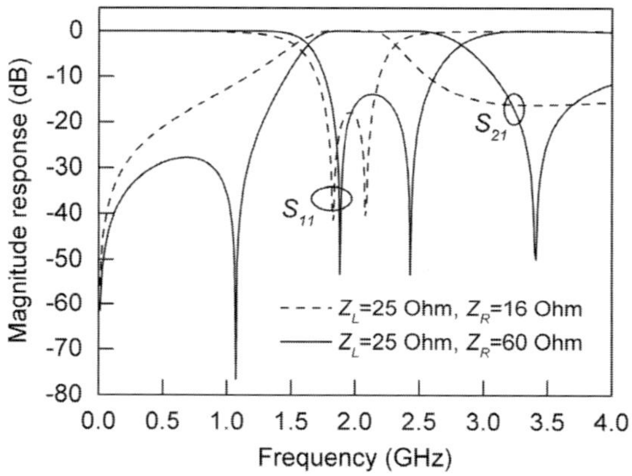

Figure 8.17 Typical frequency response of the proposed filter when Z_R>Z_L (solid lines) and Z_R<Z_L (dashed lines).

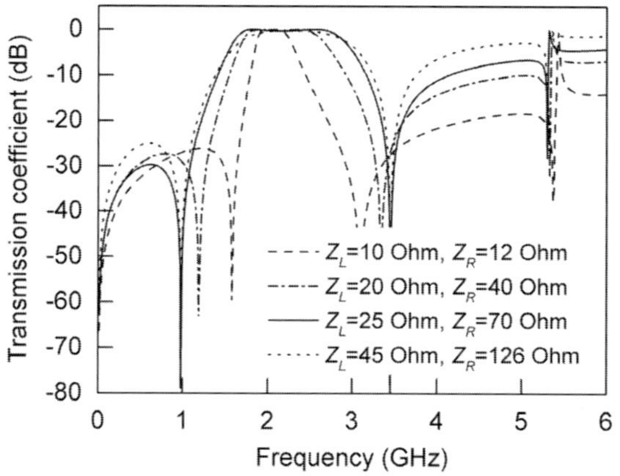

Figure 8.18 Transmission coefficient versus frequency for filters with different fractional bandwidths corresponding to different impedance ratio Z_R/Z_L.

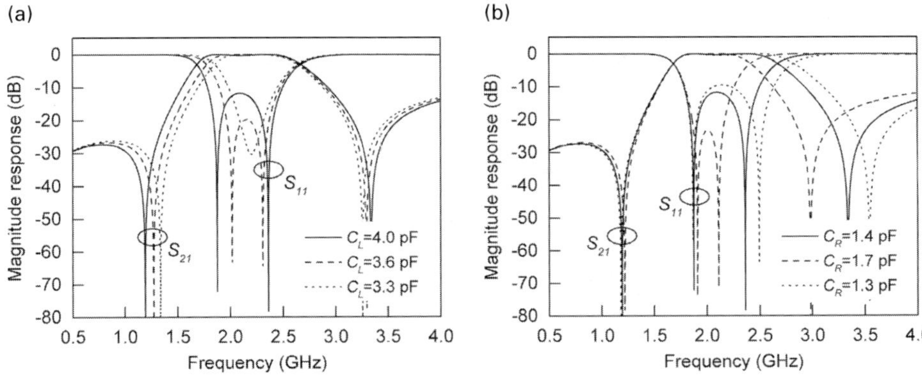

Figure 8.19 Simulated frequency response of the filter shown in Figure 8.16 (b) by adjusting the capacitances C_L(a) and C_R (b).

for the filter with Z_R = 12 Ohm and Z_L = 10 Ohm, whereas the use of Z_R = 70 Ohm and Z_L = 25 Ohm allows to achieve the fractional bandwidth of 61%. The two filter examples illustrated in Figure 8.18 have the same impedance ratio (Z_R/Z_L = 70/25 = 126/45 = 2.8) and are alike in frequency response. However, when increasing the impedance ratio to achieve a wider passband, the rejection level in the stopband gets worse accordingly, Figure 8.18. Hence, there is a trade-off between the bandwidth and stopband performance.

Figure 8.19 demonstrates that the filter response can be adjusted and optimized by varying the capacitances C_L and C_R. In Figure 8.19(a), the transmission zero below the passband is tuned by changing the values of C_L. Also the lower cut-off frequency and the location of the lower pole can be adjusted accordingly, while the upper ones remain almost unchanged. On the contrary, as shown in Figure 8.19(b), by changing the C_R values, the transmission zero above the passband, as well as the upper cut-off frequency and the location of the upper pole are adjusted more significantly than the lower ones. Hence, the locations of two transmission zeros and the bandwidth can be adjusted independently by modifying the capacitances C_L and C_R, with the fixed values of L_L and L_R that results in an optimized frequency response being obtained.

An experimental sample of the designed filter is fabricated using multi-layer LCP lamination technology. According to the above-mentioned design guidelines and filter equivalent circuit, the filter layout shown in Figure 8.20 is implemented as a five-metal-layer LCP structure using three types of LCP films (100 μm thick LCP core film, as well as 50 μm and 25 μm thick LCP bonding films). Specifically, the form of the parallel-plate capacitor is adopted to obtain the LH series capacitance C_L, which can be made extremely small by reducing the gap between the two plates. The LH shunt inductance L_L is generated by the stub lines. The high-impedance line linking two 50 Ohm termination ports introduces a small RH series inductance L_R and the spacing between the conductive plate and the ground enclosure introduces a RH shunt capacitance C_R.

The measured S parameters, the simulated filter performance obtained by Sonnet software [44], and the theoretical frequency response of the filter equivalent circuit are

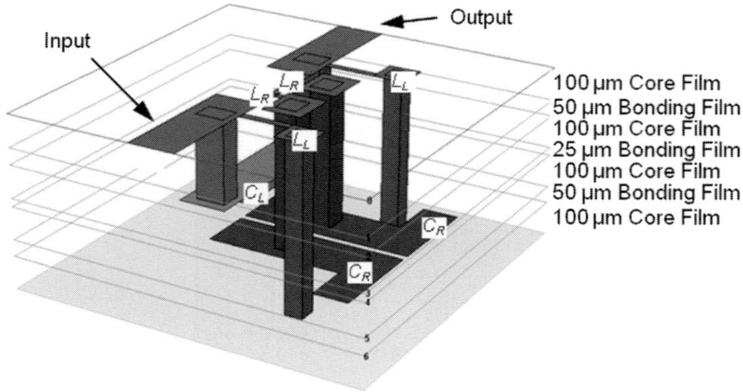

Figure 8.20 The multi-layer LCP implementation of the proposed filter based on the structure shown in Figure 8.16 (Z_L = 10 Ohm Z_R = 13 Ohm, L_L = 0.7958 nH, C_L = 7.9577 pF, L_R = 0.4285 nH, C_R = 4.3284 pF).

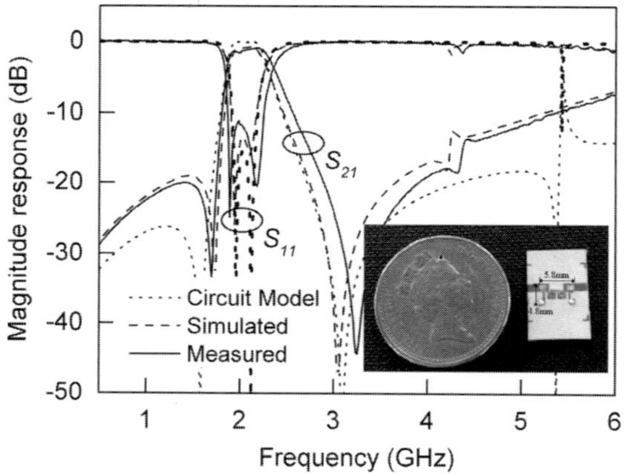

Figure 8.21 Measured S parameters of the filter compared with the results of the EM and circuit simulations.

shown in comparison in Figure 8.21. High performance is achieved for the filter with two transmission zeros designed for the centre frequency of 2.06 GHz and 3-dB bandwidth of 24%. The measured insertion loss in the passband is about 1.5 dB, and the measured transmission zeros are located at 1.69 GHz and 3.26 GHz, respectively.

The photograph of the fabricated filter prototype is shown in the inset to Figure 8.21. The size of the filter is 5.8 mm × 4.8 mm (0.06 λ_g × 0.05 λ_g), excluding the input and output microstrip lines. Furthermore, a simple performance comparison with other left-handed bandpass filters in the literature is summarized in Table 8.4 to highlight the merits of the proposed filter.

8.3.4 Ultra-wide-band filters based on composite R/LH transmission line

Since the authorization by the US Federal Communications Commission (FCC) in February 2002 for the unlicensed use of ultra-wideband (UWB) from 3.1 to 10.6 GHz

Table 8.4 Comparison with other reported left-handed bandpass filters

Ref.	Fabrication technology	f_0 (GHz)	Size (mm)	Number of transmission zeros
[42]	PCB ($\varepsilon_r = 2.2$)	4.35	6.07 × 5.14	1
[43]	PCB ($\varepsilon_r = 10.2$)	3.5	4.29 × 5.09	1
This work	**LCP ($\varepsilon_r = 3.0$)**	**2.06**	**5.80 × 4.80**	**2**

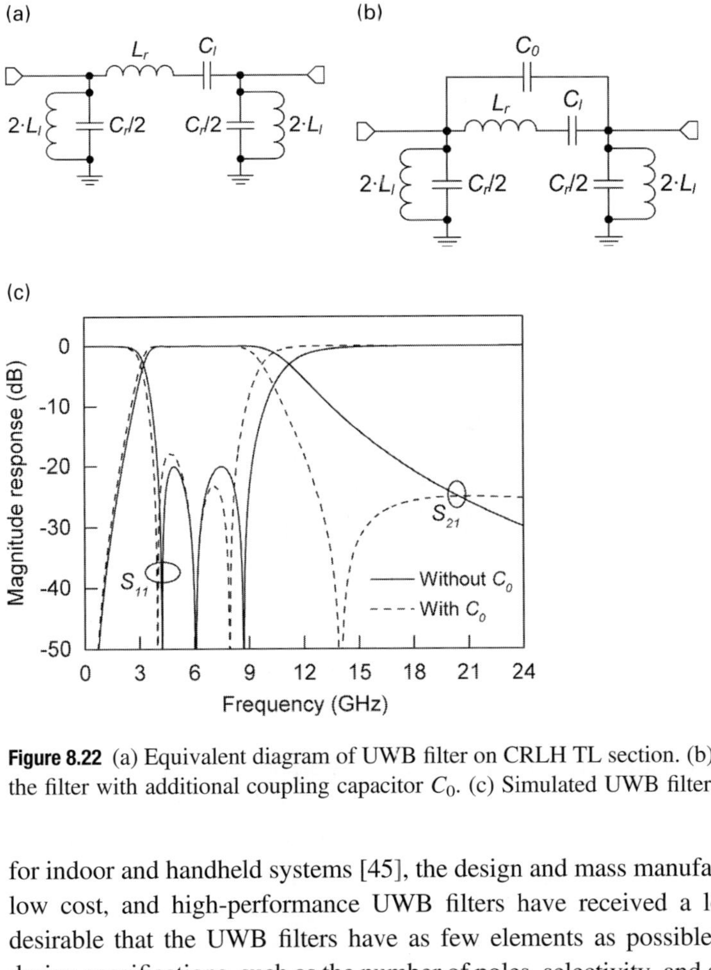

Figure 8.22 (a) Equivalent diagram of UWB filter on CRLH TL section. (b) Equivalent diagram of the filter with additional coupling capacitor C_0. (c) Simulated UWB filter performance.

for indoor and handheld systems [45], the design and mass manufacturing of small size, low cost, and high-performance UWB filters have received a lot of attention. It is desirable that the UWB filters have as few elements as possible, while still meeting design specifications, such as the number of poles, selectivity, and stopband attenuation. This does not only reduce the size and cost of the filter, but also passband insertion loss. Practical applications require that the designed filter should be easily mass manufactured with high yield. This requires that UWB filters have as simple a structure as possible to simplify the manufacturing process, reduce the cost, and improve yield.

CRLH TLs may be effectively used for UWB filter design. For this purpose, a traditional CRLH-TL [28] was suggested to be used. The filter equivalent circuit is shown in Figure 8.22(a). With this circuit, it can be easily seen, due to the high pass

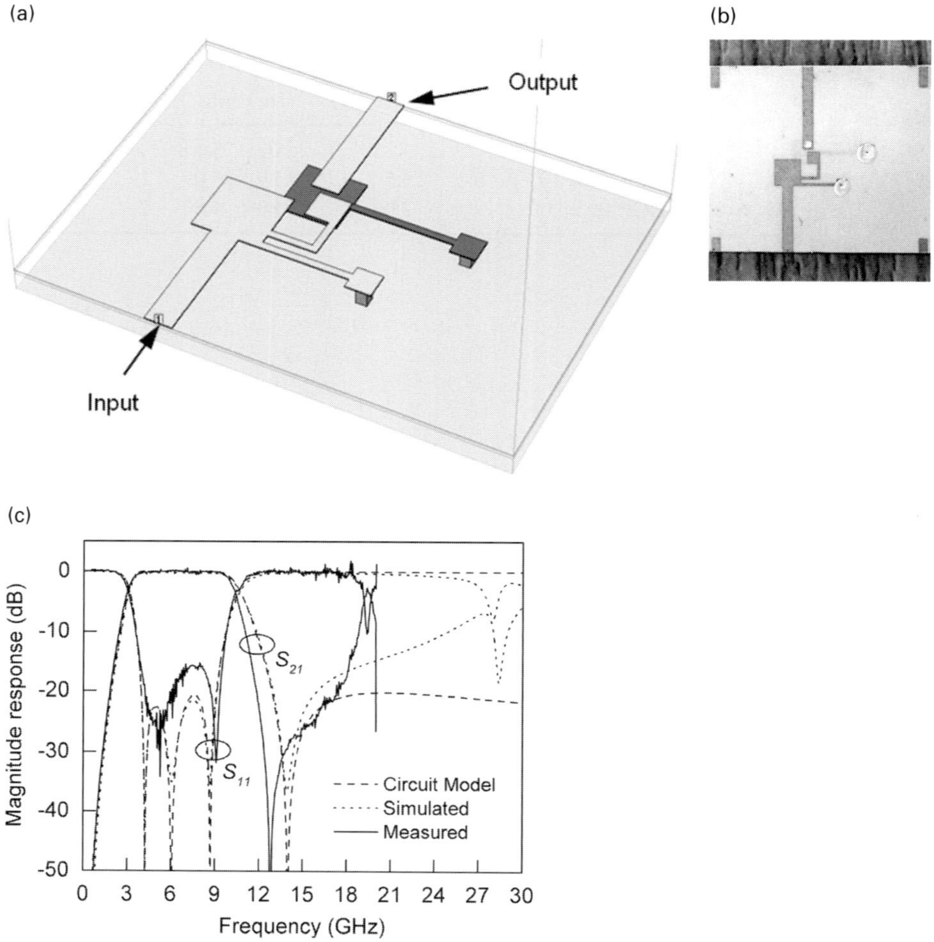

Figure 8.23. (a) LCP implementation of the UWB filter. (b) Photograph of the filter. (c) The filter performance.

nature of the LH elements (C_l and L_l) and the low pass nature of the RH elements (C_r and L_r), that a bandpass response is provided. The filter response can be improved if an additional transmission zero is produced at the upper stopband. To achieve this, a coupling capacitor C_0 between input and output ports is introduced, Figure 8.22(b) [21] [46].

The result of the circuit simulation is shown in Figure 8.22(c) (solid lines), in comparison with the result obtained without the coupling capacitor (dashed lines). The circuit efficiently produces a UWB response with additional transmission zero in the upper stopband. Only six quasi-lumped elements and a capacitive coupling are used. The filter was fabricated using LCP technology [46–47]. The multi-layer structure of the filter consists of five layers of Rogers ULTRALAM® 3000 series (ε_r = 3, tanδ = 0.0025). The LCP implementation of the filter is presented in Figure 8.23(a). The filter area is 13 mm x 10 mm, which corresponds to the linear dimensions of 0.4 λ_g at the

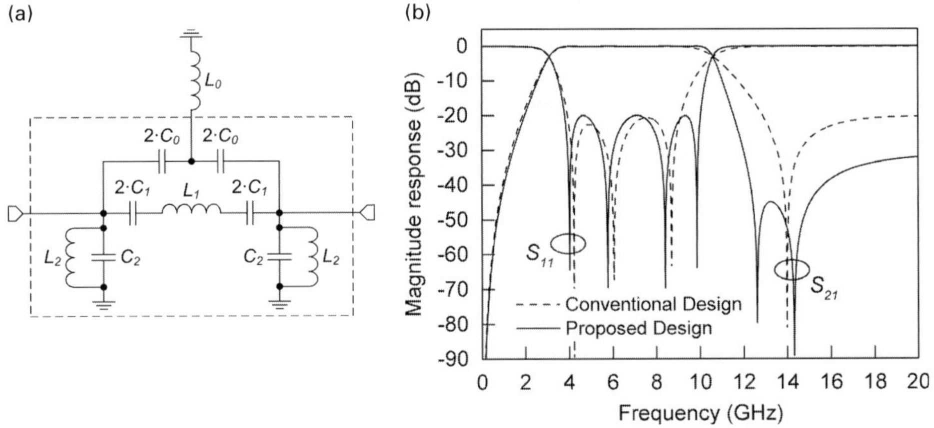

Figure 8.24 (a) Equivalent diagram of the UWB filter with resonant circuit between input and output. (b) Simulated filter performance.

central frequency f_0. The photograph of the filter is shown in Figure 8.23(b) and its experimental performance is demonstrated in Figure 8.23(c). The measured in-band insertion loss is less than 0.5 dB and in-band return loss is better than 15 dB.

The performance of the UWB filter can be improved by including resonant components between input and output for the realization of additional transmission zeros [47]. Figure 8.24(a) shows the corresponding circuit topology. The idea for this topology comes from a traditional cross-coupled lumped element bandpass filter, dash-circled in Figure 8.24(a), where the cross-coupling capacitor C_0 and series capacitor C_1 are split into two capacitors with capacitance $(2\times C_0)$ and $(2\times C_1)$, respectively, for symmetrical representation. This produces a transmission zero in the stopband as shown in Figure 8.24(b), dashed lines.

Additional inductance is added between the two cross-coupling capacitors and connected to the ground. As can be seen in Figure 8.24(b), compared to the classic cross-coupled filter, this topology gives not only an extra pole in the passband, but also provides a second transmission zero in the stopband, which gives high selectivity on the upper passband edge and a wide stopband with 31dB rejection theoretically.

The circuit topology shown in Figure 8.24(a) can be represented by the circuit model in Figure 8.25(a), and further the dash-circled T-section in Figure 8.25(a) can be transformed to the dash-circled Π-section in Figure 8.25(b) by using T- and Π-circuit equivalence. The obtained circuit can be easily described analytically in order to find S parameters and define the transmission zero locations.

To produce a UWB bandpass response with a 20 dB return loss (0.04 dB passband ripple) and 3 dB cut-off at 3.1 GHz and 10.6 GHz, a multi-layer lumped element UWB filter is designed. This design is implemented in a multi-layer LCP package (Rogers ULTRALAM® 3000).

The test structure is shown in a photo in Figure 8.26(a). The total thickness of the design is 0.4 mm and the separation between the first two metal layers is 0.05 mm to provide a strong coupling between the two layers. The size of the filter excluding the

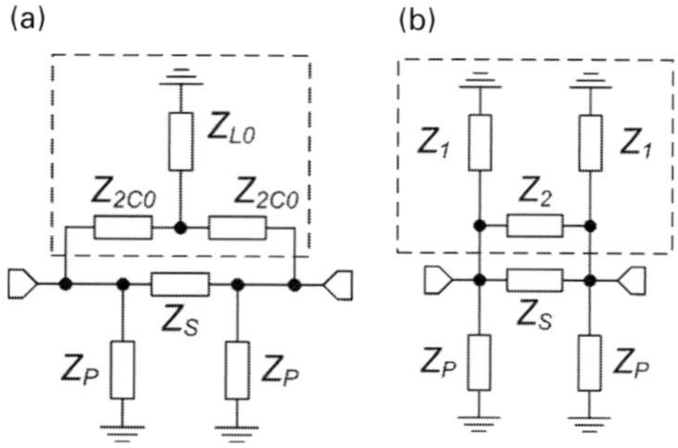

Figure 8.25 (a) Impedance representations of the proposed circuit model in Figure 24(a). (b) Simplified representations for S parameters calculations.

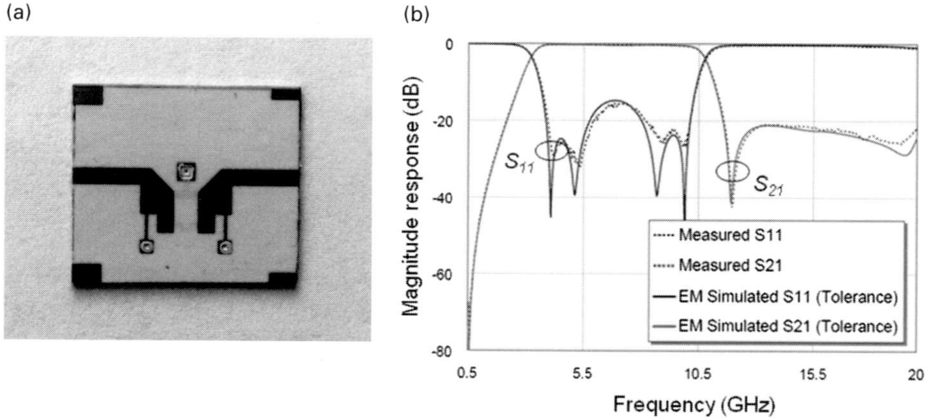

Figure 8.26 (a) Photo of the fabricated sample. (b) UWB filter performance.

50 Ohm feed line is 5.6 mm × 5.6 mm × 0.4 mm, which is 0.2 λ_g × 0.2 λ_g × 0.014 λ_g, where λ_g is the guided wavelength at the UWB centre frequency of 6.85 GHz. Compared with the previous miniature LCP UWB filter [46], it occupies 30% less planar area with similar performance. The test device connected with the Anritsu 3680 Test Fixture was measured on a HP8510 Network Analyzer. The measured result compared with the simulation is shown in Figure 8.26(b). It has less than 0.37 dB midband insertion loss, 15 dB return loss, and 21 dB stopband attenuation.

8.4 Dual-band microwave filters based on RH/LH transmission lines using multi-layer LTCC technology

In advanced multi-band wireless systems, microwave bandpass filters with dual-band operation are highly demanded for their potential for size and complexity reduction. As a rule, the multi-mode stub-loaded stepped-impedance resonators are designed to reduce the number of the assembled resonators and, thus, miniaturize the filter size [48].

In this section, multi-layer filters exhibiting dual-band response are considered. The filters are based on two-mode resonators designed as a cascaded connection of the LH TL and RH TL sections. Besides, composite right/left-handed (CRLH) transmission lines have been used to design dual-band immitance inverters. Additionally, a tunable counterpart of the dual-mode resonator is presented, which has been used for the design of a dual-band filter that exhibits a wide tunable range and high flexibility. All the devices are implemented in the multi-layer LTCC technology.

Band-pass filters designed based on a combination of artificial RH/LH TL sections using multi-layer LTCC technology demonstrate the following advantages:

- miniaturization is provided by the 3D design of the resonators and filters using a large number of ceramic layers (up to 16) of different thicknesses for the realization of passive components;
- a high degree of integrations is realized; the devices fabricated with LTCC technology represent a chip element, which can be used for surface mounting and is compatible with semiconductor monolithic integrated circuits (MIC); such a ready-made module makes it possible to not only reduce its weight and size, but also to reduce the cost of production;
- due to using lumped LC components for the design of resonators for sections of RH and LH TLs, the number of resonances is limited and no spurious passbands are observed;
- non-equidistant spectrum of the resonances is a key feature of the resonant structures using a combination of artificial TL sections with positive and negative dispersion, as well as the possibility to control the position of the resonance frequencies, which adds flexibility to the resonator (filter) synthesis procedure;
- introduction of additional redundant elements, such as additional transmission zeros provides increased frequency selectivity;
- with the help of control devices, such as semiconductor varactor diodes or MEMS devices used as surface mount components, tunable and reconfigurable frequency selective devices may be easily designed.

8.4.1 Two-mode resonators and dual-band filters based on a combination of R/LH TL sections

The design of dual-mode resonators, which find application in both single-band and multi-band filters is a challenging problem in many practical applications [49]–[51]. Using the RH/LH SIR for this purpose is beneficial because of the possibility to design a resonator with an arbitrary number of resonant responses at arbitrary frequencies, providing effective suppression of higher harmonics.

Let us consider the main approach to the design of a dual-mode resonator as a cascaded connection of the LH TL and RH TL sections. The dual-mode resonator can be designed using three unit cells (R-L-R) as shown in Figure 8.15. Different combinations of the unit cells may be used: L-R-L or R-L-R, T-Π-T or Π-T-Π, etc. The number of cells $N \geq 2$ can also be arbitrarily selected. For the open-ended resonator, the characteristics of the TL sections can be estimated for any arbitrary pair of resonant

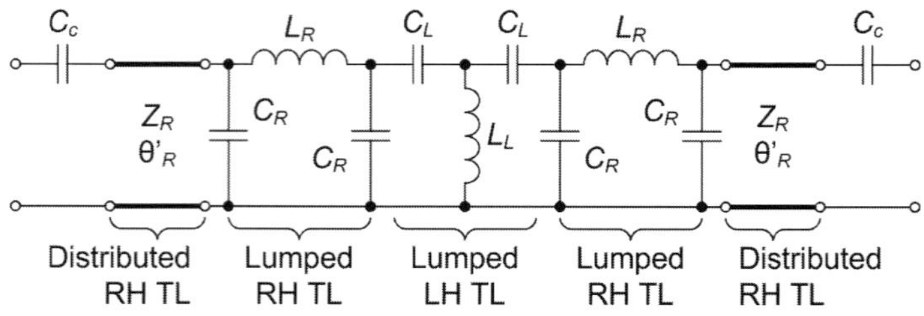

Figure 8.27 Equivalent diagram of the dual-mode resonator based on a combination of RH and LH TL sections.

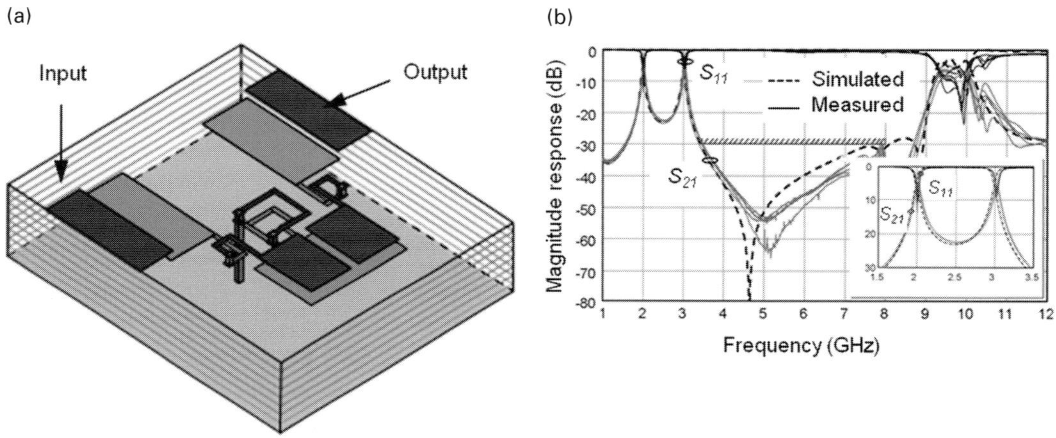

Figure 8.28 (a) Two-mode resonator with a highly suppressed spurious response: the multi-layer structure of the resonator. (b) Scattering parameters versus frequency.

frequencies solving the system of equations (8.25)–(8.28). The L-C components of the unit cells are to be found using (8.18)–(8.21).

Let us consider an example of the dual-mode resonator design. The equivalent diagram of the R-L-R resonator shown in Figure 8.27 is used as the dual-mode resonator with suppressed higher spurious responses. The frequencies ω_0 and ω_1 correspond to the lower and the higher resonant responses of the resonator. To decrease the resonator dimensions, the RH TL is designed as a cascaded connection of a Π-section of an RH TL and a distributed TL section with electrical length θ_R' and characteristic impedance Z_R. The LH TL section is designed as the T-section.

The structure of the dual-mode resonator designed for the resonant frequencies 2 and 3 GHz is presented in Figure 8.28(a) [52–53].The advanced multi-layer LTCC technology was used for resonator fabrication [54–55]. Grounded capacitances C_R are realized as parallel-plate capacitors. Series coupling capacitances C_L are produced by overlapping the electrodes of the capacitors placed in different layers. Turned stacked inductors L_R and L_L are used. The natural RH TL section is realized as a microstrip TL section with the corresponding characteristic impedance and electrical length.

The test structure of the resonator was fabricated using eight layers of the 95 μm thick LTCC DuPont Green Tape™ 951 (ε_r = 7.8, tan δ = 0.002 at 2 GHz) [53]. For conducting layer deposition, the silver ink DP6145 was used. The fired thickness of the metallization layer is 9–11 μm, while the resistivity is less than 6 mOhm/square. The paste was printed using 400 mesh metal screens with 15 μm film coating. The size of the LTCC multi-layer substrate is 22 mm × 20 mm × 0.76 mm, whereas the actual area occupied by the dual-mode resonator is 10 mm × 8 mm.

The results of electromagnetic simulation of the dual-mode resonator using Sonnet Software are shown in Figure 8.28(b) in a comparison with the measured data (Hewlett-Packard 8510C Network Analyzer) for five tested structures. In a wide frequency range, the spurious response is not observed up to 8 GHz.

A possibility to control arbitrarily the resonant frequencies of the chosen modes of the SIR designed as cascaded RH and LH TL sections can be successfully used for the design of miniature dual-band filters. The dual-band filter designed on the R-L-R SIRs implemented in the LTCC package is shown in Figure 8.29(a). The area occupied by the filter is 20 mm × 8.5 mm. Photograph of the dual-band filter is given in Figure 8.29(b).

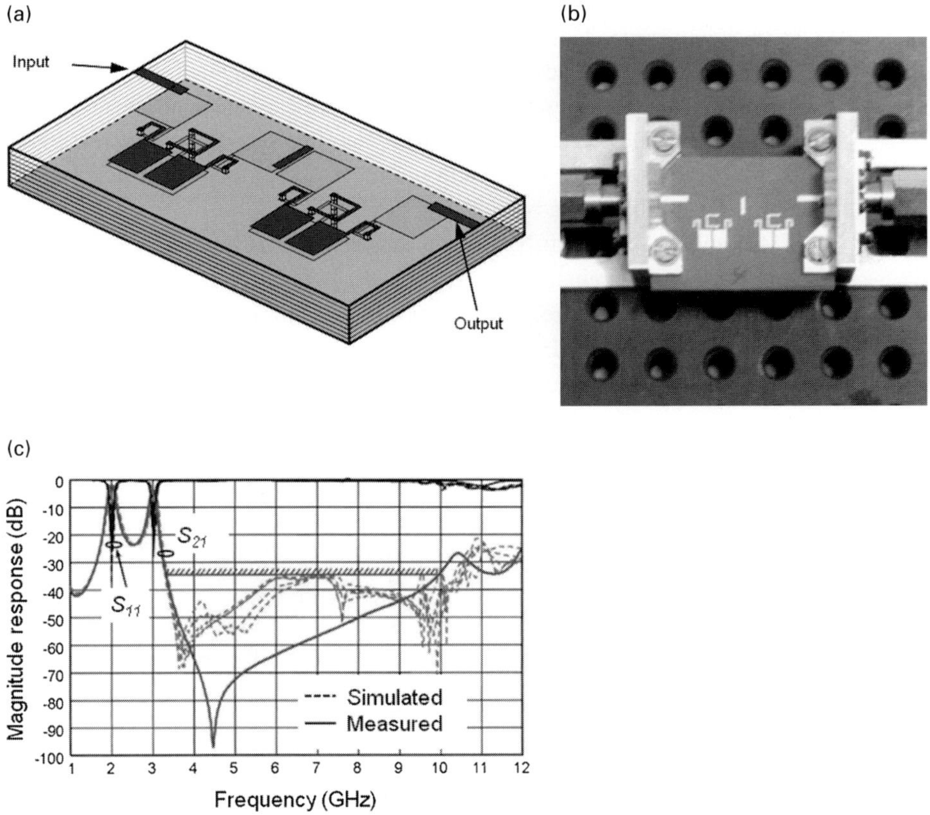

Figure 8.29 (a) LTCC implementation of the two-pole filter on the resonators in Figure 8.28(a). (b) Photo of the filter. (c) Transmission (S_{21}) and reflection (S_{11}) characteristics for five test samples.

The performance of the filter is presented in Figure 8.29(c). The filter bandwidth is 150 MHz for the operational frequencies $f_1 = 2$ GHz and $f_2 = 3$ GHz, which are non-multiple. The simulated and measured return loss is better than 20 dB and 16 dB for the frequencies f_1 and f_2 correspondingly. The measured isolation between the operational passbands is better than 20 dB. The measured insertion loss is not higher than 2.3 dB at f_1 and 3.6 dB at f_2. In the wide frequency range shown in Figure 8.29(c), the spurious response is not observed up to 10 GHz.

8.4.2 Dual-band immittance inverter

Design of dual-band filters requires using dual-mode resonators having resonances at both the central frequencies of interest and dual-band immittance inverters providing the transmission coefficient phase of ±90° within the both passbands. Conventional lumped-element immittance inverters [56] can be applied to design of dual-band filters with two narrow passbands, which are located close to each other. If this is not the case, dual-band inverters are required.

Different designs of dual-band immittance inverters have been recently reported. They are mainly based on quarter-wavelength transmission line sections of different characteristic impedance [57–58]. A significant drawback of these designs is a long electrical length of the component. Besides, depending on the given central frequencies separation and the necessary inverter parameter (inversion coefficient), the values of the characteristic impedance of the transmission line sections can be too high or too low to be practically realized. Dual-band immittance inverters on artificial transmission lines are free from these disadvantages.

A composite right/left-handed transmission line (CRLH TL) shown in Figure 8.30(a) operates as an immittance inverter at two frequencies f_1 and f_2 which can be chosen almost arbitrarily. A frequency response of such a dual-band inverter is presented in Figure 8.30(b). At the frequency f_1, the inverter behaves as a LH TL providing the 90° phase lead. Meanwhile, at the second operating frequency f_2, the inverter exhibits the

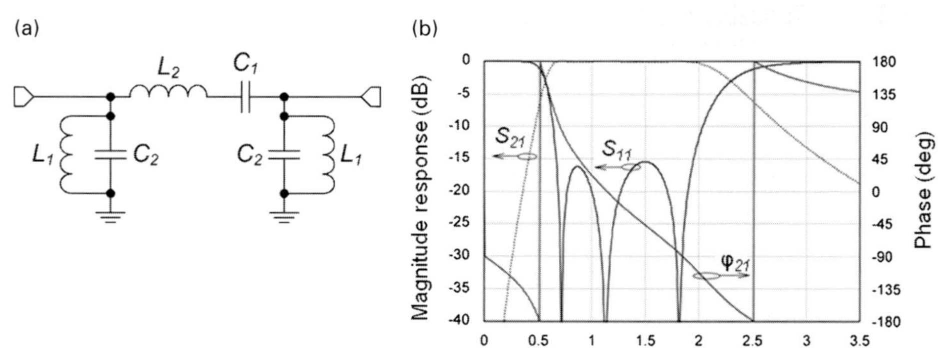

Figure 8.30 (a) Dual-band immittance inverter on CRLH TL (equivalent diagram). (b) Frequency response.

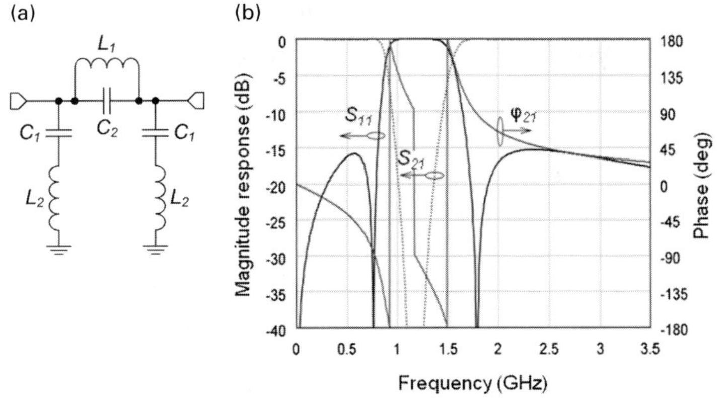

Figure 8.31 (a) Dual-band immittance inverter on D-CRLH TL (equivalent diagram). (b) Frequency response.

phase lag of –90°corresponding to a quarter-wavelength RH TL section. The CRLH TL characteristic impedance corresponds to the inverter parameter. The resonant frequency of the three LC tanks shown in Figure 8.30(a) is $f_0 = \sqrt{f_1 \cdot f_2}$.

A weak point limiting applications of the dual-bandpass filters with the CRLH TL based inverters is a low rejection level between the passbands. In this regard, dual-band immittance inverters providing attenuation zeros between the frequencies of operation seem very promising.

Figure 8.31(a) shows an equivalent diagram of a dual-CRLH TL (D-CRLH TL) [59]. An immittance inverter based on D-CRLH TL has a stopband between its two operational frequencies as illustrated in Figure 8.31(b). The stopband is formed by three transmission zeros imposed at f_0 by the corresponding LC tanks. In contrast to the CRLH TL case, a D-CRLH TL based immittance inverter behaves as a RH TL at the frequency f_1 and as a LH TL at f_2, Figure 8.31(b). The inverter is advantageous for the design of dual-bandpass filters with high rejection between the passbands.

8.4.3 Tuneable two-mode resonators and tuneable dual-band filters based on a combination of RH/LH TL sections

There is an increasing interest in developing electronically tunable or reconfigurable filters for applications in the multi-band communication systems. In the past few years, different tuneable bandpass filters have been proposed using various tuning devices, such as MEMS-devices, ferroelectric varactors, piezoelectric actuators, p-i-n diodes, and semiconductor varactor diodes. These tuning devices can tune the resonant frequencies of the resonators and the coupling coefficients between resonators, thus providing high flexibility of bandpass filters. Different types of tunable dual-band and triple-band planar filters on microstrip resonators have been proposed and realized [61–68]. Tunable dual-band bandpass filters (BPF) with independently tunable passband center frequencies and bandwidths is considered in [61–62]. The filter comprises two dual-mode single-band filters using common input/output lines. Each single BPF is realized using a varactor-loaded transmission-line dual-mode resonator. The proposed filter also offers switchable characteristics to select either of the passbands (either the first or the second passband only). To suppress the harmonics over a broad bandwidth, defected ground structures are used at input/output feeding lines without degrading the passbands characteristics. In [67], a series of multi-stub-loaded resonators, including two quad-mode resonators and one six-mode resonator, are considered for their applications in multi-band filter design. The first quad-mode resonator is utilized to realize a dual-band BPF with tunable lower passband and fixed upper passband. Meanwhile, two transmission zeros are synchronously changed with the tunable band, ensuring high selectivity. This design resulted in compact size. The second quad-mode resonator is used to realize a dual-band filter with very high skirt selectivity. The six-mode resonator is suitable to design a tri-band filter since the six modes can be easily controlled. Using this characteristic, a tri-band BPF is designed with controllable frequencies and bandwidths. In [68], a novel dual-mode dual-band microstrip filter has been investigated and designed for independently tuned dual passbands. The dual-mode resonator loaded with the varactor diodes is proposed and analyzed to independently determine the odd- and even-mode resonant frequencies. Based on the presented design equations and procedures, a two-pole dual-band filter with the tunable first-passband fractional bandwidth and constant second-passband absolute bandwidth has been designed. The filter topology presented has an attractive possibility to adjust the tuning range of the two passbands by choosing different values of the capacitances and varactors diodes. Another control component, namely MEMS-capacitors, are used to design switchable filters [64–65].

All the filters considered are designed using distributed transmission lines in planar technology. In this section, we consider tunable filters implemented in a multi-layer package. The multi-layer technology as applied to the 3D-design of the multi-band filters based on quasi-lumped LC components opens up a new approach to tunable and non-tunable filter design exhibiting miniature dimensions and enlarged functionality.

Let us consider a tuneable TL section presented by equivalent RH and LH TL sections designed as a π-network, Figure 8.32. Variable capacitances are used instead of fixed ones in the artificial TL sections. The equivalency of the TL parameters is provided at the frequency f_0. Both the TL parameters are changed while tuning capacitances: electrical length θ and characteristic impedance Z_0. Tuneable electrical length is followed by tuneable resonant frequency of the resonator containing such

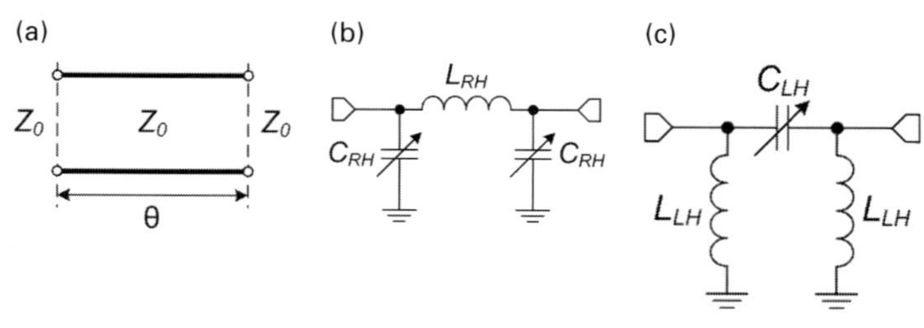

Figure 8.32 (a) Transmission line section. (b) Equivalent representation of the TL section as a Π-network of RH TL section with variable capacitances. (c) Equivalent representation of the TL section as a Π-network of LH TL section with a variable capacitance.

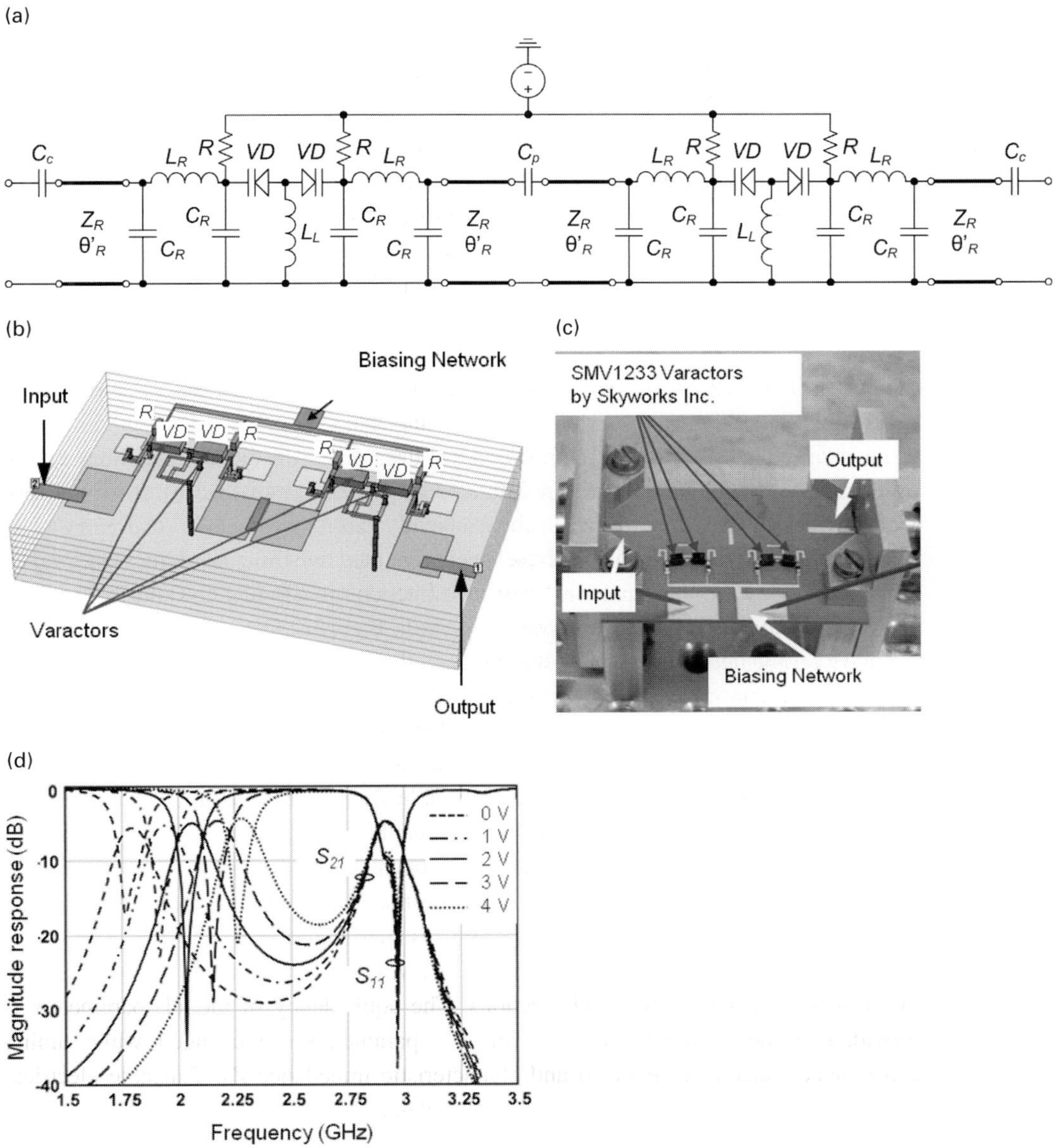

Figure 8.33 (a) Equivalent diagram of the tuneable version of the dual-band filter. (b) Layout of the dual-band filter by employing varactor diodes instead of series capacitances of LH TL and bias resistors. (c) Photo of the device under test. (d) Experimental performance of the filter tuned by the series capacitances of the LH TL section.

sections. Variation of the characteristic impedance is small enough to be neglected in the resonator and filter design. As a tuneable TL section, the T circuit can also be used.

To design a filter with a highly flexible frequency response, the possibility to control the frequency location of the passbands is considered using variable capacitances in the RH and LH TL sections of the dual-mode SIR, Figure 8.33. The simulation revealed a distinct difference between controlling the capacitances of the LH and the RH TL sections [52] [53] [69]. By replacing the series capacitances of the LH TL section by tuneable components (varactors diodes), the position of the lower passband is tuned, whilst the position of the higher passband is invariable. Controlling the shunt capacitances of both RH TL sections results in a shift of both passbands to higher frequencies. However, the tuning mechanisms influence the widths of the passbands only to a small extent. Using the RH TL section in the centre of the filter structure makes it possible to change the influence of tunable capacitances on the passbands. Moreover, using different combinations of Π or T circuits for the LC section realization also gives more possibilities to tune both passbands. This new approach to the tunable filter design is very flexible.

As an example, the equivalent diagram of the two-pole dual-band tuneable filter is depicted in Figure 8.33(a). As the voltage-dependent capacitances, the varactor diodes have been used. The C(V)-values have been taken as specified for Skyworks hyper abrupt silicon varactors. The multi-layer LTCC structure of the tuneable dual-band filter with variable capacitors and bias resistors is shown in Figure 8.33(b). The tuneable dual-band filter was manufactured and experimentally investigated. The SMV1232 varactors and 1 MOhm resistors have been mounted on the top of the multi-layer substrate, while the other components were embedded in the LTCC multi-layer structure. The photograph of the prototype is shown in Figure 8.33(c).

The experimental results are shown in Figure 8.33(d). The return loss is better than 16 dB for all values of control voltage. The insertion loss remains below 3.2 dB and 4 dB at the frequencies f_0 and f_1, respectively. As expected, the lower passband is tuned while the upper passband remains fixed. The measured tunability of the lower passband is 28%.

Additional control of the shunt capacitances of both RH TL sections by SMV1231 varactors results in a shift of both passbands to higher frequencies, Figure 8.34(a). The widths of the passbands were only slightly affected by the tuning mechanism. The tunability range of the first passband is measured to be 20%, while that of the second passband is 42%.

The simultaneous tuning of the RH and LH TL sections by varactor diodes opens up the possibility to tune the both passbands. In this case, either the ratio of the two centre frequencies or their difference could be kept constant, as illustrated by Figures 8.34(b) and 8.34(c), correspondingly.

8.5 Miniature single-band and dual-band filters using capacitively loaded cavities

LTCC technology provides an effective design for very compact microwave devices as 3D highly integrated modules. A number of miniature LTCC filters for L- and S-bands based on L-C-lumped components have been designed up to date. The main problem of the designs is a rather high insertion loss due to a rather low Q-factor of quasi-lumped-element resonators.

At the same time, low-loss LTCC filters based on high-Q cavity resonators were designed for millimetre-wave applications [70–72]. Furthermore, it is well known that

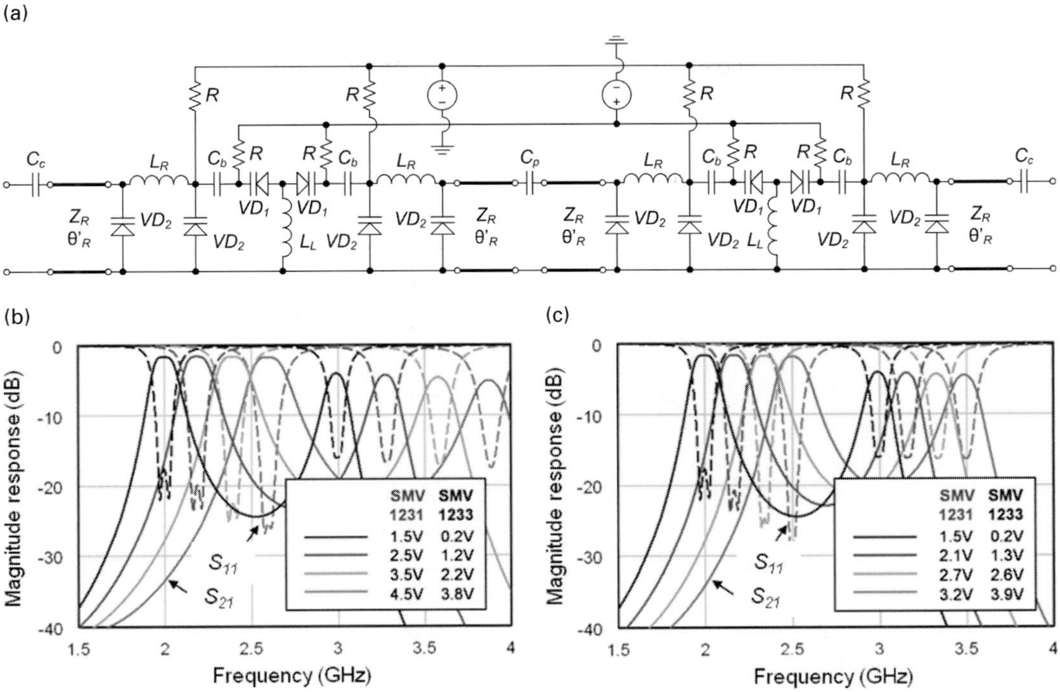

Figure 8.34 (a) Equivalent diagram of the tuneable dual-band filter. (b) Results of circuit simulation of the dual-band filter with tuneable both RH and LH TL sections in case of the ratio of the two centre frequencies is kept constant. (c) Results of circuit simulation of the dual-band filter with tuneable both RH and LH TL sections in case of the difference between the two centre frequencies is kept constant.

by loading a cavity resonator with a capacitive post, we can reduce the resonant frequency of the cavity exhibiting a relatively high unloaded Q-factor [73]. Various designs of compact microwave filters based on highly loaded evanescent-mode cavities using different fabrication technologies (LTCC, PCB, silicon micromachining process), have been recently presented [73–75]. A further miniaturization of capacitively loaded cavity resonators to design small-size and low-loss narrowband LTCC filters for S-band is discussed. The size of a highly loaded LTCC cavity can be as low as one-eighth of the guided wavelength ($\lambda_g/8$), while the Q-factor of such a miniaturized resonator still remains much higher as compared to a quasi-lumped-element LTCC resonator.

Resonators based on capacitively loaded cavities (CLC) can be redesigned to exhibit dual-mode behavior and employed to design dual-band filters. Here, we present two different dual-mode CLCs – one with capacitively loaded cavities of different size nested one into another and another with CLC, in which a complementary split-ring resonator (CSRR) is formed on the cavity top cover. The proposed resonators are used for a design of dual-band filters exhibiting small dimensions and high performance.

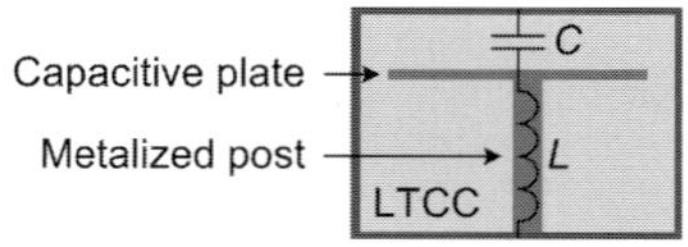

Figure 8.35 Schematic representation of LTCC capacitively loaded cavity.

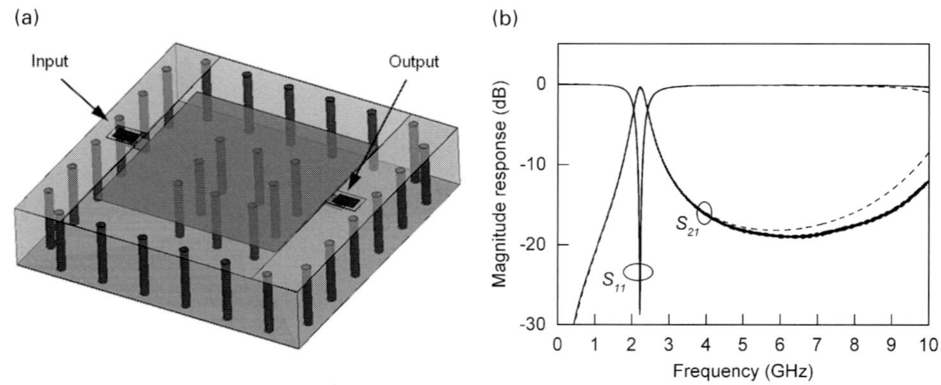

Figure 8.36 (a) Structure of the LTCC capacitively loaded cavity. (b) Characteristics obtained by electromagnetic simulation (dashed lines) and measured characteristics of four experimental samples (solid lines).

8.5.1 Single- and multi-mode resonators on capacitively loaded cavities

A schematic representation of a cavity, which is filled with LTCC and loaded by a conductive post with a capacitive plate at one end, is shown in Figure 8.35. The height of the LTCC cavity resonator cannot be greater than a few millimetres, due to technological reasons. Such a resonator operates in the TM_{110} mode [70]. The higher the capacitive loading, the lower is the resonant frequency of the cavity. On the other hand, for a given resonant frequency, the dimensions of the cavity decrease while increasing capacitive loading. Since the post inductance L is rather low, a high capacitive loading is required for an efficient size reduction. The resonator size is determined by the area of the capacitive plate, which can be decreased by making the upper LTCC layer as thin as possible.

A highly loaded cavity can be considered as a quasi-lumped-element resonator because the electric field is concentrated between the capacitive plate and the top of the cavity. However, the magnetic field distribution remains nearly unchanged and, consequently, the power dissipation in the metallic parts does not change significantly. As a result, the unloaded Q-factor of the resonator does not drop as strongly as the resonant frequency does [73]. This leaves room to design highly loaded LTCC cavities, which are comparable in size with quasi-lumped-element LTCC resonators but have a two to three times higher unloaded Q-factor.

Figure 8.36(a) presents the LTCC capacitively loaded square shaped cavity [76]. In order to demonstrate the internal structure of the resonator, the middle part of the top ground plane is not shown in the figure. Sidewalls and the square post are formed by rows of stacked via holes. The resonator is connected to external circuits by inductive coupling elements. Input and output coplanar feed lines are arranged on the top of the LTCC structure.

The resonator was implemented with six layers of DuPont Green Tape™ 951 LTCC (ε_r = 7.8), each of 210 μm thickness after sintering, two layers with the thickness of 42 μm, and one layer with the thickness of 95 μm. The total height of the LTCC structure is about 1.44 mm. The internal size of the cavity with a resonant frequency of 2.22 GHz was reduced to 6 mm × 6 mm (6 mm corresponds to $\lambda_g/8$). The post is of 2 mm × 2 mm and is covered with the capacitive plate with an area of 4.4 mm × 4.4 mm. The resonator was manufactured using DuPont 6148 Ag co-fireable conductor paste and DuPont 6141 AgPd co-fireable via fill paste. The resonator was designed with the aid of the Ansoft HFSS 3D electromagnetic field solver.

Experimental investigations were carried out using a vector network analyzer and a probe station with coplanar probes for on-wafer measurements. The TRL calibration technique was used. Figure 8.36(b) presents the measured characteristics of four manufactured samples of the resonator. Very good repeatability of the characteristics is observed. The estimated unloaded Q-factor of the resonator is about 150, which is at least three times higher than for a LTCC resonator based on quasi-lumped elements.

Capacitively loaded cavities (CLCs) of different size, having different resonant frequencies can be nested one into another in such a way that the inner cavity serves as the capacitive load for the outer cavity [76]. In Figure 8.37(a), the structure of such a dual-mode resonator consisting of two nested CLCs is depicted. An equivalent diagram of the resonator is shown in Figure 8.37(b).

Advantages of the dual-mode resonator of this kind are as follows: (i) the resonant frequencies can be chosen in an arbitrary way; (ii) a transmission zero is provided between the resonant frequencies; (iii) the overall size is small; and (iv) absence of a spurious response over a wide frequency range. A transmission zero below each resonant frequency can be easily introduced by surrounding the dual-mode resonator with an external conducting box as shown in Figure 8.37(c). The corresponding equivalent diagram is presented in Figure 8.37(d).

The frequency behavior of the resonators of both types was analysed by numerical simulations and confirmed with an experimental investigation (Figure 8.38). Test structures of the dual-mode resonators with non-multiple resonant frequencies f_{01} = 748 MHz and f_{02} = 1748 MHz were designed and fabricated in the multi-layer LTCC technology, Figure 8.38(a). The area of the test sample of the dual-mode resonator with a transmission zero embedded in 16 layers of DuPont Green Tape™ 951 is 7.6 mm × 5.2 mm. The height of the LTCC structure is 2.9 mm. The linear size of the resonator does not exceed $\lambda_g/18$ at f_{01} and $\lambda_g/8$ at f_{02}. Simulated and experimentally obtained characteristics of the dual-mode resonators are compared in Figure 8.38(b) and (c).

8.5.2 Miniature microwave filters using capacitively loaded cavities

Miniature bandpass filters can be designed using capacitively loaded cavities. A two-pole bandpass filter with Chebyshev response was designed for UMTS/LTE-2100 applications (2110–2170 MHz) on the basis of the resonator shown in Figure 8.36(a)

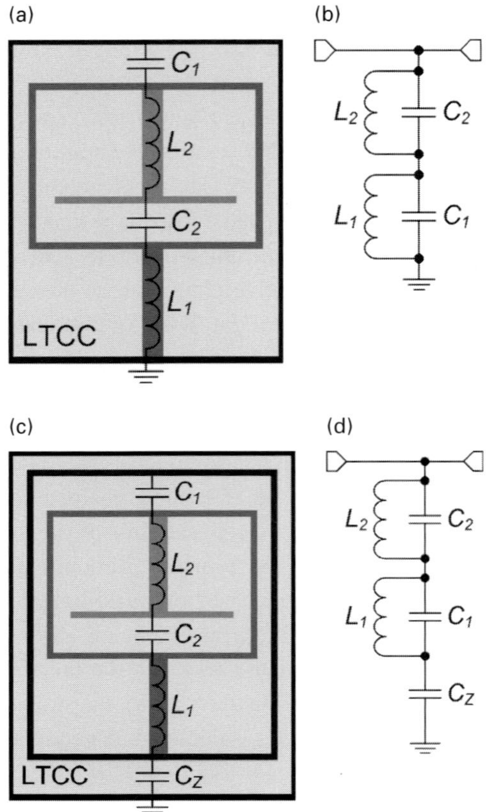

Figure 8.37 (a) Two nested CLCs: cross-section of the dual-mode resonator. (b) The corresponding equivalent diagram. (c) The cross-section of the dual-mode resonator with an additional transmission zero. (d) The corresponding equivalent diagram.

[76]. The filter consists of two capacitively loaded cavities coupled to each other by means of an iris in the mutual sidewall, Figure 8.39(a). Though the reactive nature of the iris coupling is in general mixed, the main contribution comes from the electric field (i.e., capacitive coupling). The value of the coupling depends on the size b of the iris, Figure 8.39(a). The larger the size chosen, the stronger becomes the coupling. Since $b < a$, the cavity size a cannot be chosen arbitrary. The entire multi-layer LTCC structure of the filter designed occupies an area of 13 mm $\times$ 7 mm.

The filter characteristics obtained by full-wave simulations and measurements are displayed in Figure 8.39(b). The experimental investigation of fabricated test sample revealed 1 dB of insertion loss at the central frequency, and a return loss of about 15 dB. The measured data confirm the ability to design low-loss and small-size LTCC filters based on cavity resonators with a high capacitive loading

A four-pole bandpass filter with a cross-coupling was also designed for the same frequency band [76]. The multi-layer LTCC structure of the filter is shown in Figure 8.40(a). The filter employs three inductive couplings, which are formed by

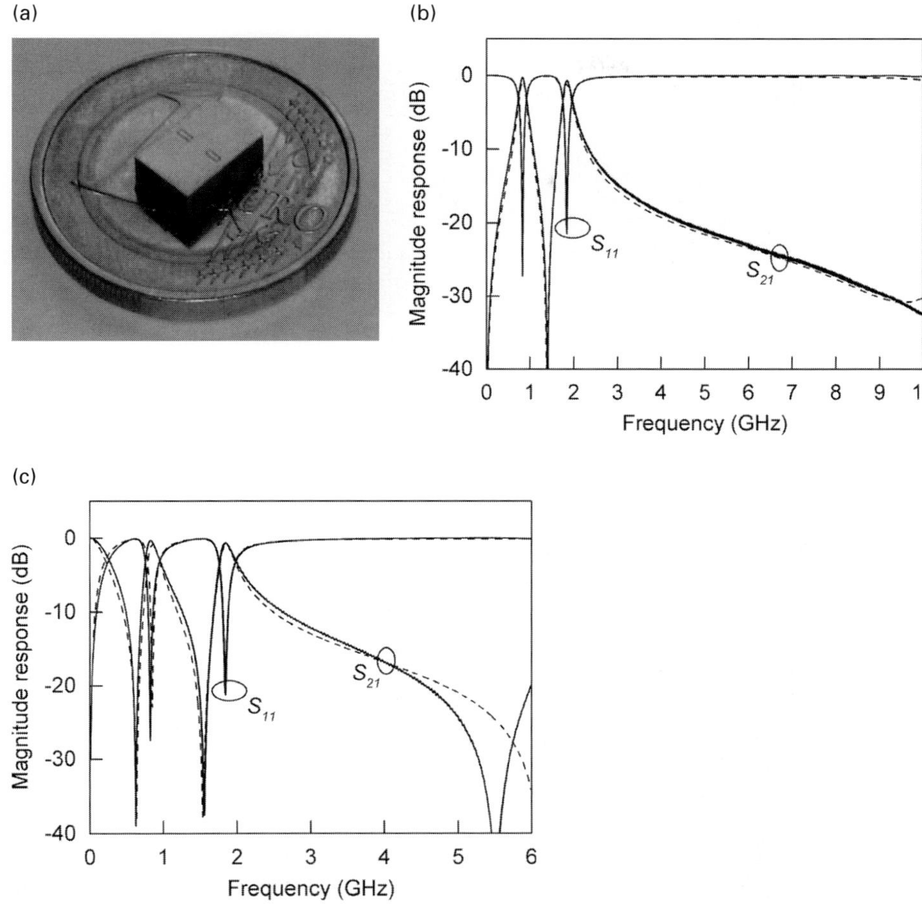

Figure 8.38 (a) Photograph of the test sample of dual-mode LTCC resonators on nested CLCs. (b) Measured (solid lines) and simulated (dashed lines) characteristics of the resonator. (c) Measured (solid lines) and simulated (dashed lines) characteristics of the resonator providing an additional transmission zero, Figure 8.37(c).

narrow strip lines connected to the capacitive plates of adjacent cavities. A capacitive cross-coupling between the outermost resonators is achieved with a rather wide strip line overlaying the capacitive plate of both resonators. The filter area is 13 mm × 13 mm.

Simulated characteristics of the filter are presented in Figure 8.40(b). The dielectric loss tangent specified by DuPont (tan δ = 0.002) and a typical value of the conductive paste surface resistance (R_{sur} = 5 mOhm/square) was taken into account by the electromagnetic simulation. The cross-coupled four-pole filter provides drastically improved selectivity as compared with the previous two-pole filter design. The midband insertion loss of this filter is 2.5 dB.

Another filter based on cavities was proposed in [77]. The filter presents a three-pole highly selective LTCC filter realized using three coupled low-profile cavities. Each

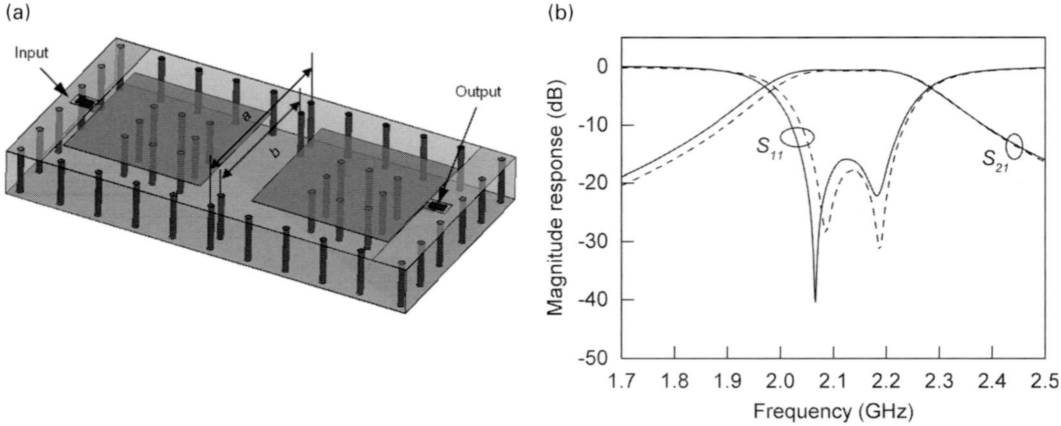

Figure 8.39 (a) Structure of the two-pole bandpass LTCC filter based on iris-coupled capacitively loaded cavities. (b) Characteristics obtained by electromagnetic simulation (dashed lines) and measured characteristics of the experimental sample (solid lines).

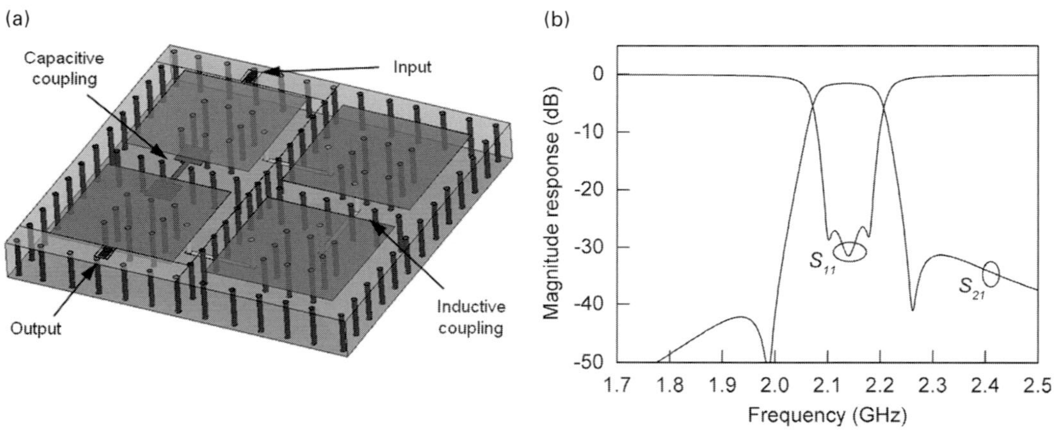

Figure 8.40 (a) Structure of four-pole bandpass LTCC filter with a cross-coupling. (b) Characteristics obtained by electromagnetic simulations.

cavity is loaded by a conductive post with a capacitive plate at one end, Figure 8.41(a). Sidewalls and the square post are formed by rows of stacked via holes. The filter resonators are coupled to each other by means of an iris in the mutual sidewall. The outermost resonators are connected to external circuits by inductive coupling elements. An equivalent diagram of the filter is shown in Figure 8.41(b). The filter with a Chebyshev response was embedded in nine layers of DuPont Green Tape™ 951 LTCC (ε_r = 7.8, tan δ = 0.002). Six layers are 210 μm thick after sintering, two other layers have the thickness of 42 μm, and one layer has the thickness of 95 μm. The total height of the LTCC structure is about 1.44 mm. The filter occupies an area of 19 mm × 7 mm (0.38 λ_g × 0.14 λ_g).

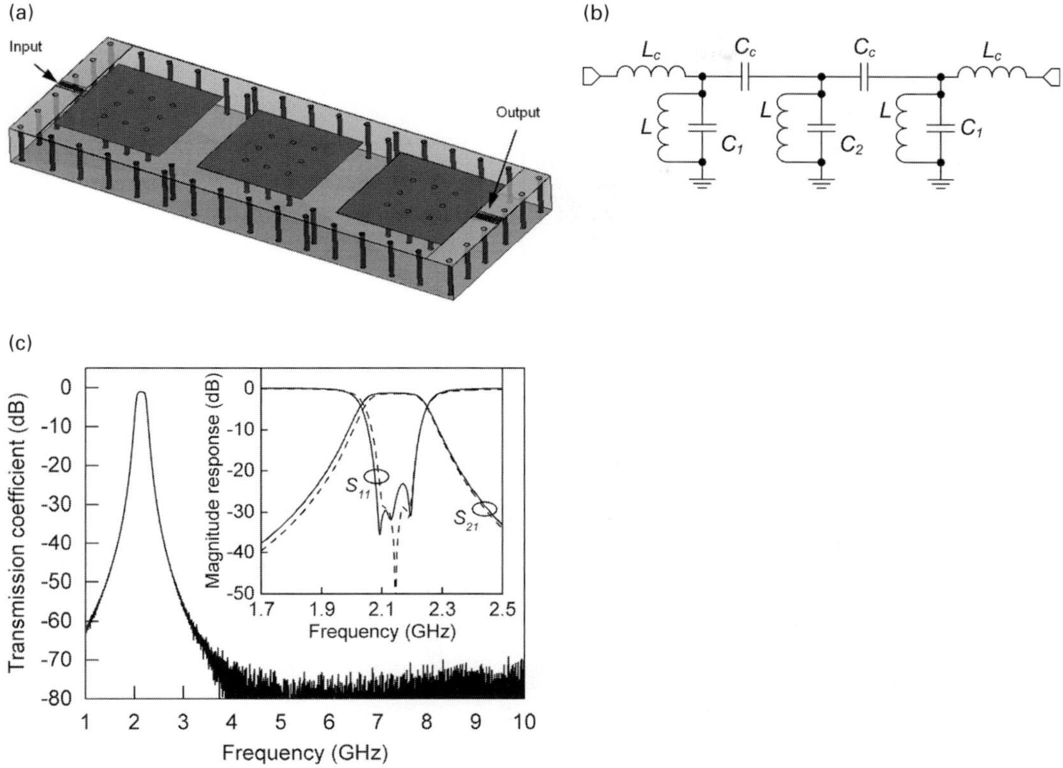

Figure 8.41 (a) Structure of the three-pole Chebyshev bandpass LTCC filter based on iris coupled resonators. (b) Equivalent diagram. (c) Measured (solid lines) and simulated (dashed lines) characteristics.

The simulated (Ansoft HFSS 3D electromagnetic solver) and measured filter performance are plotted in Figure 8.41(c). The in-band insertion loss does not exceed 1.2 dB, which corresponds to an average unloaded Q-factor of the resonators equal to 150. Furthermore, the filter exhibits no spurious response over a wide frequency range (at least up to 10 GHz).

8.5.3 Two-mode resonators and dual-band filters based on capacitively loaded cavities

Due to their behavior, coupled nested CLCs are good candidates for the design of dual-band filters. The structure of coupled dual-mode resonators based on two nested CLCs is schematically shown in Figure 8.42(a). The outer cavities are coupled to each other through an iris in the mutual side wall. Additional irises in the adjacent side walls of the inner cavities are used to provide a coupling between them.

An equivalent diagram of the coupled resonators is presented in Figure 8.42(b). The coupling between the tanks with the resonant frequencies f_{01} and $f_{02} > f_{01}$ is represented by the capacitances C_{c1} and C_{c2} of the different branches of the circuit. At the frequency f_{01}, the two coupling capacitances are connected in parallel and the total capacitance

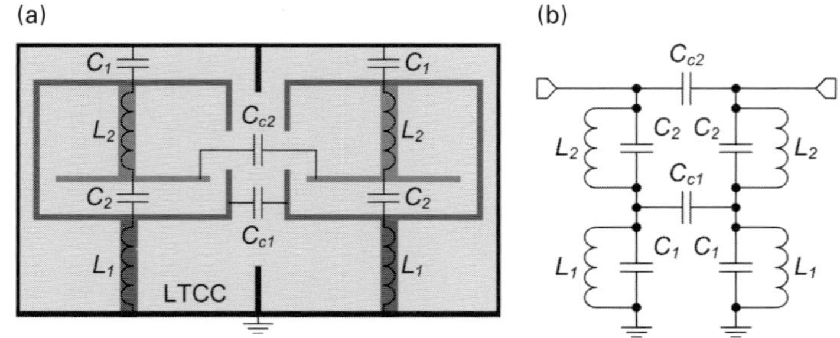

Figure 8.42 (a) Coupled dual-mode resonators based on nested CLCs. (b) Equivalent diagram of the coupled CLCs.

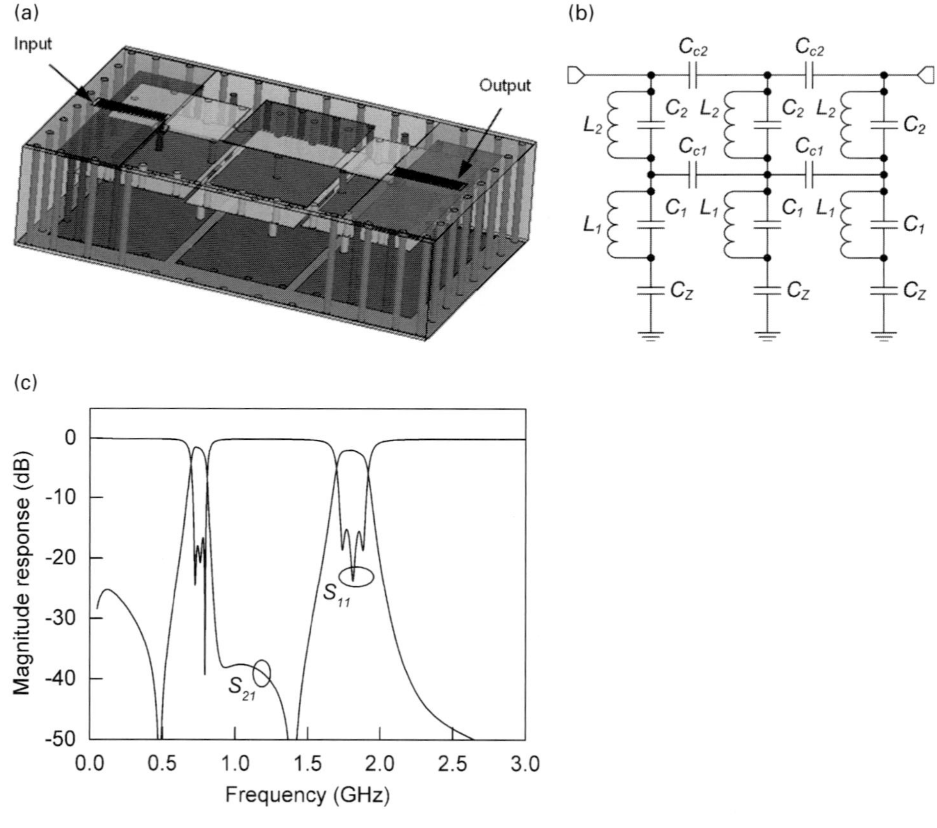

Figure 8.43 (a) Structure of the 3-pole dual-band LTCC filter on nested CLCs. (b) Equivalent diagram of the filter. (c) Filter characteristics obtained by the full-wave simulation.

of the coupling is $C_c(f_{01}) = C_{c1} + C_{c2}$. Meanwhile, at f_{02} the coupling capacitance is $C_c(f_{02}) = C_{c2}$.

The following design method is suggested for the filters based on nested CLCs. As the first step, two single-band filters with the central frequencies f_{01} and f_{02} are designed separately using a conventional synthesis technique. Then the coupling values of the

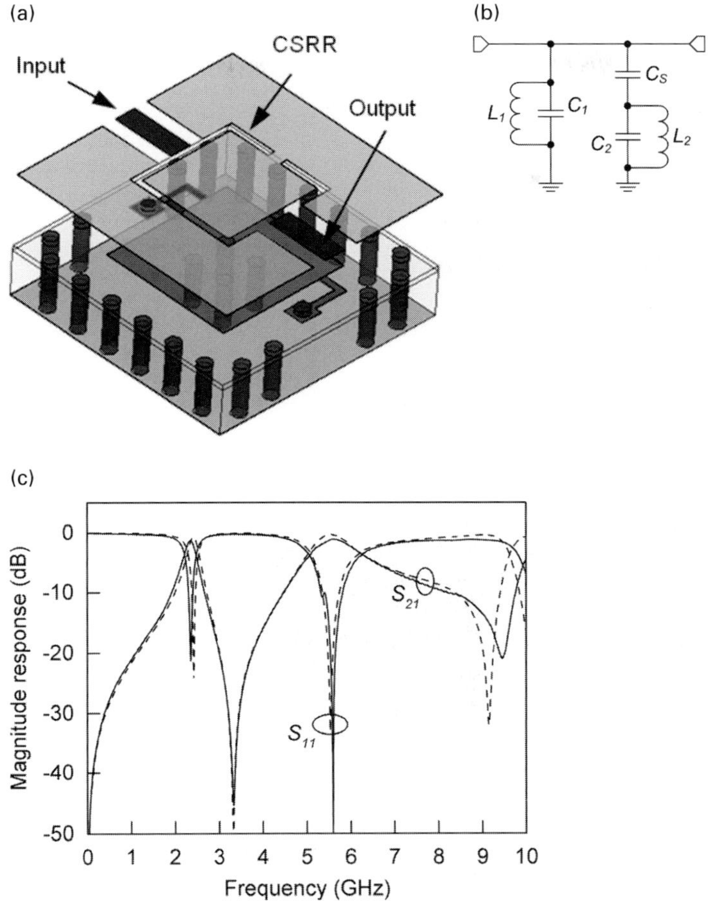

Figure 8.44 (a) Dual-mode resonator on a combination of CLC and CSRR. (b) Equivalent diagram. (c) Measured (solid lines) and simulated (dashed lines) frequency responses.

filter designed for f_{01} are corrected as $C'_{c1} = C_{c1} - C_{c2}$. After this, the filters are integrated, as shown in Figure 8.42(b). The method can be also used to design dual-band filters with additional transmission zeros based on the resonator structure in Figure 8.37(c).

A design example of the 3-pole dual-band filter based on CLCs for application at 698–798 MHz and 1710–1788 MHz is demonstrated in Figure 8.43. The filter was implemented in the LTCC technology using the above-mentioned structure of the dual-mode resonator, Figure 8.37(c). The filter is as small as 12 mm × 7.6 mm, corresponding to the linear dimension $\lambda_g/12$ at the lower frequency. The filter exhibits a low insertion loss (1.4 dB and 2.1 dB in the lower and higher passbands, respectively) and high selectivity.

A dual-mode resonator can also be designed as a combination of a CLC and a complementary split-ring resonator (CSRR) formed on the cavity top cover, Figure 8.44(a) [77]. The cavity operates at the lower resonant frequency, while the upper resonance is

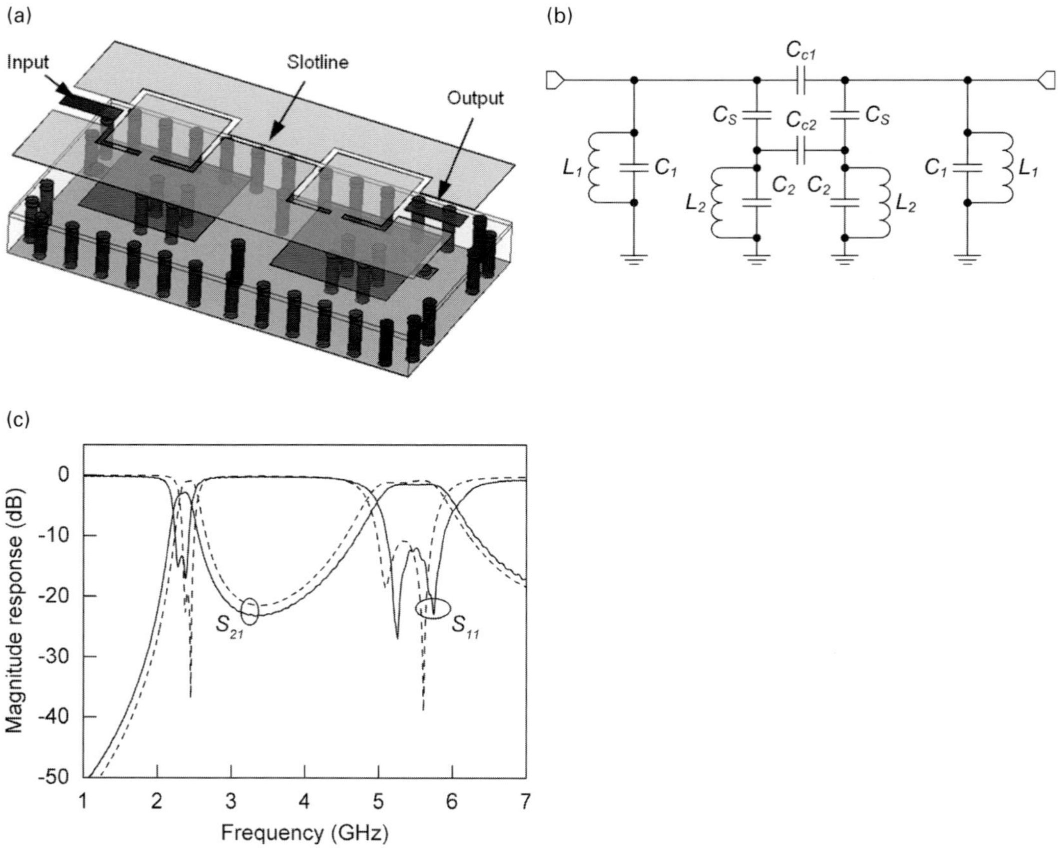

Figure 8.45 (a) Two-pole dual-band filter based on CLCs with integrated CSRRs. (b) Equivalent diagram. (c) Measured (solid lines) and simulated (dashed lines) characteristics.

provided by the CSRR. An equivalent diagram of the dual-mode resonator is shown in Figure 8.45(b). The size of such a dual-mode resonator is limited by the dimensions of the single-mode cavity. Since the CSRR is a planar structure, its integration with the cavity does not result in an increase in height of the structure as for the dual-mode resonator of nested CLCs. However, the unloaded Q-factor of the CSRR is lower in comparison with that of the CLC. Hence, dual-band resonators of this type can be favorably used when the insertion loss in the upper filter passband is not of a crucial importance, e.g. to design filters with the upper fractional bandwidth being wider than the lower one.

Characteristics of the dual-mode resonator with the resonant frequencies of 2.4 GHz and 5.6 GHz are presented in Figure 8.44(c). The resonator implemented on 1.7 mm thick two layer Rogers RO4003 (ε_r = 3.55, tan δ = 0.0027) occupies the area of 10 mm $\times$ 10 mm corresponding to $\lambda_g/8$ at the lower resonant frequency. The unloaded Q-factor is 130 at the both resonant frequencies.

A coupling between the dual-mode resonators may be provided in the same way as for the single-mode cavities. Changing the iris size influences the coupling between the

CLCs, as well as the coupling between the CSSRs. In order to adjust the coupling between the CSRRs independently, they are connected by a slot line (see Figure 8.45(a)) providing a tapered coupling whose value depends on the connection point position. An equivalent diagram is illustrated by Figure 8.45(b).

A two-pole dual-band filter for WLAN application at 2.3–2.4 GHz and 5.1–5.8 GHz was designed. The filter area is 20 mm × 10 mm. The simulated and measured performance of the filter is presented in Figure 8.45(c). The measured insertion loss in the two passbands is 2.5 dB and 1.4 dB, correspondingly.

8.6 Conclusion

In this chapter, the great potential of multi-layer technologies for the design of miniature high-performance filters has been presented.

Multilayer single-band filters based on different design approaches have been proposed and discussed. Beside filter architecture with quasi-lumped elements, filters based on a combination of RH and LH TL sections and capacitively loaded cavities have been analysed and their features presented.

A combination of RH and LH sections has also been used for the design of dual-band filters, as well as a dual-band immittance inverter. Capacitively loaded cavities have been redesigned to exhibit dual-mode behavior and as such have been employed for dual-band filter design. The proposed dual-band filters show compact size and good performance.

References

[1] A. Djaiz and T. A. Denidni, "A new compact microstrip two-layer bandpass filter using aperture-coupled SIR-hairpin resonators with transmission zeros," *IEEE Transactions on Microwave Theory and Techniques,* vol. 54, no. 5, pp. 1929–1936, May 2006.

[2] W. Schwab and W. Menzel, "Compact bandpass filters with improved stop-band characteristics using planar multilayer structures," *Proceedings of IEEE MTT-S International Microwave Symposium Digest*, Albuquerque, NM, 1992, pp. 1207–1209.

[3] C. Cho and K. C. Gupta, "Design methodology for multilayer coupled line filters," *Proceedings of IEEE MTT-S International Microwave Symposium Digest*, Denver, CO, June 1997, pp. 785–788.

[4] D. Ahn, J. S. Park,C. S. Kim, J. N. Kin, Y. Qian, and T. Itoh, "A design of the low-pass filter using the novel microstrip defected ground structure," *IEEE Transactions on Microwave Theory and Techniques,* vol. 49, no. 1, pp. 86–93, Jan. 2001.

[5] N. Karmakar, S. M. Roy, and I. Balbin, "Quasi-static modeling of defected ground structure," *IEEE Transactions on Microwave Theory and Techniques,* vol. 54, no. 5, pp. 2160–2168, May 2006.

[6] PILKOR CND Co., "LTCC technology," www.pilkorcnd.co.kr. [Online]. Available: http://www.pilkorcnd.co.kr/sub/sub3_1.asp?mNum=3 [Accessed: June 18, 2014].

[7] Y. Imanaka, *Multilayered Low Temperature Cofired Ceramics (LTCC) Technology*, Springer, New York, 2005.

[8] IIT Ltd, "What is the LTCC?," www.ltcc-consulting.com.

[9] C.-W. Tang, C.-W. Shen, and P.-J. Hsieh, "Design of low-temperature co-fired ceramic bandpass filters with modified coupled inductors," *IEEE Transactions on Microwave Theory and Techniques,* vol. 57, no. 1, pp. 172–178, Jan. 2009.
[10] L. K. Yeung and K. L. Wu, "A compact second-order LTCC bandpass filter with two finite transmission zero," *IEEE Transactions on Microwave Theory and Techniques,* vol. 51, no. 2, pp. 337–341, Feb. 2003.
[11] W. Tung, Y. Chiang, and J. Cheng, "A new compact LTCC bandpass filter using negative coupling," *IEEE Microwave and Wireless Components Letters*, vol. 15, no. 10, pp. 641–643, Oct. 2005.
[12] G. Brzezina, L. Roy, and L. MacEachern, "A miniature LTCC filter using novel resonators for GPS applications," *Proceeding of the 37th European Microwave Conference*, Munich, Oct. 2007, pp. 536–539.
[13] C.W. Tang, Y.C. Lin, and C.Y. Chang, "Realization of transmission zeros in combine filters using an auxiliary inductively coupled ground plane," *IEEE Transactions on Microwave Theory and Techniques*, vol. 51, no. 10, pp. 2112–2118, Oct. 2003.
[14] A. Sutono, J. Laskar, and W. R. Smith, "Development of three dimensional integrated Bluetooth image reject filter," *Proceedings of IEEE MTT-S International Microwave Symposium Digest*, Boston, MA, May 2000, pp. 339–342.
[15] Y. Zhang and K. A. Zaki, "LTCC multi-layer coupled strip-resonator filters," *Proceedings of IEEE MTT-S International Microwave Symposium Digest*, Honolulu, HI, June 2007, pp. 1039–1042.
[16] A. Simine, D.V. Kholodnyak, P. Turalchuk, V. Piatnitsa, Heli Jantunen, and I.B. Vendik, "Enhancement of inductance Q-factor for LTCC filter design," *Proceedings of 35th European Microwave Conference*, Paris, France, Oct. 2005, pp. 1319–1322.
[17] A. Simin, D. Kholodnyak, and I. B. Vendik, "A novel approach to synthesis of bandpass filters with minimized insertion loss," *Proceedings of 37th European Microwave Conference*, Munich, Germany, Oct. 2007, pp. 336–339.
[18] D. Thompson, "Characterization and design of liquid crystal polymer (LCP) based multi-layer RF components and packages," Ph.D. dissertation, School of Elec. and Comp. Eng., Georgia Inst. of Tech., Atlanta, GA, 2006.
[19] J.-S. Hong, *Microstrip Filters for RF/Microwave Application*, 2nd edn. New York: Wiley, 2010.
[20] C. Cho and K. C. Gupta, "Design methodology for multilayer coupled line filters," *Proceedings of IEEE MTT-S International Microwave Symposium Digest*, Denver, CO, June 1997, pp. 785–788.
[21] S. Qian and J.-S. Hong, "Miniature quasi-lumped element wideband bandpass filter at 0.5–2 GHz band using multilayer liquid crystal polymer technology," *IEEE Transactions on Microwave Theory and Techniques*, vol. 60, no. 9, pp. 2799–2807, Sept. 2012.
[22] Z.-C. Hao and J.-S. Hong, "Ultra-wideband bandpass filter using multilayer liquid-crystal-polymer technology," *IEEE Transactions on Microwave Theory and Techniques,* vol. 56, no. 9, pp. 2095–2100, Sept. 2008.
[23] Z.-C. Hao and J.-S. Hong, "UWB bandpass filter using cascaded miniature high-pass and low-pass filters with multilayer liquid crystal polymer technology," *IEEE Transactions on Microwave Theory and Techniques,* vol. 58, no. 4, pp. 941–948, April 2010.
[24] Z.-C. Hao and J.-S. Hong, "Ultra wideband bandpass filter using embedded stepped impedance resonators on multilayer liquid crystal polymer substrate," *IEEE Microwave and Wireless Components Letters*, vol. 18, no. 9, pp. 581–583, Sept. 2008.

[25] R.Bairava Subramanian, "Development of V-band integrated front-ends on liquid crystal polymer (LCP) technology," *IEEE Antenna and Wireless Propag. Lett.*, vol. 7, pp. 134–137, July 2008.

[26] G. V. Eleftheriades, O. Siddiqui, and A. K. Iyer, "Transmission line models for negative refractive index media and associated implementations without excess resonators," *IEEE Microwave and Wireless Components Letters*, vol. 13, pp. 51–53, Feb. 2003.

[27] I. B. Vendik, D. V. Kholodnyak, and P. V. Kapitanova, "Microwave phase sifters and filters based on a combination of left-handed and right-handed transmission lines," *Applications of Metamaterials, Metamaterials Handbook*, ed. F. Capolino, CRC Press, Taylor & Francis Group, Boca Raton, FL, 2009, pp. 13-1–13-21.

[28] C. Caloz and T. Itoh, *Electromagnetic Metamaterials: Transmission Line Theory and Microwave Applications*, John Wiley, New Jersey, 2006.

[29] R. Marques, F. Martin, and M. Sorolla, *Metamaterials with Negative Parameters: Theory, Design and Microwave Applications,* Wiley-Blackwell, New Jersey, 2008.

[30] Au Jin, *Kong, Electromagnetic Wave Theory,* EMW Publishing, Cambridge, MA, 2005.

[31] L. Brillouin and M. Parodi, *Propagation des ondes dans les milieux periodiques,* Paris: Masson et cie éditeurs, Dunod éditeurs, 1956.

[32] I. B. Vendik, O. G. Vendik, D. V. Kholodnyak, E. V. Serebryakova, and P. V. Kapitanova, "Digital phase shifters based on right- and left-handed transmission lines," *Proceedings of the European Microwave Association*, vol. 2, no. 2, March 2006, pp. 30–37.

[33] A. Lai, C. Caloz, and T. Itoh, "Composite right/left-handed transmission line metamaterials," *IEEE Microwave Magazine*, vol. 5, no. 3, pp. 34–50, 2004.

[34] I. Vendik, D. Kholodnyak, P. Kapitanova, and K. Zemlyakov, "Multifunctional microwave devices based on metamaterial transmission lines", *Proceedings of the IEEE International Conference on Microwaves, Communications, Antennas and Electronics System (COMCAS 2009)*, Tel Aviv, Israel, Nov. 2009, pp. 195–199.

[35] K. Zemlyakov and I. Vendik, "Tuneable microwave resonators and filters on combination of right/left handed transmission line sections for multiband applications," *Metamaterials 2010*, Karlsruhe, Germany, Sept. 2010, pp. 423–425.

[36] I. B. Vendik, D. V. Kholodnyak, I. V. Kolmakova, E. V. Serbryakova, P. V. Kapitanova, F. Martín, J. Bonache, J. García, I. Gil, and M. Gil, "Applications of right/left handed and resonant left handed transmission lines for microwave circuit design," *Proceedings of 36th European Microwave Conference*, Manchester, UK, Oct. 2006, pp. 955–958.

[37] R. Gómez-García and J. I. Alonso, "Design of Sharp-rejection and low-loss wide-band planar filters using signal-interference techniques," *IEEE Microwave and Wireless Components Letters*, vol. 15, no. 8, pp. 530–532, Aug. 2005.

[38] R. Gómez-García and M. Sánchez-Renedo, "A class of microwave transversal signal-interference dual-passband planar filters," *IEEE Microwave and Wireless Components Letters*, vol. 19, no. 3, pp. 158–160, March 2009.

[39] M. Á. Sánchez-Soriano and E. Bronchalo, "Compact UWB bandpass filter based on signal interference techniques," *IEEE Microwave and Wireless Components Letters*, vol. 19, no. 11, pp. 692–694, Nov. 2009.

[40] R. Gómez-García, J. Muñoz-Ferreras, and M. Sánchez-Renedo, "Signal-interference stepped-impedance-line microstrip filters and application to duplexers," *IEEE Microwave and Wireless Components Letters*, vol. 21, no. 8, pp. 421–423, Aug. 2011.

[41] Y. Horii, C. Caloz, and T. Itoh, "Super-compact multilayered left-handed transmission line and diplexer application," *IEEE Transactions on Microwave Theory and Techniques,* vol. 53, no. 4, pp. 1527–1534, April 2005.

[42] N. Jia, I. Vendik, D. Kholodnyak, and J.-S. Hong, "A compact bandpass filter based on right- and left-handed transmission line sections," *IEEE Microwave and Wireless Components Letters*, vol. 23, no. 6, pp. 279–281, June 2013.

[43] H. Wen, L. Jin, and H. Chen, "Improved composite right/left handed transmission line metamaterials and its application in bandpass filters," *International Workshop on Microwave and Millimeter Wave Circuits and System Technology*, Chengdu, China, April 2012, pp. 1–3.

[44] Sonnet Software, Sonnet Em, version 13.54 [CD-ROM], North Syracuse, NY, 2011.

[45] *Revision of part* 15 *of the Commision's rules regarding ultra-wideband transmission system*, FCC, Washington, DC, Tech. Rep. ET-Docket 98–153 FCC02–48, 2002.

[46] A. S. Rusakov, I. B. Vendik, and S. Qian, "A direct synthesis method for UWB bandpass filters based on metamaterial transmission lines," *Metamaterials'2012*, St. Petersburg, Russia, Sept. 2012, pp. 387–389.

[47] S. Qian, J.-S. Hong, A. Rusakov,and I. B. Vendik, "A novel compact ultra-wideband bandpass filter," *Proceedings of 43rd European Microwave Conference*, Oct. 2013, pp. 896–899.

[48] A. Rusakov, N. Jankovic, and V. Crnojevic-Bengin, "A compact tri-band bandpass filter based on grounded tri-mode stepped-impedance stub-loaded resonator," *Proceedings of 11th International Conference on Telecommunications in Modern Satellite, Cable and Broadcasting Services (TELSIKS'2013), 2013*, Nis, Serbia, 2013, pp. 273–276.

[49] H. Joshi and W. J. Chappel, "Dual-band lumped-element bandpass filter," *IEEE Transactions on Microwave Theory and Techniques,*, vol. 54, no.12, pp. 4169–4177, Dec. 2006.

[50] X. Q. Lin, R. P. Liu, and X. M. Yang, "Arbitrarily dual-band components using simplified structures of conventional CRLH TLs," *IEEE Transactions on Microwave Theory and Techniques*, vol. 54, no. 7, pp. 2902–2909, July 2006.

[51] A. Görür and C. Karpuz, "Miniature dual-mode microstrip filters," *IEEE Microwave and Wireless Components Letters*, vol. 17, no. 1, pp. 37–39, Jan. 2007.

[52] P. Kapitanova, D. Kholodnyak, S. Humbla, R. Perrone, J. Mueller, M. A. Hein, and I. B. Vendik, "Multi-band and tunable multi-band microwave resonators and filters on cascaded left/right – handed transmission line sections," *EUROCON 2009*, Saint Petersburg, Russia, May 2009, pp. 60–66.

[53] P. Kapitanova, D. Kholodhyak, S. Humbla, R. Perrone, J. Mueller, M. A. Hein, and I. Vendik, "Right- and left-handed transmission line resonators and filters for dual band application," *Microwave and Optical Technology Letters*, vol. 51, no. 3, pp. 629–633, March 2009.

[54] P. V. Kapitanova, A. V. Simine, D. V. Kholodnyak, and I. B. Vendik, "Multilayer thick-film technology as applied to design of microwave device," *Journal of the European Ceramic Society*, vol. 27, no. 8–9, pp. 2941–2944, 2007.

[55] J. Müller, R. Perrone, H. Thust, K. H. Drüe, C. Kutscher, R. Stephan, J. Trabert, M. A. Hein, D. Schwanke, J. Pohlner, G. Reppe, R. Kulke, P. Uhlig, A. F. Jacob, T. Baras, and A. Molke, "Technology benchmarking of high-resolution structures on LTCC for microwave circuits," *Proceeding of the Electronics System Integration Technology Conference*, ESTC, Dresden, Germany, Sept. 2006, pp. 5–7.

[56] G. L. Matthaei, L. Young, and E. M. T. Jones, *Microwave Filters, Impedance-Matching Networks, and Coupling Structure*, 2nd edn, Artech House, Dedham, MA, 1980.

[57] C. Monzon, "A small dual-frequency transformer in two sections," *IEEE Transactions on Microwave Theory and Techniques*, vol. 51, no. 4, pp. 1157–1161, April 2003.

[58] H.-Y. A. Yim and K.-K. M. Cheng, "Novel dual-band planar resonator and admittance inverter for filter design and applications," *Proceedings of IEEE MTT-S International Microwave Symposium Digest*, pp. 2187–2190, June 2005.

[59] C. Caloz, "Dual composite right/left-handed (D-CRLH) transmission line metamaterial," *IEEE Microwave and Wireless Components Letters*, vol. 16, no. 11, pp. 585–587, Nov. 2006.

[60] D. V. Kholodnyak, P. A. Turalchuk, and E. Yu. Zameshaeva, "Dual-frequency immittance inverter with a stopband between the operation frequencies based on one-dimensional metamaterials," RU Patent 2 463 690 C2, Oct. 10, 2012 (in Russian).

[61] G. Chaudhary, Y. Jeong, and J. Lim, "Harmonic suppressed dual-band bandpass filter with tunable passbands," *IEEE Transactions on Microwave Theory and Techniques,* vol. 60, no. 7, pp. 2115–2123, July 2012.

[62] G. Chaudhary, Y. Jeong, and J. Lim, "Dual-band bandpass filter with independently tunable center frequencies and bandwidths," *IEEE Transactions on Microwave Theory and Techniques,* vol. 61, no. 1, pp. 107–116, Jan. 2013.

[63] Q. Y. Xiang, Q. Y. Feng, X. G. Huang, and D. H. Jia, "Electrical tunable microstrip LC bandpass filters with constant bandwidth," *IEEE Transactions on Microwave Theory and Techniques,* vol. 61, no. 3, pp. 1124–1130, March 2013.

[64] S. J. Park, K.Y. Lee, and G. M. Rebeiz, "Low-loss 5.15–5.70-GHz RF MEMS switchable filter for wireless LAN application," *IEEE Transactions on Microwave Theory and Techniques*, vol. 54, no. 11, pp. 3931–3939, Nov. 2006.

[65] S. J. Park, M. A. El-Tanani, I. Reines, and G. M. Rebeiz, "Low-loss 4–6-GHz tunable filter with 3-bit high-orthogonal bias RF-MEMS capacitance network," *IEEE Transactions on Microwave Theory and Techniques,* vol. 56, no. 10, pp. 2348–2355, Oct. 2008.

[66] Chi-Feng Chen, "A compact reconfigurable microstrip dual-band filter using varactor-tuned stub-loaded stepped-impedance resonators," *IEEE Microwave and Wireless Components Letters*, vol. 23, no. 1, pp. 16–18, Jan. 2013.

[67] Li Gao, Xiu Yin Zhang, Bin-Jie Hu, and Quan Xue, "Novel multi-stub loaded resonators and their applications to various bandpass filters," *IEEE Transactions on Microwave Theory and Techniques*, vol. 62, no. 5, pp. 1162–1172, May 2014.

[68] X. Huang, Lei Zhu, Q. Feng, Q. Xiang, and D. Jia, "Tunable bandpass filter with independently controllable dual passbands," *IEEE Transactions on Microwave Theory and Techniques*,vol. 61, no. 9, pp. 3200–3208, May 2014.

[69] P. Kapitanova, D. Kholodnyak, S. Humbla, R. Perrone, J. Mueller, M. A. Hein, and I. Vendik, "Tunable microwave devices based on left/right-handed transmission line sections in multilayer implementation," *International Journal of Microwave and Wireless Technologies*, vol. 1, no. 4, pp. 323–329, Aug. 2009.

[70] P. Ferrand, D. Baillargeat, S. Verdeyme, J. Puech, M. Lahti, and T. Jaakola, "LTCC reduced-size bandpass filters based on capacitively loaded cavities for Q band application," *Proceedings of IEEE MTT-S International Microwave Symposium Digest*, vol. 2, pp. 1789–1792, June 2005.

[71] J.-H. Lee, S. Pinel, J. Papapolymerou, J. Laskar, and M. M. Tentzeris, "Low loss LTCC cavity filters using system-on-package technology at 60 GHz," *IEEE Transactions on Microwave Theory and Techniques,* vol. 53, no. 12, pp. 3817–3824, Dec. 2005.

[72] L. Rigaudeau, P. Ferrand, D. Baillargeat, S. Bila, S. Verdeyme, M. Lahti, and T. Jaakola, "LTCC 3-D resonators applied to the design of very compact filters for Q-band applications," *IEEE Transactions on Microwave Theory and Techniques,* vol. 54, no. 6, pp. 2620–2627, June 2006.

[73] X. Gong, A. Margomenos, B. Liu, W. J. Chappell, and L. P. B. Katehi, "High-Q evanescent-mode filters using silicon micromachining and polymer stereo-lithography (SL) processing," *Proceedings of IEEE MTT-S International Microwave Symposium Digest*, vol. 2, pp. 433–436, June 2004.

[74] H. Joshi, H. H. Sigmarsson, D. Peroulis, and W. J. Chappell, "Highly loaded evanescent cavities for widely tunable high-Q filters,"*Proceedings of IEEE MTT-S International Microwave Symposium Digest,*June 2007, pp. 2133–2136.

[75] T. Kushta and T. Harada, "New 3-D design of filtering components using multilayer board technologies," *Proceedings of 38th European Microwave Conference*, Amsterdam, Nederlands, Oct. 2008, pp. 218–221.

[76] V. Turgaliev, D. Kholodnyak, I. Vendik, D. Stöpel, S. Humbla, J. Müller, and M. A. Hein, "LTCC Highly loaded cavities for the design of single- and dual-band low-loss miniature filters," *Proceedings of 40th European Microwave Conference*, Paris, France, Oct. 2010, pp. 180–183.

[77] V. Turgaliev, D. Kholodnyak, and I. Vendik, "Small-size dual-band filters on capacitively loaded cavities," *Proceedings of 43th European Microwave Conference*, Nuremberg, Germany, Oct. 2013, pp. 660–663.

Index